Research Software Engineering with Python

Building software that makes research possible

Damien Irving
Kate Hertweck
Luke Johnston
Joel Ostblom
Charlotte Wickham
Greg Wilson

CRC Press
Taylor & Francis Group
Boca Raton London New York

CRC Press is an imprint of the
Taylor & Francis Group, an **informa** business
A CHAPMAN & HALL BOOK

First edition published 2022
by CRC Press
6000 Broken Sound Parkway NW, Suite 300, Boca Raton, FL 33487-2742

and by CRC Press
2 Park Square, Milton Park, Abingdon, Oxon, OX14 4RN

CRC Press is an imprint of Taylor & Francis Group, LLC

Library of Congress Cataloging-in-Publication Data

Names: Irving, Damien, author.
Title: Research software engineering with Python : building software that makes research possible / Damien Irving, Kate Hertweck, Luke William Johnston, Joel Ostblom, Charlotte Wickham, and Greg Wilson.
Description: First edition. | Boca Raton, FL : CRC Press, 2021. | Includes bibliographical references and index.
Identifiers: LCCN 2021006032 | ISBN 9780367698348 (hardback) | ISBN 9780367698324 (paperback) | ISBN 9781003143482 (ebook)
Subjects: LCSH: Research--Data processing. | Computer software--Development. | Python (Computer program language)
Classification: LCC Q180.55.E4 I76 2021 | DDC 001.40285/5133--dc23
LC record available at https://lccn.loc.gov/2021006032

ISBN: 978-0-367-69834-8 (hbk)
ISBN: 978-0-367-69832-4 (pbk)
ISBN: 978-1-003-14348-2 (ebk)

DOI: 10.1201/9781003143482

Publisher's note: This book has been prepared from camera-ready copy provided by the authors.

To David Flanders
who taught me so much about growing and sustaining coding communities.
— Damien

To the UofT Coders Group
who taught us much more than we taught them.
— Luke and Joel

To my parents Judy and John
who taught me to love books and everything I can learn from them.
— Kate

To Joshua.
— Charlotte

To Brent Gorda
without whom none of this would have happened.
— Greg

All royalties from this book are being donated to The Carpentries,
an organization that teaches foundational coding and data science skills
to researchers worldwide.

Contents

Welcome

> It's still magic even if you know how it's done.
>
> — Terry Pratchett

Software is now as essential to research as telescopes, test tubes, and reference libraries. This means that researchers *need* to know how to build, check, use, and share programs. However, most introductions to programming focus on developing commercial applications, not on exploring problems whose answers aren't yet known. Our goal is to show you how to do that, both on your own and as part of a team.

We believe every researcher should know how to write short programs that clean and analyze data in a reproducible way and how to use version control to keep track of what they have done. But just as some astronomers spend their careers designing telescopes, some researchers focus on building the software that makes research possible. People who do this are called **research software engineers**; the aim of this book is to get you ready for this role by helping you go from writing code for yourself to creating tools that help your entire field advance.

0.1 The Big Picture

Our approach to research software engineering is based on three related concepts:

- **Open science**: Making data, methods, and results freely available to all by publishing them under **open licenses**.
- **Reproducible research**: Ensuring that anyone with access to the data and software can feasibly reproduce results, both to check them and to build on them.
- **Sustainable software**: The ease with which to maintain and extend it

rather than to replace it. Sustainability isn't just a property of the software: it also depends on the skills and culture of its users.

People often conflate these three ideas, but they are distinct. For example, if you share your data and the programs that analyze it, but don't document what steps to take in what order, your work is open but not reproducible. Conversely, if you completely automate your analysis, but your data is only available to people in your lab, your work is reproducible but not open. Finally, if a software package is being maintained by a couple of post-docs who are being paid a fraction of what they could earn in industry and have no realistic hope of promotion because their field doesn't value tool building, then sooner or later it will become **abandonware**, at which point openness and reproducibility become less relevant.

Nobody argues that research should be irreproducible or unsustainable, but "not against it" and actively supporting it are very different things. Academia doesn't yet know how to reward people for writing useful software, so while you may be thanked, the effort you put in may not translate into academic job security or decent pay.

Some people worry that if they make their data and code publicly available, someone else will use it and publish a result they could have come up with themselves. This is almost unheard of in practice, but that doesn't stop it being used as a scare tactic. Other people are afraid of looking foolish or incompetent by sharing code that might contain bugs. This isn't just **impostor syndrome**: members of marginalized groups are frequently judged more harshly than others, so being wrong in public is much riskier for them.

With this course, we hope to give researchers the tools and knowledge to be better research software developers, to be more efficient in their work, make less mistakes, and work more openly and reproducibly. We hope that by having more researchers with these skills and knowledge, research culture can improve to address the issues raised above.

0.2 Intended Audience

This book is written for researchers who are already using Python for their data analysis, but who want to take their coding and software development to the next level. You don't have to be highly proficient with Python, but you should already be comfortable doing things like reading data from files and writing loops, conditionals, and functions. The following personas are examples of the types of people that are our target audience.

Amira Khan completed a master's in library science five years ago and has since worked for a small aid organization. She did some statistics during her degree, and has learned some R and Python by doing data science courses online, but has no formal training in programming. Amira would like to tidy up the scripts, datasets, and reports she has created in order to share them with her colleagues. These lessons will show her how to do this.

Jun Hsu completed an Insight Data Science[1] fellowship last year after doing a PhD in geology and now works for a company that does forensic audits. He uses a variety of machine learning and visualization packages, and would now like to turn some of his own work into an open source project. This book will show him how such a project should be organized and how to encourage people to contribute to it.

Sami Virtanen became a competent programmer during a bachelor's degree in applied math and was then hired by the university's research computing center. The kinds of applications they are being asked to support have shifted from fluid dynamics to data analysis; this guide will teach them how to build and run data pipelines so that they can pass those skills on to their users.

0.3 What You Will Learn

Rather than simply providing reference material about good coding practices, the book follows Amira and Sami as they work together to write an actual software package to address a real research question. The data analysis task that we focus on relates to a fascinating result in the field of quantitative linguistics. Zipf's Law[2] states that the second most common word in a body of text appears half as often as the most common, the third most common appears a third as often, and so on. To test whether Zipf's Law holds for a collection of classic novels that are freely available from Project Gutenberg[3], we write a software package that counts and analyzes the word frequency distribution in any arbitrary body of text.

In the process of writing and publishing a Python package to verify Zipf's Law, we will show you how to do the following:

- Organize small and medium-sized data science projects.
- Use the Unix shell to efficiently manage your data and code.

[1] `https://www.insightdatascience.com/`
[2] `https://en.wikipedia.org/wiki/Zipf%27s_law`
[3] `https://www.gutenberg.org/`

- Write Python programs that can be used on the command line.
- Use Git and GitHub to track and share your work.
- Work productively in a small team where everyone is welcome.
- Use Make to automate complex workflows.
- Enable users to configure your software without modifying it directly.
- Test your software and know which parts have not yet been tested.
- Find, handle, and fix errors in your code.
- Publish your code and research in open and reproducible ways.
- Create Python packages that can be installed in standard ways.

0.4 Using this Book

This book was written to be used as the material for a (potentially) semester-long course at the university level, although it can also be used for independent self-study. Participatory live-coding is the anticipated style for teaching the material, rather than lectures simply talking about the code presented (N. C. C. Brown and Wilson 2018; Wilson 2019a). The chapters and their content are generally designed to be used in the order given.

Chapters are structured with the introduction at the start, content in the middle, and exercises at the end. Callout boxes are interspersed throughout the content to be used as a supplement to the main text, but not a requirement for the course overall. Early chapters have many small exercises; later chapters have fewer but larger exercises. In order to break up long periods of live-coding while teaching, it may be preferable to stop and complete some of the exercises at key points throughout the chapter, rather than waiting until the end. Possible exercise solutions are provided (Appendix A), in addition to learning objectives (Appendix B) and key points (Appendix C) for each chapter.

0.5 Contributing and Re-Use

The source for the book can be found at the `py-rse` GitHub repository[4] and any corrections, additions, or contributions are very welcome. Everyone whose work is included will be credited in the acknowledgments. Check out our

[4] `https://github.com/merely-useful/py-rse`

contributing guidelines[5] as well as our Code of Conduct[6] for more information on how to contribute.

The content and code of this book can be freely re-used as it is licensed[7] under a Creative Commons Attribution 4.0 International License[8] (CC-BY 4.0) and a MIT License[9], so the material can be used, re-used, and modified, as long as there is attribution to this source.

0.6 Acknowledgments

This book owes its existence to everyone we met through The Carpentries[10]. We are also grateful to Insight Data Science[11] for sponsoring the early stages of this work, to the authors of Noble (2009), Haddock and Dunn (2010), Wilson et al. (2014), Scopatz and Huff (2015), Taschuk and Wilson (2017), Wilson et al. (2017), N. C. C. Brown and Wilson (2018), Devenyi et al. (2018), Sholler et al. (2019), Wilson (2019b) and to everyone who has contributed, including Madeleine Bonsma-Fisher, Jonathan Dursi, Christina Koch, Sara Mahallati, Brandeis Marshall, and Elizabeth Wickes.

- Many of the explanations and exercises in Chapters 2–4 have been adapted from Software Carpentry's lesson *The Unix Shell*[12].
- Many of the explanations and exercises in Chapters 6 and 7 have been adapted from Software Carpentry's lesson *Version Control with Git*[13] and an adaptation/extension of that lesson[14] that is maintained by the University of Wisconsin-Madison Data Science Hub.
- Chapter 9 is based on Software Carpentry's lesson *Automation and Make*[15] and on Jonathan Dursi's *Introduction to Pattern Rules*[16].
- Chapter 14 is based in part on *Python 102*[17] by Ashwin Srinath.

[5]`https://github.com/merely-useful/py-rse/blob/book/CONTRIBUTING.md`
[6]`https://github.com/merely-useful/py-rse/blob/book/CONDUCT.md`
[7]`https://github.com/merely-useful/py-rse/blob/book/LICENSE.md`
[8]`https://creativecommons.org/licenses/by/4.0/`
[9]`https://github.com/merely-useful/py-rse/blob/book/LICENSE-MIT.md`
[10]`https://carpentries.org/`
[11]`https://www.insightdatascience.com/`
[12]`http://swcarpentry.github.io/shell-novice/`
[13]`http://swcarpentry.github.io/git-novice/`
[14]`https://uw-madison-datascience.github.io/git-novice-custom/`
[15]`http://swcarpentry.github.io/make-novice/`
[16]`https://github.com/ljdursi/make_pattern_rules`
[17]`https://python-102.readthedocs.io/`

1

Getting Started

> Everything starts somewhere, though many physicists disagree.
>
> — Terry Pratchett

As with many research projects, the first step in our Zipf's Law analysis is to download the research data and install the required software. Before doing that, it's worth taking a moment to think about how we are going to organize everything. We will soon have a number of books from Project Gutenberg[1] in the form of a series of text files, plots we've produced showing the word frequency distribution in each book, as well as the code we've written to produce those plots and to document and release our software package. If we aren't organized from the start, things could get messy later on.

1.1 Project Structure

Project organization is like a diet: everyone has one, it's just a question of whether it's healthy or not. In the case of a project, "healthy" means that people can find what they need and do what they want without becoming frustrated. This depends on how well organized the project is and how familiar people are with that style of organization.

As with good coding style, small pieces in predictable places with readable names are easier to find and use than large chunks that vary from project to project and have names like "stuff." We can be messy while we are working and then tidy up later, but experience teaches that we will be more productive if we make tidiness a habit.

In building the Zipf's Law project, we'll follow a widely used template for organizing small and medium-sized data analysis projects (Noble 2009). The project will live in a directory called `zipf`, which will also be a Git repository

[1]`https://www.gutenberg.org/`

stored on GitHub (Chapter 6). The following is an abbreviated version of the project directory tree as it appears toward the end of the book:

```
zipf/
├── .gitignore
├── CITATION.md
├── CONDUCT.md
├── CONTRIBUTING.md
├── LICENSE.md
├── README.md
├── Makefile
├── bin
│   ├── book_summary.sh
│   ├── collate.py
│   ├── countwords.py
│   └── ...
├── data
│   ├── README.md
│   ├── dracula.txt
│   ├── frankenstein.txt
│   └── ...
├── docs
│   └── ...
├── results
│   ├── collated.csv
│   ├── dracula.csv
│   ├── dracula.png
│   └── ...
└── ...
```

The full, final directory tree is documented in Appendix D.

1.1.1 Standard information

Our project will contain a few standard files that should be present in every research software project, open source or otherwise:

- `README` includes basic information on our project. We'll create it in Chapter 7, and extend it in Chapter 14.
- `LICENSE` is the project's license. We'll add it in Section 8.4.
- `CONTRIBUTING` explains how to contribute to the project. We'll add it in Section 8.11.

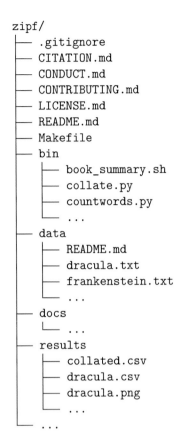

- `CONDUCT` is the project's Code of Conduct. We'll add it in Section 8.3.
- `CITATION` explains how to cite the software. We'll add it in Section 14.7.

Some projects also include a `CONTRIBUTORS` or `AUTHORS` file that lists everyone who has contributed to the project, while others include that information in the `README` (we do this in Chapter 7) or make it a section in `CITATION`. These files are often called **boilerplate**, meaning they are copied without change from one use to the next.

1.1.2 Organizing project content

Following Noble (2009), the directories in the repository's root are organized according to purpose:

- Runnable programs go in `bin/` (an old Unix abbreviation for "binary," meaning "not text"). This will include both shell scripts, e.g., `book_summary.sh` developed in Chapter 4, and Python programs, e.g., `countwords.py`, developed in Chapter 5.
- Raw data goes in `data/` and is never modified after being stored. You'll set up this directory and its contents in Section 1.2.
- Results are put in `results/`. This includes cleaned-up data, figures, and everything else created using what's in `bin` and `data`. In this project, we'll describe exactly how `bin` and `data` are used with `Makefile` created in Chapter 9.
- Finally, documentation and manuscripts go in `docs/`. In this project, `docs` will contain automatically generated documentation for the Python package, created in Section 14.6.2.

This structure works well for many computational research projects and we encourage its use beyond just this book. We will add some more folders and files not directly addressed by Noble (2009) when we talk about testing (Chapter 11), provenance (Chapter 13), and packaging (Chapter 14).

1.2 Downloading the Data

The data files used in the book are archived at an online repository called Figshare (which we discuss in detail in Section 13.1.2) and can be accessed at:

https://doi.org/10.6084/m9.figshare.13040516

We can download a zip file containing the data files by clicking "download all" at this URL and then unzipping the contents into a new `zipf/data` directory (also called a **folder**) that follows the project structure described above. Here's how things look once we're done:

```
zipf/
└── data
    ├── README.md
    ├── dracula.txt
    ├── frankenstein.txt
    ├── jane_eyre.txt
    ├── moby_dick.txt
    ├── sense_and_sensibility.txt
    ├── sherlock_holmes.txt
    └── time_machine.txt
```

1.3 Installing the Software

In order to conduct our analysis, we need to install the following software:

1. A **Bash shell**
2. **Git** version control
3. A text editor
4. Python 3[2] (via the Anaconda distribution)
5. GNU Make[3]

Comprehensive software installation instructions for Windows, Mac, and Linux operating systems (with video tutorials) are maintained by The Carpentries[4] as part of their workshop website template at:

https://carpentries.github.io/workshop-template/#setup

We can follow those instructions to install the Bash shell, Git, a text editor and Anaconda. We recommend Anaconda as the method for installing Python, as it includes Conda as well as many of the packages we'll use in this book.

You can check if Make is already on your computer by typing `make -v` into the Bash shell. If it is not, you can install it as follows:

[2]https://www.python.org/
[3]https://www.gnu.org/software/make/
[4]https://carpentries.org/

- *Linux (Debian/Ubuntu)*: Install it from the Bash shell using `sudo apt-get install make`.
- *Mac*: Install Xcode[5] (via the App Store).
- *Windows*: Follow the installation instructions[6] maintained by the Master of Data Science program at the University of British Columbia.

conda in the Shell on Windows

If you are using Windows and the `conda` command isn't available at the Bash shell, you'll need to open the Anaconda Prompt program (via the Windows start menu) and run the command `conda init bash` (this only needs to be done once). After that, your shell will be configured to use conda going forward.

Software Versions

Throughout the book, we'll be showing you examples of the output you can expect to see. This output is derived from running a Mac with: Git version 2.29.2, Python version 3.7.6, GNU bash version 3.2.57(1)-release (x86_64-apple-darwin19), GNU Make 3.81, and conda 4.9.2. In some cases, what you see printed to the screen may differ slightly based on software version. We'll help you understand how to interpret the output so you can keep working and troubleshoot regardless of software version.

1.4 Summary

Now that our project structure is set up, our data is downloaded, and our software is installed, we are ready to start our analysis.

[5]`https://developer.apple.com/xcode/`
[6]`https://ubc-mds.github.io/resources_pages/install_ds_stack_windows/#make`

1.5 Exercises

1.5.1 Getting ready

Make sure you've downloaded the required data files (following Section 1.2) and installed the required software (following Section 1.3) before progressing to the next chapter.

1.6 Key Points

- Make tidiness a habit, rather than cleaning up your project files later.
- Include a few standard files in all your projects, such as README, LICENSE, CONTRIBUTING, CONDUCT and CITATION.
- Put runnable code in a `bin/` directory.
- Put raw/original data in a `data/` directory and never modify it.
- Put results in a `results/` directory. This includes cleaned-up data and figures (i.e., everything created using what's in `bin` and `data`).
- Put documentation and manuscripts in a `docs/` directory.
- Refer to The Carpentries software installation guide[7] if you're having trouble.

[7] `https://carpentries.github.io/workshop-template/#setup`

2

The Basics of the Unix Shell

Ninety percent of most magic merely consists of knowing one extra fact.

— Terry Pratchett

Computers do four basic things: store data, run programs, talk with each other, and interact with people. They do the interacting in many different ways, of which **graphical user interfaces** (GUIs) are the most widely used. The computer displays icons to show our files and programs, and we tell it to copy or run those by clicking with a mouse. GUIs are easy to learn but hard to automate, and don't create a record of what we did.

In contrast, when we use a **command-line interface** (CLI) we communicate with the computer by typing commands, and the computer responds by displaying text. CLIs existed long before GUIs; they have survived because they are efficient, easy to automate, and automatically record what we have done.

The heart of every CLI is a **read-evaluate-print loop** (REPL). When we type a command and press Return (also called Enter) the CLI **r**eads the command, **e**valuates it (i.e., executes it), **p**rints the command's output, and **l**oops around to wait for another command. If you have used an interactive console for Python, you have already used a simple CLI.

This lesson introduces another CLI that lets us interact with our computer's operating system. It is called a "command shell," or just **shell** for short, and in essence is a program that runs other programs on our behalf (Figure 2.1). Those "other programs" can do things as simple as telling us the time or as complex as modeling global climate change; as long as they obey a few simple rules, the shell can run them without having to know what language they are written in or how they do what they do.

```
amiras-computer$ ls
Applications    Downloads       Music           todo.txt
Desktop         Library         Pictures        zipf
Documents       Movies          Public
amiras-computer$ cd Documents/
amiras-computer$ ls
personal_writing        work_files
amiras-computer$ 
```

FIGURE 2.1: The Bash shell.

What's in a Name?

Programmers have written many different shells over the last forty years, just as they have created many different text editors and plotting packages. The most popular shell today is called Bash (an acronym of **B**ourne **A**gain **SH**ell, and a weak pun on the name of its predecessor, the Bourne shell). Other shells may differ from Bash in minor ways, but the core commands and ideas remain the same. In particular, the most recent versions of MacOS use a shell called the Z Shell or `zsh`; we will point out a few differences as we go along.

Please see Section 1.3 for instructions on how to install and launch the shell on your computer.

2.1 Exploring Files and Directories

Our first shell commands will let us explore our folders and files, and will also introduce us to several conventions that most Unix tools follow. To start, when Bash runs it presents us with a **prompt** to indicate that it is waiting for us to type something. This prompt is a simple dollar sign by default:

```
$
```

However, different shells may use a different symbol: in particular, the `zsh` shell, which is the default on newer versions of MacOS, uses `%`. As we'll see in Section 4.6, we can customize the prompt to give us more information.

Don't Type the Dollar Sign

We show the `$` prompt so that it's clear what you are supposed to type, particularly when several commands appear in a row, but you should *not* type it yourself.

Let's run a command to find out who the shell thinks we are:

```
$ whoami
```

```
amira
```

Learn by Doing

Amira is one of the learners described in Section 0.2. For the rest of the book, we'll present code and examples from her perspective. You should follow along on your own computer, though what you see might deviate in small ways because of differences in operating system (and because your name probably isn't Amira).

Now that we know who we are, we can explore where we are and what we have. The part of the operating system that manages files and directories (also called **folders**) is called the **filesystem**. Some of the most commonly used commands in the shell create, inspect, rename, and delete files and directories. Let's start exploring them by running the command `pwd`, which stands for **p**rint **w**orking **d**irectory. The "print" part of its name is straightforward; the "working directory" part refers to the fact that the shell keeps track of our **current working directory** at all times. Most commands read and write files in the current working directory unless we tell them to do something else, so knowing where we are before running a command is important.

```
$ pwd
```

```
/Users/amira
```

Here, the computer's response is `/Users/amira`, which tells us that we are in a directory called `amira` that is contained in a top-level directory called `Users`. This directory is Amira's **home directory**; to understand what that

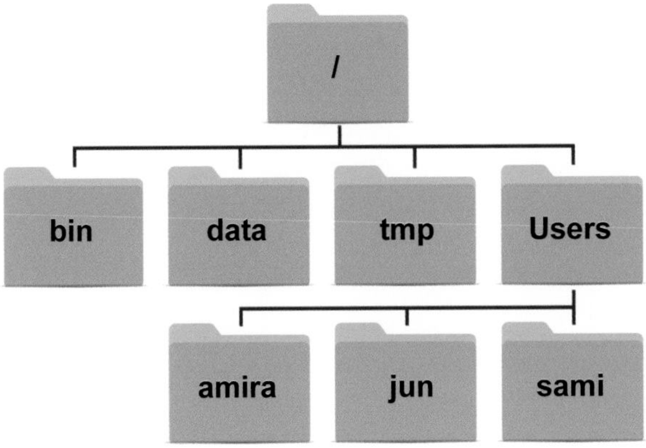

FIGURE 2.2: A sample filesystem.

means, we must first understand how the filesystem is organized. On Amira's computer it looks like Figure 2.2.

At the top is the **root directory** that holds everything else, which we can refer to using a slash character `/` on its own. Inside that directory are several other directories, including `bin` (where some built-in programs are stored), `data` (for miscellaneous data files), `tmp` (for temporary files that don't need to be stored long-term), and `Users` (where users' personal directories are located). We know that `/Users` is stored inside the root directory `/` because its name begins with `/`, and that our current working directory `/Users/amira` is stored inside `/Users` because `/Users` is the first part of its name. A name like this is called a **path** because it tells us how to get from one place in the filesystem (e.g., the root directory) to another (e.g., Amira's home directory).

Slashes

The `/` character means two different things in a path. At the front of a path or on its own, it refers to the root directory. When it appears inside a name, it is a separator. Windows uses backslashes (\\) instead of forward slashes as separators.

Underneath `/Users`, we find one directory for each user with an account on this machine. Jun's files are stored in `/Users/jun`, Sami's in `/Users/sami`, and Amira's in `/Users/amira`. This is where the name "home directory" comes from: when we first log in, the shell puts us in the directory that holds our files.

Home Directory Variations

Our home directory will be in different places on different operating systems. On Linux it may be `/home/amira`, and on Windows it may be `C:\Documents and Settings\amira` or `C:\Users\amira` (depending on the version of Windows). Our examples show what we would see on MacOS.

Now that we know where we are, let's see what we have using the command `ls` (short for "listing"), which prints the names of the files and directories in the current directory:

```
$ ls
```

```
Applications    Downloads    Music        todo.txt
Desktop         Library      Pictures     zipf
Documents       Movies       Public
```

Again, our results may be different depending on our operating system and what files or directories we have.

We can make the output of `ls` more informative using the `-F` **option** (also sometimes called a **switch** or a **flag**). Options are exactly like arguments to a function in Python; in this case, `-F` tells `ls` to decorate its output to show what things are. A trailing `/` indicates a directory, while a trailing `*` tells us something is a runnable program. Depending on our setup, the shell might also use colors to indicate whether each entry is a file or directory.

```
$ ls -F
```

```
Applications/   Downloads/    Music/        todo.txt
Desktop/        Library/      Pictures/     zipf/
Documents/      Movies/       Public/
```

Here, we can see that almost everything in our home directory is a **subdirectory**; the only thing that isn't is a file called `todo.txt`.

Spaces Matter

`1+2` and `1 + 2` mean the same thing in mathematics, but `ls -F` and `ls-F` are very different things in the shell. The shell splits whatever we type into pieces based on spaces, so if we forget to separate `ls` and `-F` with at least one space, the shell will try to find a program called `ls-F` and (quite sensibly) give an error message like `ls-F: command not found`.

Some options tell a command how to behave, but others tell it what to act on. For example, if we want to see what's in the `/Users` directory, we can type:

```
$ ls /Users
```

```
amira   jun     sami
```

We often call the file and directory names that we give to commands **arguments** to distinguish them from the built-in options. We can combine options and arguments:

```
$ ls -F /Users
```

```
amira/  jun/    sami/
```

but we must put the options (like `-F`) before the names of any files or directories we want to work on, because once the command encounters something that *isn't* an option it assumes there aren't any more:

```
$ ls /Users -F
```

```
ls: -F: No such file or directory
amira   jun     sami
```

Command Line Differences

Code can sometimes behave in unexpected ways on different computers, and this applies to the command line as well. For example, the following code actually *does* work on some Linux operating systems:

```
$ ls /Users -F
```

Some people think this is convenient; others (including us) believe it is confusing, so it's best to avoid doing this.

2.2 Moving Around

Let's run `ls` again. Without any arguments, it shows us what's in our current working directory:

```
$ ls -F
```

```
Applications/    Downloads/    Music/       todo.txt
Desktop/         Library/      Pictures/    zipf/
Documents/       Movies/       Public/
```

If we want to see what's in the `zipf` directory we can ask `ls` to list its contents:

```
$ ls -F zipf
```

```
data/
```

Notice that `zipf` doesn't have a leading slash before its name. This absence tells the shell that it is a **relative path**, i.e., that it identifies something starting from our current working directory. In contrast, a path like `/Users/amira` is an **absolute path**: it is always interpreted from the root directory down, so it always refers to the same thing. Using a relative path is like telling someone to go two kilometers north and then half a kilometer east; using an absolute path is like giving them the latitude and longitude of their destination.

We can use whichever kind of path is easiest to type, but if we are going to do a lot of work with the data in the `zipf` directory, the easiest thing would be to change our current working directory so that we don't have to type `zipf` over and over again. The command to do this is `cd`, which stands for **c**hange **d**irectory. This name is a bit misleading because the command doesn't change the directory; instead, it changes the shell's idea of what directory we are in. Let's try it out:

```
$ cd zipf
```

`cd` doesn't print anything. This is normal: many shell commands run silently unless something goes wrong, on the theory that they should only ask for our attention when they need it. To confirm that `cd` has done what we asked, we can use `pwd`:

```
$ pwd
```

```
/Users/amira/zipf
```

```
$ ls -F
```

```
data/
```

Missing Directories and Unknown Options

If we give a command an option that it doesn't understand, it will usually print an error message, and (if we're lucky) tersely remind us of what we should have done:

```
$ cd -j
```

```
-bash: cd: -j: invalid option
cd: usage: cd [-L|-P] [dir]
```

On the other hand, if we get the syntax right but make a mistake in the name of a file or directory, it will tell us that:

```
$ cd whoops
```

```
-bash: cd: whoops: No such file or directory
```

We now know how to go down the directory tree, but how do we go up? This doesn't work:

```
$ cd amira
```

```
cd: amira: No such file or directory
```

because `amira` on its own is a relative path meaning "a file or directory called `amira` *below our current working directory.*" To get back home, we can either use an absolute path:

```
$ cd /Users/amira
```

or a special relative path called `..` (two periods in a row with no spaces), which always means "the directory that contains the current one." The directory that contains the one we are in is called the **parent directory**, and sure enough, `..` gets us there:

```
$ cd ..
$ pwd
```

```
/Users/amira
```

`ls` usually doesn't show us this special directory—since it's always there, displaying it every time would be a distraction. We can ask `ls` to include it using the `-a` option, which stands for "all." Remembering that we are now in `/Users/amira`:

```
$ ls -F -a
```

```
./                Documents/     Music/       zipf/
../               Downloads/     Pictures/
Applications/     Library/       Public/
Desktop/          Movies/        todo.txt
```

The output also shows another special directory called `.` (a single period), which refers to the current working directory. It may seem redundant to have a name for it, but we'll see some uses for it soon.

Combining Options

You'll occasionally need to use multiple options in the same command. In most command-line tools, multiple options can be combined with a single `-` and no spaces between the options:

```
$ ls -Fa
```

This command is synonymous with the previous example. While you may see commands written like this, we don't recommend you use this approach in your own work. This is because some commands take **long options** with multi-letter names, and it's very easy to mistake `--no` (meaning "answer 'no' to all questions") with `-no` (meaning `-n -o`).

The special names `.` and `..` don't belong to `cd`: they mean the same thing to every program. For example, if we are in `/Users/amira/zipf`, then `ls ..` will display a listing of `/Users/amira`. When the meanings of the parts are the same no matter how they're combined, programmers say they are **orthogonal**. Orthogonal systems tend to be easier for people to learn because there are fewer special cases to remember.

Other Hidden Files

In addition to the hidden directories `..` and `.`, we may also come across files with names like `.jupyter`. These usually contain settings or other data for particular programs; the prefix `.` is used to prevent `ls` from cluttering up the output when we run `ls`. We can always use the `-a` option to display them.

`cd` is a simple command, but it allows us to explore several new ideas. First, several `..` can be joined by the path separator to move higher than the parent directory in a single step. For example, `cd ../..` will move us up two directories (e.g., from `/Users/amira/zipf` to `/Users`), while `cd ../Movies` will move us up from `zipf` and back down into `Movies`.

What happens if we type `cd` on its own without giving a directory?

```
$ pwd
```

```
/Users/amira/Movies
```

```
$ cd
$ pwd
```

```
/Users/amira
```

No matter where we are, `cd` on its own always returns us to our home directory. We can achieve the same thing using the special directory name `~`, which is a shortcut for our home directory:

```
$ ls ~
```

```
Applications   Downloads   Music      todo.txt
Desktop        Library     Pictures   zipf
Documents      Movies      Public
```

(`ls` doesn't show any trailing slashes here because we haven't used `-F`.) We can use `~` in paths, so that (for example) `~/Downloads` always refers to our download directory.

Finally, `cd` interprets the shortcut `-` (a single dash) to mean the last directory we were in. Using this is usually faster and more reliable than trying to remember and type the path, but unlike `~`, it only works with `cd`: `ls -` tries to print a listing of a directory called `-` rather than showing us the contents of our previous directory.

2.3 Creating New Files and Directories

We now know how to explore files and directories, but how do we create them? To find out, let's go back to our `zipf` directory:

```
$ cd ~/zipf
$ ls -F
```

```
data/
```

To create a new directory, we use the command `mkdir` (short for **make** **dir**ectory):

```
$ mkdir docs
```

Since `docs` is a relative path (i.e., does not have a leading slash) the new directory is created below the current working directory:

```
$ ls -F
```

```
data/  docs/
```

Using the shell to create a directory is no different than using a graphical tool. If we look at the current directory with our computer's file browser we will see the `docs` directory there too. The shell and the file explorer are two different ways of interacting with the files; the files and directories themselves are the same.

Naming Files and Directories

Complicated names of files and directories can make our life painful. Following a few simple rules can save a lot of headaches:

1. **Don't use spaces.** Spaces can make a name easier to read, but since they are used to separate arguments on the command line, most shell commands interpret a name like `My Thesis` as two names `My` and `Thesis`. Use `-` or `_` instead, e.g., `My-Thesis` or `My_Thesis`.
2. **Don't begin the name with `-` (dash)** to avoid confusion with command options like `-F`.
3. **Stick with letters, digits, `.` (period or 'full stop'), `-` (dash) and `_` (underscore).** Many other characters mean special things in the shell. We will learn about some of those special characters during this lesson, but the characters cited here are always safe.

If we need to refer to files or directories that have spaces or other special characters in their names, we can surround the name in quotes (`""`). For example, `ls "My Thesis"` will work where `ls My Thesis` does not.

Since we just created the `docs` directory, `ls` doesn't display anything when we ask for a listing of its contents:

```
$ ls -F docs
```

Let's change our working directory to `docs` using `cd`, then use a very simple text editor called **Nano** to create a file called `draft.txt` (Figure 2.3):

```
$ cd docs
$ nano draft.txt
```

When we say "Nano is a text editor" we really do mean "text": it can only work with plain character data, not spreadsheets, images, Microsoft Word files, or anything else invented after 1970. We use it in this lesson because it runs everywhere, and because it is as simple as something can be and still be called an editor. However, that last trait means that we *shouldn't* use it for larger tasks like writing a program or a paper.

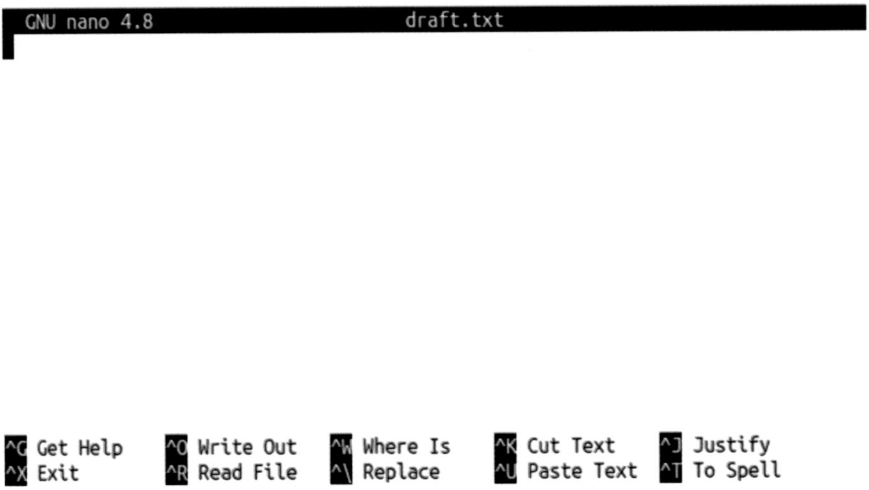

FIGURE 2.3: The Nano editor.

Recycling Pixels

Unlike most modern editors, Nano runs *inside* the shell window instead of opening a new window of its own. This is a holdover from an era when graphical terminals were a rarity and different applications had to share a single screen.

Once Nano is open we can type in a few lines of text, then press Ctrl+O (the Control key and the letter 'O' at the same time) to save our work. Nano will ask us what file we want to save it to; press Return to accept the suggested default of `draft.txt`. Once our file is saved, we can use Ctrl+X to exit the editor and return to the shell.

Control, Ctrl, or ^ Key

The Control key, also called the "Ctrl" key, can be described in a bewildering variety of ways. For example, Control plus X may be written as:

- `Control-X`
- `Control+X`
- `Ctrl-X`
- `Ctrl+X`
- `C-x`
- `^X`

When Nano runs, it displays some help in the bottom two lines of the screen using the last of these notations: for example, `^G Get Help` means "use Ctrl+G to get help" and `^O WriteOut` means "use Ctrl+O to write out the current file."

Nano doesn't leave any output on the screen after it exits, but `ls` will show that we have indeed created a new file `draft.txt`:

```
$ ls
```

```
draft.txt
```

Dot Something

All of Amira's files are named "something dot something." This is just a convention: we can call a file `mythesis` or almost anything else. However, both people and programs use two-part names to help them tell different kinds of files apart. The part of the filename after the dot is called the **filename extension** and indicates what type of data the file holds: `.txt` for plain text, `.pdf` for a PDF document, `.png` for a PNG image, and so on. This is just a convention: saving a PNG image of a whale as `whale.mp3` doesn't somehow magically turn it into a recording of whalesong, though it *might* cause the operating system to try to open it with a music player when someone double-clicks it.

2.4 Moving Files and Directories

Let's go back to our `zipf` directory:

```
cd ~/zipf
```

The `docs` directory contains a file called `draft.txt`. That isn't a particularly informative name, so let's change it using `mv` (short for **move**):

```
$ mv docs/draft.txt docs/prior-work.txt
```

The first argument tells `mv` what we are "moving," while the second is where it's to go. "Moving" `docs/draft.txt` to `docs/prior-work.txt` has the same effect as renaming the file:

```
$ ls docs
```

```
prior-work.txt
```

We must be careful when specifying the destination because `mv` will overwrite existing files without warning. An option `-i` (for "interactive") makes `mv` ask us for confirmation before overwriting. `mv` also works on directories, so `mv analysis first-paper` would rename the directory without changing its contents.

Now suppose we want to move `prior-work.txt` into the current working directory. If we don't want to change the file's name, just its location, we can provide `mv` with a directory as a destination and it will move the file there. In this case, the directory we want is the special name `.` that we mentioned earlier:

```
$ mv docs/prior-work.txt .
```

`ls` now shows us that `docs` is empty:

```
$ ls docs
```

and that our current directory now contains our file:

```
$ ls
```

```
data/  docs/  prior-work.txt
```

If we only want to check that the file exists, we can give its name to `ls` just like we can give the name of a directory:

```
$ ls prior-work.txt
```

```
prior-work.txt
```

2.5 Copying Files and Directories

The `cp` command **cop**ies files. It works like `mv` except it creates a file instead of moving an existing one:

```
$ cp prior-work.txt docs/section-1.txt
```

We can check that `cp` did the right thing by giving `ls` two arguments to ask it to list two things at once:

```
$ ls prior-work.txt docs/section-1.txt
```

```
docs/section-1.txt  prior-work.txt
```

Notice that `ls` shows the output in alphabetical order. If we leave off the second filename and ask it to show us a file and a directory (or multiple directories) it lists them one by one:

```
$ ls prior-work.txt docs
```

```
prior-work.txt

docs:
section-1.txt
```

Copying a directory and everything it contains is a little more complicated. If we use `cp` on its own, we get an error message:

```
$ cp docs backup
```

```
cp: analysis is a directory (not copied).
```

If we really want to copy everything, we must give `cp` the `-r` option (meaning **recursive**):

```
$ cp -r docs backup
```

Once again we can check the result with `ls`:

```
$ ls docs backup
```

```
docs/:
section-1.txt

backup/:
section-1.txt
```

Copying Files to and from Remote Computers

For many researchers, a motivation for learning how to use the shell is that it's often the only way to connect to a remote computer (e.g., located at a supercomputing facility or in a university department).

Similar to the `cp` command, there exists a **s**ecure **c**o**p**y (`scp`) command for copying files between computers. See Appendix E for details, including how to set up a secure connection to a remote computer via the shell.

2.6 Deleting Files and Directories

Let's tidy up by removing the `prior-work.txt` file we created in our `zipf` directory. The command to do this is `rm` (for **r**e**m**ove):

```
$ rm prior-work.txt
```

We can confirm the file is gone using `ls`:

```
$ ls prior-work.txt
```

```
ls: prior-work.txt: No such file or directory
```

Deleting is forever: unlike most GUIs, the Unix shell doesn't have a trash bin that we can recover deleted files from. Tools for finding and recovering deleted files do exist, but there is no guarantee they will work, since the computer may recycle the file's disk space at any time. In most cases, when we delete a file it really is gone.

In a half-hearted attempt to stop us from erasing things accidentally, `rm` refuses to delete directories:

```
$ rm docs
```

```
rm: docs: is a directory
```

We can tell `rm` we really want to do this by giving it the recursive option `-r`:

```
$ rm -r docs
```

`rm -r` should be used with great caution: in most cases, it's safest to add the `-i` option (for **i**nteractive) to get `rm` to ask us to confirm each deletion. As a halfway measure, we can use `-v` (for **v**erbose) to get `rm` to print a message for each file it deletes. This option works the same way with `mv` and `cp`.

2.7 Wildcards

`zipf/data` contains the text files for several ebooks from Project Gutenberg[1]:

```
$ ls data
```

```
README.md          moby_dick.txt
dracula.txt        sense_and_sensibility.txt
frankenstein.txt   sherlock_holmes.txt
jane_eyre.txt      time_machine.txt
```

The `wc` command (short for **w**ord **c**ount) tells us how many lines, words, and letters there are in one file:

```
$ wc data/moby_dick.txt
```

```
 22331  215832 1276222 data/moby_dick.txt
```

What's in a Word?

`wc` only considers spaces to be word breaks: if two words are connected by a long dash—like "dash" and "like" in this sentence—then `wc` will count them as one word.

[1] `https://www.gutenberg.org/`

We could run `wc` more times to find out how many lines there are in the other files, but that would be a lot of typing and we could easily make a mistake. We can't just give `wc` the name of the directory as we do with `ls`:

```
$ wc data
```

```
wc: data: read: Is a directory
```

Instead, we can use **wildcards** to specify a set of files at once. The most commonly used wildcard is `*` (a single asterisk). It matches zero or more characters, so `data/*.txt` matches all of the text files in the `data` directory:

```
$ ls data/*.txt
```

```
data/dracula.txt        data/sense_and_sensibility.txt
data/frankenstein.txt   data/sherlock_holmes.txt
data/jane_eyre.txt      data/time_machine.txt
data/moby_dick.txt
```

while `data/s*.txt` only matches the two whose names begin with an 's':

```
$ ls data/s*.txt
```

```
data/sense_and_sensibility.txt  data/sherlock_holmes.txt
```

Wildcards are expanded to match filenames *before* commands are run, so they work exactly the same way for every command. This means that we can use them with `wc` to (for example) count the number of words in the books with names that contain an underscore:

```
$ wc data/*_*.txt
```

```
  21054  188460 1049294 data/jane_eyre.txt
  22331  215832 1253891 data/moby_dick.txt
  13028  121593  693116 data/sense_and_sensibility.txt
  13053  107536  581903 data/sherlock_holmes.txt
   3582   35527  200928 data/time_machine.txt
  73048  668948 3779132 total
```

or the number of words in Frankenstein:

```
$ wc data/frank*.txt
```

```
  7832  78100 442967 data/frankenstein.txt
```

The exercises will introduce and explore other wildcards. For now, we only need to know that it's possible for a wildcard expression to *not* match anything. In this case, the command will usually print an error message:

```
$ wc data/*.csv
```

```
wc: data/*.csv: open: No such file or directory
```

2.8 Reading the Manual

`wc` displays lines, words, and characters by default, but we can ask it to display only the number of lines:

```
$ wc -l data/s*.txt
```

```
  13028 sense_and_sensibility.txt
  13053 sherlock_holmes.txt
  26081 total
```

`wc` has other options as well. We can use the `man` command (short for **man**ual) to find out what they are:

```
$ man wc
```

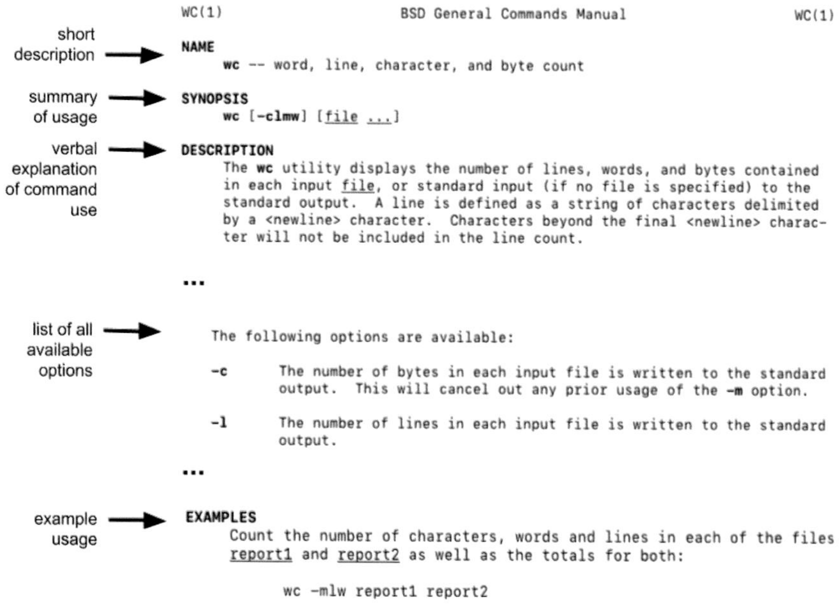

FIGURE 2.4: Key features of Unix manual pages.

Paging through the Manual

If our screen is too small to display an entire manual page at once, the shell will use a **paging program** called `less` to show it piece by piece. We can use ↑ and ↓ to move line-by-line or Ctrl+Spacebar and Spacebar to skip up and down one page at a time. (B and F also work.)

To search for a character or word, use / followed by the character or word to search for. If the search produces multiple hits, we can move between them using N (for "next"). To quit, press Q.

Manual pages contain a lot of information—often more than we really want. Figure 2.3 includes excerpts from the manual on your screen, and highlights a few of features useful for beginners.

Some commands have a `--help` option that provides a succinct summary of possibilities, but the best place to go for help these days is probably the TLDR[2] website. The acronym stands for "too long, didn't read," and its help for `wc` displays this:

[2] `https://tldr.sh/`

```
wc
Count words, bytes, or lines.

Count lines in file:
wc -l {{file}}

Count words in file:
wc -w {{file}}

Count characters (bytes) in file:
wc -c {{file}}

Count characters in file (taking multi-byte character sets into
account):
wc -m {{file}}

edit this page on github
```

As the last line suggests, all of its examples are in a public GitHub repository so that users like you can add the examples you wish it had. For more information, we can search on Stack Overflow[3] or browse the GNU manuals[4] (particularly those for the core GNU utilities[5], which include many of the commands introduced in this lesson). In all cases, though, we need to have some idea of what we're looking for in the first place: someone who wants to know how many lines there are in a data file is unlikely to think to look for `wc`.

2.9 Summary

The original Unix shell is celebrating its fiftieth anniversary. Its commands may be cryptic, but few programs have remained in daily use for so long. The next chapter will explore how we can combine and repeat commands in order to create powerful, efficient workflows.

[3] https://stackoverflow.com/questions/tagged/bash
[4] https://www.gnu.org/manual/manual.html
[5] https://www.gnu.org/software/coreutils/manual/coreutils.html

2.10 Exercises

The exercises below involve creating and moving new files, as well as considering hypothetical files. Please note that if you create or move any files or directories in your Zipf's Law project, you may want to reorganize your files following the outline at the beginning of the next chapter. If you accidentally delete necessary files, you can start with a fresh copy of the data files by following the instructions in Section 1.2.

2.10.1 Exploring more `ls` flags

What does the command `ls` do when used with the `-l` option?

What happens if you use two options at the same time, such as `ls -l -h`?

2.10.2 Listing recursively and by time

The command `ls -R` lists the contents of directories recursively, which means the subdirectories, sub-subdirectories, and so on at each level are listed. The command `ls -t` lists things by time of last change, with most recently changed files or directories first.

In what order does `ls -R -t` display things? Hint: `ls -l` uses a long listing format to view timestamps.

2.10.3 Absolute and relative paths

Starting from `/Users/amira/data`, which of the following commands could Amira use to navigate to her home directory, which is `/Users/amira`?

1. `cd .`
2. `cd /`
3. `cd /home/amira`
4. `cd ../..`
5. `cd ~`
6. `cd home`
7. `cd ~/data/..`
8. `cd`
9. `cd ..`
10. `cd ../.`

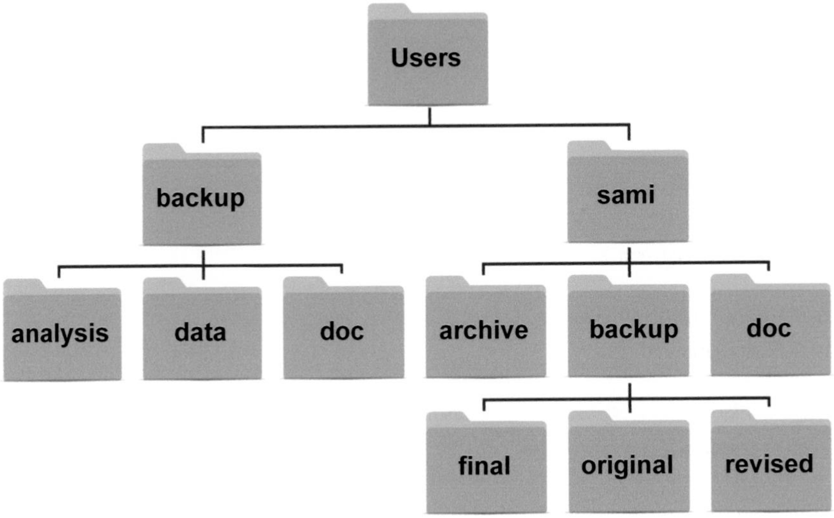

FIGURE 2.5: Filesystem for exercises.

2.10.4 Relative path resolution

Using the filesystem shown in Figure 2.5, if `pwd` displays `/Users/sami`, what will `ls -F ../backup` display?

1. `../backup: No such file or directory`
2. `final original revised`
3. `final/ original/ revised/`
4. `data/ analysis/ doc/`

2.10.5 `ls` reading comprehension

Using the filesystem shown in Figure 2.5, if `pwd` displays `/Users/backup`, and `-r` tells `ls` to display things in reverse order, what command(s) will result in the following output:

```
doc/ data/ analysis/
```

1. `ls pwd`
2. `ls -r -F`
3. `ls -r -F /Users/backup`

2.10.6 Creating files a different way

What happens when you execute `touch my_file.txt`? (Hint: use `ls -l` to find information about the file)

When might you want to create a file this way?

2.10.7 Using `rm` safely

What would happen if you executed `rm -i my_file.txt` on the file created in the previous exercise? Why would we want this protection when using `rm`?

2.10.8 Moving to the current folder

After running the following commands, Amira realizes that she put the (hypothetical) files `chapter1.txt` and `chapter2.txt` into the wrong folder:

```
$ ls -F
```

```
 data/  docs/
```

```
$ ls -F data
```

```
README.md          frankenstein.txt              sherlock_holmes.txt
chapter1.txt       jane_eyre.txt                 time_machine.txt
chapter2.txt       moby_dick.txt
dracula.txt        sense_and_sensibility.txt
```

```
$ cd docs
```

Fill in the blanks to move these files to the current folder (i.e., the one she is currently in):

```
$ mv ___/chapter1.txt  ___/chapter2.txt ___
```

2.10.9 Renaming files

Suppose that you created a plain-text file in your current directory to contain a list of the statistical tests you will need to do to analyze your data, and named it: `statstics.txt`

After creating and saving this file you realize you misspelled the filename! You want to correct the mistake, which of the following commands could you use to do so?

1. `cp statstics.txt statistics.txt`
2. `mv statstics.txt statistics.txt`
3. `mv statstics.txt .`
4. `cp statstics.txt .`

2.10.10 Moving and copying

Assuming the following hypothetical files, what is the output of the closing `ls` command in the sequence shown below?

```
$ pwd
```

```
/Users/amira/data
```

```
$ ls
```

```
books.dat
```

```
$ mkdir doc
$ mv books.dat doc/
$ cp doc/books.dat ../books-saved.dat
$ ls
```

1. `books-saved.dat doc`
2. `doc`
3. `books.dat doc`
4. `books-saved.dat`

2.10.11 Copy with multiple filenames

This exercise explores how `cp` responds when attempting to copy multiple things.

What does `cp` do when given several filenames followed by a directory name?

```
$ mkdir backup
$ cp dracula.txt frankenstein.txt backup/
```

What does `cp` do when given three or more filenames?

```
$ cp dracula.txt frankenstein.txt jane_eyre.txt
```

2.10.12 List filenames matching a pattern

When run in the `data` directory of your project directory, which `ls` command(s) will produce this output?

```
jane_eyre.txt   sense_and_sensibility.txt
```

1. `ls ??n*.txt`
2. `ls *e_*.txt`
3. `ls *n*.txt`
4. `ls *n?e*.txt`

2.10.13 Organizing directories and files

Amira is working on a project and she sees that her files aren't very well organized:

```
$ ls -F
```

```
books.txt    data/    results/   titles.txt
```

The `books.txt` and `titles.txt` files contain output from her data analysis. What command(s) does she need to run to produce the output shown?

```
$ ls -F
```

```
data/   results/
```

```
$ ls results
```

```
books.txt    titles.txt
```

2.10.14 Reproduce a directory structure

You're starting a new analysis, and would like to duplicate the directory structure from your previous experiment so you can add new data.

Assume that the previous experiment is in a folder called `2016-05-18`, which contains a `data` folder that in turn contains folders named `raw` and `processed` that contain data files. The goal is to copy the folder structure of `2016-05-18/data` into a folder called `2016-05-20` so that your final directory structure looks like this:

```
2016-05-20/
└── data
    ├── processed
    └── raw
```

Which of the following commands would achieve this objective? What would the other commands do?

```
# Set 1
$ mkdir 2016-05-20
$ mkdir 2016-05-20/data
$ mkdir 2016-05-20/data/processed
$ mkdir 2016-05-20/data/raw
```

```
# Set 2
$ mkdir 2016-05-20
$ cd 2016-05-20
$ mkdir data
$ cd data
$ mkdir raw processed
```

```
# Set 3
$ mkdir 2016-05-20/data/raw
$ mkdir 2016-05-20/data/processed
```

```
# Set 4
$ mkdir 2016-05-20
$ cd 2016-05-20
$ mkdir data
$ mkdir raw processed
```

2.10.15 Wildcard expressions

Wildcard expressions can be very complex, but you can sometimes write them in ways that only use simple syntax, at the expense of being a bit more verbose. In your `data/` directory, the wildcard expression `[st]*.txt` matches all files beginning with `s` or `t` and ending with `.txt`. Imagine you forgot about this.

1. Can you match the same set of files with basic wildcard expressions that do not use the `[]` syntax? *Hint*: You may need more than one expression.
2. Under what circumstances would your new expression produce an error message where the original one would not?

2.10.16 Removing unneeded files

Suppose you want to delete your processed data files, and only keep your raw files and processing script to save storage. The raw files end in `.txt` and the processed files end in `.csv`. Which of the following would remove all the processed data files, and *only* the processed data files?

1. `rm ?.csv`
2. `rm *.csv`
3. `rm * .csv`
4. `rm *.*`

2.10.17 Other wildcards

The shell provides several wildcards beyond the widely used `*`. To explore them, explain in plain language what (hypothetical) files the expression `novel-????-[ab]*.{txt,pdf}` matches and why.

2.11 Key Points

- A **shell** is a program that reads commands and runs other programs.
- The **filesystem** manages information stored on disk.
- Information is stored in files, which are located in directories (folders).
- Directories can also store other directories, which forms a directory tree.
- `pwd` prints the user's **current working directory**.
- `/` on its own is the **root directory** of the whole filesystem.
- `ls` prints a list of files and directories.
- An **absolute path** specifies a location from the root of the filesystem.
- A **relative path** specifies a location in the filesystem starting from the current directory.
- `cd` changes the current working directory.
- `..` means the **parent directory**.
- `.` on its own means the current directory.
- `mkdir` creates a new directory.
- `cp` copies a file.
- `rm` removes (deletes) a file.
- `mv` moves (renames) a file or directory.
- `*` matches zero or more characters in a filename.
- `?` matches any single character in a filename.
- `wc` counts lines, words, and characters in its inputs.
- `man` displays the manual page for a given command; some commands also have a `--help` option.

3

Building Tools with the Unix Shell

> Wisdom comes from experience. Experience is often a result of lack of wisdom.
>
> — Terry Pratchett

The shell's greatest strength is that it lets us combine programs to create pipelines that can handle large volumes of data. This lesson shows how to do that, and how to repeat commands to process as many files as we want automatically.

We'll be continuing to work in the `zipf` project, which should contain the following files after the previous chapter:

```
zipf/
└── data
    ├── README.md
    ├── dracula.txt
    ├── frankenstein.txt
    ├── jane_eyre.txt
    ├── moby_dick.txt
    ├── sense_and_sensibility.txt
    ├── sherlock_holmes.txt
    └── time_machine.txt
```

3.1 Combining Commands

To see how the shell lets us combine commands, let's go into the `zipf/data` directory and count the number of lines in each file once again:

```
$ cd ~/zipf/data
$ wc -l *.txt
```

```
  15975 dracula.txt
   7832 frankenstein.txt
  21054 jane_eyre.txt
  22331 moby_dick.txt
  13028 sense_and_sensibility.txt
  13053 sherlock_holmes.txt
   3582 time_machine.txt
  96855 total
```

Which of these books is shortest? We can check by eye when there are only 16 files, but what if there were eight thousand?

Our first step toward a solution is to run this command:

```
$ wc -l *.txt > lengths.txt
```

The greater-than symbol `>` tells the shell to **redirect** the command's output to a file instead of printing it. Nothing appears on the screen; instead, everything that would have appeared has gone into the file `lengths.txt`. The shell creates this file if it doesn't exist, or overwrites it if it already exists.

We can print the contents of `lengths.txt` using `cat`, which is short for con**cat**enate (because if we give it the names of several files it will print them all in order):

```
$ cat lengths.txt
```

```
  15975 dracula.txt
   7832 frankenstein.txt
  21054 jane_eyre.txt
  22331 moby_dick.txt
  13028 sense_and_sensibility.txt
  13053 sherlock_holmes.txt
   3582 time_machine.txt
  96855 total
```

We can now use `sort` to sort the lines in this file:

```
$ sort lengths.txt -n
```

```
   3582 time_machine.txt
   7832 frankenstein.txt
  13028 sense_and_sensibility.txt
  13053 sherlock_holmes.txt
  15975 dracula.txt
  21054 jane_eyre.txt
  22331 moby_dick.txt
  96855 total
```

Just to be safe, we use `sort`'s `-n` option to specify that we want to sort numerically. Without it, `sort` would order things alphabetically so that `10` would come before `2`.

`sort` does not change `lengths.txt`. Instead, it sends its output to the screen just as `wc` did. We can therefore put the sorted list of lines in another temporary file called `sorted-lengths.txt` using `>` once again:

```
$ sort lengths.txt > sorted-lengths.txt
```

Redirecting to the Same File

It's tempting to send the output of `sort` back to the file it reads:

```
$ sort -n lengths.txt > lengths.txt
```

However, all this does is wipe out the contents of `lengths.txt`. The reason is that when the shell sees the redirection, it opens the file on the right of the `>` for writing, which erases anything that file contained. It then runs `sort`, which finds itself reading from a newly empty file.

Creating intermediate files with names like `lengths.txt` and `sorted-lengths.txt` works, but keeping track of those files and cleaning them up when they're no longer needed is a burden. Let's delete the two files we just created:

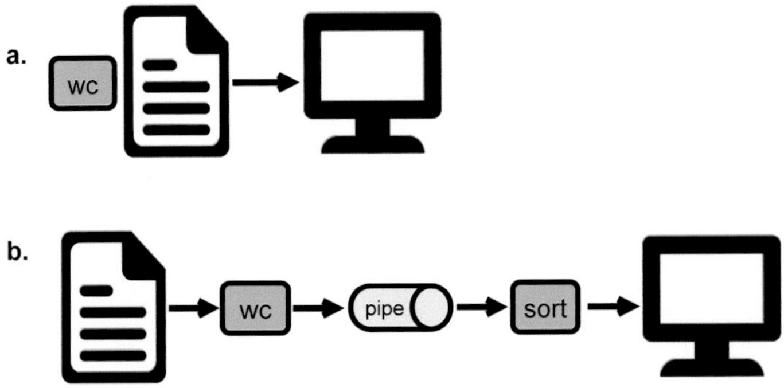

FIGURE 3.1: Piping commands.

```
rm lengths.txt sorted-lengths.txt
```

We can produce the same result more safely and with less typing using a **pipe**:

```
$ wc -l *.txt | sort -n
```

```
   3582 time_machine.txt
   7832 frankenstein.txt
  13028 sense_and_sensibility.txt
  13053 sherlock_holmes.txt
  15975 dracula.txt
  21054 jane_eyre.txt
  22331 moby_dick.txt
  96855 total
```

The vertical bar `|` between the `wc` and `sort` commands tells the shell that we want to use the output of the command on the left as the input to the command on the right.

Running a command with a file as input has a clear flow of information: the command performs a task on that file and prints the output to the screen (Figure 3.1a). When using pipes, however, the information flows differently after the first (upstream) command. The downstream command doesn't read from a file. Instead, it reads the output of the upstream command (Figure 3.1b).

We can use `|` to build pipes of any length. For example, we can use the command `head` to get just the first three lines of sorted data, which shows us the three shortest books:

```
$ wc -l *.txt | sort -n | head -n 3
```

```
 3582 time_machine.txt
 7832 frankenstein.txt
13028 sense_and_sensibility.txt
```

Options Can Have Values

When we write `head -n 3`, the value 3 is not input to `head`. Instead, it is associated with the option `-n`. Many options take values like this, such as the names of input files or the background color to use in a plot. Some versions of `head` may allow you to use `head -3` as a shortcut, though this can be confusing if other options are included.

We could always redirect the output to a file by adding `> shortest.txt` to the end of the pipeline, thereby retaining our answer for later reference.

In practice, most Unix users would create this pipeline step by step, just as we have: by starting with a single command and adding others one by one, checking the output after each change. The shell makes this easy by letting us move up and down in our **command history** with the ↑ and ↓ keys. We can also edit old commands to create new ones, so a very common sequence is:

- Run a command and check its output.
- Use ↑ to bring it up again.
- Add the pipe symbol `|` and another command to the end of the line.
- Run the pipe and check its output.
- Use ↑ to bring it up again.
- And so on.

3.2 How Pipes Work

In order to use pipes and redirection effectively, we need to know a little about how they work. When a computer runs a program—any program—it creates a **process** in memory to hold the program's instructions and data. Every process in Unix has an input channel called **standard input** and an output channel called **standard output**. (By now you may be surprised that their names are so memorable, but don't worry: most Unix programmers call them "stdin" and "stdout," which are pronounced "stuh-Din" and "stuh-Dout").

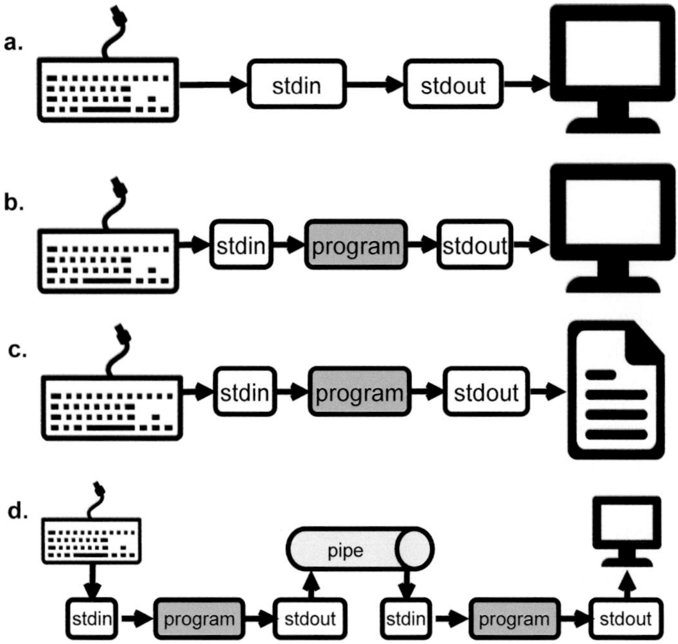

FIGURE 3.2: Standard I/O.

The shell is a program like any other, and like any other, it runs inside a process. Under normal circumstances its standard input is connected to our keyboard and its standard output to our screen, so it reads what we type and displays its output for us to see (Figure 3.2a). When we tell the shell to run a program, it creates a new process and temporarily reconnects the keyboard and stream to that process's standard input and output (Figure 3.2b).

If we provide one or more files for the command to read, as with `sort lengths.txt`, the program reads data from those files. If we don't provide any filenames, though, the Unix convention is for the program to read from standard input. We can test this by running `sort` on its own, typing in a few lines of text, and then pressing Ctrl+D to signal the end of input . `sort` will then sort and print whatever we typed:

```
$ sort
one
two
three
four
^D
```

```
four
one
three
two
```

Redirection with > tells the shell to connect the program's standard output to a file instead of the screen (Figure 3.2c).

When we create a pipe like `wc *.txt | sort`, the shell creates one process for each command so that `wc` and `sort` will run simultaneously, and then connects the standard output of `wc` directly to the standard input of `sort` (Figure 3.2d).

`wc` doesn't know whether its output is going to the screen, another program, or to a file via `>`. Equally, `sort` doesn't know if its input is coming from the keyboard or another process; it just knows that it has to read, sort, and print.

Why Isn't It Doing Anything?

What happens if a command is supposed to process a file but we don't give it a filename? For example, what if we type:

```
$ wc -l
```

but don't type `*.txt` (or anything else) after the command? Since `wc` doesn't have any filenames, it assumes it is supposed to read from the keyboard, so it waits for us to type in some data. It doesn't tell us this; it just sits and waits.

This mistake can be hard to spot, particularly if we put the filename at the end of the pipeline:

```
$ wc -l | sort moby_dick.txt
```

In this case, `sort` ignores standard input and reads the data in the file, but `wc` still just sits there waiting for input.

If we make this mistake, we can end the program by typing Ctrl+C. We can also use this to interrupt programs that are taking a long time to run or are trying to connect to a website that isn't responding.

Just as we can redirect standard output with `>`, we can connect standard input to a file using `<`. In the case of a single file, this has the same effect as providing the file's name to the command:

```
$ wc < moby_dick.txt
```

```
   22331  215832 1276222
```

If we try to use redirection with a wildcard, though, the shell *doesn't* concatenate all of the matching files:

```
$ wc < *.txt
```

```
-bash: *.txt: ambiguous redirect
```

It also doesn't print the error message to standard output, which we can prove by redirecting:

```
$ wc < *.txt > all.txt
```

```
-bash: *.txt: ambiguous redirect
```

```
$ cat all.txt
```

```
cat: all.txt: No such file or directory
```

Instead, every process has a second output channel called **standard error** (or **stderr**). Programs use it for error messages so that their attempts to tell us something has gone wrong don't vanish silently into an output file. There are ways to redirect standard error, but doing so is almost always a bad idea.

3.3 Repeating Commands on Many Files

A loop is a way to repeat a set of commands for each item in a list. We can use them to build complex workflows out of simple pieces, and (like wildcards) they reduce the typing we have to do and the number of mistakes we might make.

Let's suppose that we want to take a section out of each book whose name starts with the letter "s" in the `data` directory. More specifically, suppose we want to get the first 8 lines of each book *after* the 9 lines of license information that appear at the start of the file. If we only cared about one file, we could write a pipeline to take the first 17 lines and then take the last 8 of those:

```
$ head -n 17 sense_and_sensibility.txt | tail -n 8
```

```
Title: Sense and Sensibility

Author: Jane Austen
Editor:
Release Date: May 25, 2008 [EBook #161]
Posting Date:
Last updated: February 11, 2015
Language: English
```

If we try to use a wildcard to select files, we only get 8 lines of output, not the 16 we expect:

```
$  head -n 17 s*.txt | tail -n 8
```

```
Title: The Adventures of Sherlock Holmes

Author: Arthur Conan Doyle
Editor:
Release Date: April 18, 2011 [EBook #1661]
Posting Date: November 29, 2002
Latest Update:
Language: English
```

The problem is that `head` is producing a single stream of output containing 17 lines for each file (along with a header telling us the file's name):

```
$ head -n 17 s*.txt


==> sense_and_sensibility.txt <==
The Project Gutenberg EBook of Sense and Sensibility, by ...

This eBook is for the use of anyone anywhere at no cost and with
almost no restrictions whatsoever.  You may copy it, give it ...
re-use it under the terms of the Project Gutenberg License ...
with this eBook or online at www.gutenberg.net



Title: Sense and Sensibility

Author: Jane Austen
Editor:
Release Date: May 25, 2008 [EBook #161]
Posting Date:
Last updated: February 11, 2015
Language: English

==> sherlock_holmes.txt <==
Project Gutenberg's The Adventures of Sherlock Holmes, by Arthur
Conan Doyle

This eBook is for the use of anyone anywhere at no cost and ...
almost no restrictions whatsoever.  You may copy it, give ...
re-use it under the terms of the Project Gutenberg License ...
with this eBook or online at www.gutenberg.net



Title: The Adventures of Sherlock Holmes

Author: Arthur Conan Doyle
Editor:
Release Date: April 18, 2011 [EBook #1661]
Posting Date: November 29, 2002
Latest Update:
Language: English
```

Let's try this instead:

```
$ for filename in sense_and_sensibility.txt sherlock_holmes.txt
> do
>   head -n 17 $filename | tail -n 8
> done
```

```
Title: Sense and Sensibility

Author: Jane Austen
Editor:
Release Date: May 25, 2008 [EBook #161]
Posting Date:
Last updated: February 11, 2015
Language: English
Title: The Adventures of Sherlock Holmes

Author: Arthur Conan Doyle
Editor:
Release Date: April 18, 2011 [EBook #1661]
Posting Date: November 29, 2002
Latest Update:
Language: English
```

As the output shows, the loop runs our pipeline once for each file. There is a lot going on here, so we will break it down into pieces:

1. The keywords `for`, `in`, `do`, and `done` create the loop, and must always appear in that order.
2. `filename` is a variable just like a variable in Python. At any moment it contains a value, but that value can change over time.
3. The loop runs once for each item in the list. Each time it runs, it assigns the next item to the variable. In this case, `filename` will be `sense_and_sensibility.txt` the first time around the loop and `sherlock_holmes.txt` the second time.
4. The commands that the loop executes are called the **body** of the loop and appear between the keywords `do` and `done`. Those commands use the current value of the variable `filename`, but to get it, we must put a dollar sign `$` in front of the variable's name. If we forget and use `filename` instead of `$filename`, the shell will think that we are referring to a file that is actually called `filename`.

5. The shell prompt changes from `$` to a **continuation prompt** `>` as we type in our loop to remind us that we haven't finished typing a complete command yet. We don't type the `>`, just as we don't type the `$`. The continuation prompt `>` has nothing to do with redirection; it's used because there are only so many punctuation symbols available.

Continuation Prompts May Differ Too

As mentioned in Chapter 2, there is variation in how different shells look and operate. If you noticed the second, third, and fourth code lines in your for loop were prefaced with `for`, it's not because you did something wrong! That difference is one of the ways in which `zsh` differs from `bash`.

It is very common to use a wildcard to select a set of files and then loop over that set to run commands:

```
$ for filename in s*.txt
> do
>   head -n 17 $filename | tail -n 8
> done
```

```
Title: Sense and Sensibility

Author: Jane Austen
Editor:
Release Date: May 25, 2008 [EBook #161]
Posting Date:
Last updated: February 11, 2015
Language: English



Title: The Adventures of Sherlock Holmes

Author: Arthur Conan Doyle
Editor:
Release Date: April 18, 2011 [EBook #1661]
Posting Date: November 29, 2002
Latest Update:
Language: English
```

3.4 Variable Names

We should always choose meaningful names for variables, but we should remember that those names don't mean anything to the computer. For example, we have called our loop variable `filename` to make its purpose clear to human readers, but we could equally well write our loop as:

```
$ for x in s*.txt
> do
>   head -n 17 $x | tail -n 8
> done
```

or as:

```
$ for username in s*.txt
> do
>   head -n 17 $username | tail -n 8
> done
```

Don't do this. Programs are only useful if people can understand them, so meaningless names like `x` and misleading names like `username` increase the odds of misunderstanding.

3.5 Redoing Things

Loops are useful if we know in advance what we want to repeat, but we can also repeat commands that we have run recently. One way is to use ↑ and ↓ to go up and down in our command history as described earlier. Another is to use `history` to get a list of the last few hundred commands we have run:

```
$ history
```

```
  551  wc -l *.txt | sort -n
  552  wc -l *.txt | sort -n | head -n 3
  553  wc -l *.txt | sort -n | head -n 1 > shortest.txt
```

We can use an exclamation mark `!` followed by a number to repeat a recent command:

```
$ !552
```

```
wc -l *.txt | sort -n | head -n 3
```

```
   3582 time_machine.txt
   7832 frankenstein.txt
  13028 sense_and_sensibility.txt
```

The shell prints the command it is going to re-run to standard error before executing it, so that (for example) `!572 > results.txt` puts the command's output in a file *without* also writing the command to the file.

Having an accurate record of the things we have done and a simple way to repeat them are two of the main reasons people use the Unix shell. In fact, being able to repeat history is such a powerful idea that the shell gives us several ways to do it:

- `!head` re-runs the most recent command starting with `head`, while `!wc` re-runs the most recent starting with `wc`.
- If we type Ctrl+R (for **r**everse search) the shell searches backward through its history for whatever we type next. If we don't like the first thing it finds, we can type Ctrl+R again to go further back.

If we use `history`, ↑, or Ctrl+R we will quickly notice that loops don't have to be broken across lines. Instead, their parts can be separated with semi-colons:

```
$ for filename in s*.txt; do head -n 17 $filename | tail -n 8;
done
```

This is fairly readable, though it becomes more challenging if our for loop includes multiple commands. For example, we may choose to include the `echo` command, which prints its arguments to the screen, so we can keep track of progress or for debugging. Compare this:

```
$ for filename in s*.txt
> do
>   echo $filename
>   head -n 17 $filename | tail -n 8
> done
```

with this:

```
$ for filename in s*.txt; do echo $filename; head -n 17 $filename
  | tail -n 8; done
```

Even experienced users have a tendency to (incorrectly) put the semi-colon after `do` instead of before it. If our loop contains multiple commands, though, the multi-line format is much easier to read and troubleshoot. Note that (depending on the size of your shell window) the format separated by semi-colons may be printed onto more than one line, as shown in the previous code example. You can tell whether code entered into your shell is intended to be run as a single line based on the prompt: both the original command prompt (`$`) and the continuation prompt (`>`) indicate the code is on separate lines; the absence of either in shell commands indicates it is a single line of code.

3.6 Creating New Filenames Automatically

Suppose we want to create a backup copy of each book whose name ends in "e." If we don't want to change the files' names, we can do this with `cp`:

```
$ cd ~/zipf
$ mkdir backup
$ cp data/*e.txt backup
$ ls backup
```

```
jane_eyre.txt  time_machine.txt
```

Warnings

If you attempt to re-execute the code chunk above, you'll end up with an error after the second line:

```
mkdir: backup: File exists
```

This warning isn't necessarily a cause for alarm. It lets you know that the command couldn't be completed, but will not prevent you from proceeding.

But what if we want to append the extension `.bak` to the files' names? `cp` can do this for a single file:

```
$ cp data/time_machine.txt backup/time_machine.txt.bak
```

but not for all the files at once:

```
$ cp data/*e.txt backup/*e.txt.bak
```

```
cp: target 'backup/*e.txt.bak' is not a directory
```

`backup/*e.txt.bak` doesn't match anything—those files don't yet exist—so after the shell expands the `*` wildcards, what we are actually asking `cp` to do is:

```
$ cp data/jane_eyre.txt data/time_machine.txt backup/*e.bak
```

This doesn't work because `cp` only understands how to do two things: copy a single file to create another file, or copy a bunch of files into a directory. If we give it more than two names as arguments, it expects the last one to be a directory. Since `backup/*e.bak` is not, `cp` reports an error.

Instead, let's use a loop to copy files to the backup directory and append the `.bak` suffix:

```
$ cd data
$ for filename in *e.txt
> do
>   cp $filename ../backup/$filename.bak
> done
$ ls ../backup
```

```
jane_eyre.txt.bak  time_machine.txt.bak
```

3.7 Summary

The shell's greatest strength is the way it combines a few powerful ideas with pipes and loops. The next chapter will show how we can make our work more reproducible by saving commands in files that we can run over and over again.

3.8 Exercises

The exercises below involve creating and moving new files, as well as considering hypothetical files. Please note that if you create or move any files or directories in your Zipf's Law project, you may want to reorganize your files following the outline at the beginning of the next chapter. If you accidentally delete necessary files, you can start with a fresh copy of the data files by following the instructions in Section 1.2.

3.8.1 What does >> mean?

We have seen the use of `>`, but there is a similar operator `>>` which works slightly differently. We'll learn about the differences between these two operators by printing some strings. We can use the `echo` command to print strings as shown below:

```
$ echo The echo command prints text
```

```
The echo command prints text
```

Now test the commands below to reveal the difference between the two operators:

```
$ echo hello > testfile01.txt
```

and:

```
$ echo hello >> testfile02.txt
```

Hint: Try executing each command twice in a row and then examining the output files.

3.8.2 Appending data

Given the following commands, what will be included in the file `extracted.txt`:

```
$ head -n 3 dracula.txt > extracted.txt
$ tail -n 2 dracula.txt >> extracted.txt
```

1. The first three lines of `dracula.txt`
2. The last two lines of `dracula.txt`
3. The first three lines and the last two lines of `dracula.txt`
4. The second and third lines of `dracula.txt`

3.8.3 Piping commands

In our current directory, we want to find the 3 files which have the least number of lines. Which command listed below would work?

1. `wc -l * > sort -n > head -n 3`
2. `wc -l * | sort -n | head -n 1-3`
3. `wc -l * | head -n 3 | sort -n`
4. `wc -l * | sort -n | head -n 3`

3.8.4 Why does `uniq` only remove adjacent duplicates?

The command `uniq` removes adjacent duplicated lines from its input. Consider a hypothetical file `genres.txt` containing the following data:

```
science fiction
fantasy
science fiction
fantasy
science fiction
science fiction
```

Running the command `uniq genres.txt` produces:

```
science fiction
fantasy
science fiction
fantasy
science fiction
```

Why do you think `uniq` only removes *adjacent* duplicated lines? (Hint: think about very large datasets.) What other command could you combine with it in a pipe to remove all duplicated lines?

3.8.5 Pipe reading comprehension

A file called `titles.txt` contains a list of book titles and publication years:

```
Dracula,1897
Frankenstein,1818
Jane Eyre,1847
Moby Dick,1851
Sense and Sensibility,1811
The Adventures of Sherlock Holmes,1892
The Invisible Man,1897
The Time Machine,1895
Wuthering Heights,1847
```

What text passes through each of the pipes and the final redirect in the pipeline below?

```
$ cat titles.txt | head -n 5 | tail -n 3 | sort -r > final.txt
```

Hint: build the pipeline up one command at a time to test your understanding

3.8.6 Pipe construction

For the file `titles.txt` from the previous exercise, consider the following command:

```
$ cut -d , -f 2 titles.txt
```

What does the `cut` command (and its options) accomplish?

3.8.7 Which pipe?

Consider the same `titles.txt` from the previous exercises.

The `uniq` command has a `-c` option which gives a count of the number of times a line occurs in its input. If `titles.txt` was in your working directory, what command would you use to produce a table that shows the total count of each publication year in the file?

1. `sort titles.txt | uniq -c`
2. `sort -t, -k2,2 titles.txt | uniq -c`
3. `cut -d, -f 2 titles.txt | uniq -c`
4. `cut -d, -f 2 titles.txt | sort | uniq -c`
5. `cut -d, -f 2 titles.txt | sort | uniq -c | wc -l`

3.8.8 Doing a dry run

A loop is a way to do many things at once—or to make many mistakes at once if it does the wrong thing. One way to check what a loop *would* do is to `echo` the commands it would run instead of actually running them.

Suppose we want to preview the commands the following loop will execute without actually running those commands (`analyze` is a hypothetical command):

```
$ for file in *.txt
> do
>   analyze $file > analyzed-$file
> done
```

What is the difference between the two loops below, and which one would we want to run?

```
$ for file in *.txt
> do
>   echo analyze $file > analyzed-$file
> done
```

or:

```
$ for file in *.txt
> do
>   echo "analyze $file > analyzed-$file"
> done
```

3.8.9 Variables in loops

Given the files in `data/`, what is the output of the following code?

```
$ for datafile in *.txt
> do
>    ls *.txt
> done
```

Now, what is the output of the following code?

```
$ for datafile in *.txt
> do
>   ls $datafile
> done
```

Why do these two loops give different outputs?

3.8.10 Limiting sets of files

What would be the output of running the following loop in your `data/` directory?

```
$ for filename in d*
> do
>     ls $filename
> done
```

How would the output differ from using this command instead?

```
$ for filename in *d*
> do
>     ls $filename
> done
```

3.8.11 Saving to a file in a loop

Consider running the following loop in the `data/` directory:

```
for book in *.txt
> do
>       echo $book
>       head -n 16 $book > headers.txt
> done
```

Why would the following loop be preferable?

```
for book in *.txt
> do
>       head -n 16 $book >> headers.txt
> done
```

3.8.12 Why does `history` record commands before running them?

If you run the command:

```
$ history | tail -n 5 > recent.sh
```

the last command in the file is the `history` command itself, i.e., the shell has added `history` to the command log before actually running it. In fact, the shell *always* adds commands to the log before running them. Why do you think it does this?

3.9 Key Points

- `cat` displays the contents of its inputs.
- `head` displays the first few lines of its input.
- `tail` displays the last few lines of its input.
- `sort` sorts its inputs.
- Use the up-arrow key to scroll up through previous commands to edit and repeat them.
- Use `history` to display recent commands and `!number` to repeat a command by number.
- Every process in Unix has an input channel called **standard input** and an output channel called **standard output**.
- > redirects a command's output to a file, overwriting any existing content.
- >> appends a command's output to a file.
- < operator redirects input to a command.
- A **pipe** | sends the output of the command on the left to the input of the command on the right.
- A `for` loop repeats commands once for every thing in a list.
- Every `for` loop must have a variable to refer to the thing it is currently operating on and a **body** containing commands to execute.
- Use `$name` or `${name}` to get the value of a variable.

4

Going Further with the Unix Shell

> There isn't a way things should be. There's just what happens, and what we do.
>
> — Terry Pratchett

The previous chapters explained how we can use the command line to do all of the things we can do with a GUI, and how to combine commands in new ways using pipes and redirection. This chapter extends those ideas to show how we can create new tools by saving commands in files and how to use a more powerful version of **wildcards** to extract data from files.

We'll be continuing to work in the `zipf` project, which after the previous chapter should contain the following files:

```
zipf/
└── data
    ├── README.md
    ├── dracula.txt
    ├── frankenstein.txt
    ├── jane_eyre.txt
    ├── moby_dick.txt
    ├── sense_and_sensibility.txt
    ├── sherlock_holmes.txt
    └── time_machine.txt
```

Deleting Extra Files

You may have additional files if you worked through all of the exercises in the previous chapter. Feel free to delete them or move them to a separate directory. If you have accidentally deleted files you need, you can download them again by following the instructions in Section 1.2.

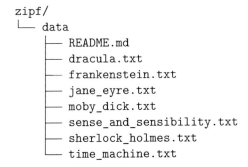

4.1 Creating New Commands

Loops let us run the same commands many times, but we can go further and save commands in files so that we can repeat complex operations with a few keystrokes. For historical reasons, a file full of shell commands is usually called a **shell script**, but it is really just another kind of program.

Let's start by creating a new directory for our runnable programs called `bin/`, consistent with the project structure described in Section 1.1.2.

```
$ cd ~/zipf
$ mkdir bin
$ cd bin
```

Edit a new file called `book_summary.sh` to hold our shell script:

```
$ nano book_summary.sh
```

and insert this line:

```
head -n 17 ../data/moby_dick.txt | tail -n 8
```

Note that we do *not* put the `$` prompt at the front of the line. We have been showing that to highlight interactive commands, but in this case we are putting the command in a file rather than running it immediately.

Empty Line at the End of a File?

You'll often see scripts from many languages that end in an empty line. What you are seeing, though, is the last line of code ending in a newline character. This indicates to the computer that the code has ended. While this newline character is not *required* for shell scripts to work, and sometimes isn't shown by coding tools, it does make it easier to view and modify scripts. When you are copying code from this book, remember to add an empty line at the end!

Once we have added this line, we can save the file with Ctrl+O and exit with Ctrl+X. `ls` shows that our file now exists:

```
$ ls
```

```
book_summary.sh
```

We can check the contents of the file using `cat book_summary.sh`. More importantly, we can now ask the shell to run this file:

```
$ bash book_summary.sh
```

```
Title: Moby Dick
       or The Whale
Author: Herman Melville
Editor:
Release Date: December 25, 2008 [EBook #2701]
Posting Date:
Last Updated: December 3, 2017
Language: English
```

Sure enough, our script's output is exactly the same text we would get if we ran the command directly. If we want, we can pipe the output of our shell script to another command to count how many lines it contains:

```
$ bash book_summary.sh | wc -l
```

```
     8
```

What if we want our script to print the name of the book's author? The command `grep` finds and prints lines that match a pattern. We'll learn more about `grep` in Section 4.4, but for now we can edit the script:

```
$ nano book_summary.sh
```

and add a search for the word "Author":

```
head -n 17 ../data/moby_dick.txt | tail -n 8 | grep Author
```

Sure enough, when we run our modified script:

```
$ bash book_summary.sh
```

we get the line we want:

```
Author: Herman Melville
```

And once again we can pipe the output of our script into other commands just as we would pipe the output from any other program. Here, we count the number of words in the author line:

```
$ bash book_summary.sh | wc -w
```

```
       3
```

4.2 Making Scripts More Versatile

Getting the name of the author for only one of the books isn't particularly useful. What we really want is a way to get the name of the author from any of our files. Let's edit `book_summary.sh` again and replace `../data/moby_dick.txt` with a special variable `$1`. Once our change is made, `book_summary.sh` should contain:

```
head -n 17 $1 | tail -n 8 | grep Author
```

Inside a shell script, `$1` means "the first argument on the command line." If we now run our script like this:

```
$ bash book_summary.sh ../data/moby_dick.txt
```

then `$1` is assigned `../data/moby_dick.txt` and we get exactly the same output as before. If we give the script a different filename:

```
$  bash book_summary.sh ../data/frankenstein.txt
```

we get the name of the author of that book instead:

```
Author: Mary Wollstonecraft (Godwin) Shelley
```

Our small script is now doing something useful, but it may take the next person who reads it a moment to figure out exactly what. We can improve our script by adding **comments** at the top:

```
# Get author information from a Project Gutenberg eBook.
# Usage: bash book_summary.sh /path/to/file.txt
head -n 17 $1 | tail -n 8 | grep Author
```

As in Python, a comment starts with a `#` character and runs to the end of the line. The computer ignores comments, but they help people (including our future self) understand and use what we've created.

Let's make one more change to our script. Instead of always extracting the author name, let's have it select whatever information the user specified:

```
# Get desired information from a Project Gutenberg eBook.
# Usage: bash book_summary.sh /path/to/file.txt what_to_look_for
head -n 17 $1 | tail -n 8 | grep $2
```

The change is very small: we have replaced the fixed string 'Author' with a reference to the special variable `$2`, which is assigned the value of the second command-line argument we give the script when we run it.

Update Your Comments

As you update the code in your script, don't forget to update the comments that describe the code. A description that sends readers in the wrong direction is worse than none at all, so do your best to avoid this common oversight.

Let's check that it works by asking for *Frankenstein*'s release date:

```
$ bash book_summary.sh ../data/frankenstein.txt Release
```

```
Release Date: June 17, 2008 [EBook #84]
```

4.3 Turning Interactive Work into a Script

Suppose we have just run a series of commands that did something useful, such as summarizing all books in a given directory:

```
$ for x in ../data/*.txt
> do
>   echo $x
>   bash book_summary.sh $x Author
> done > authors.txt
$ for x in ../data/*.txt
> do
>   echo $x
>   bash book_summary.sh $x Release
> done > releases.txt
$ ls
```

```
authors.txt         book_summary.sh       releases.txt
```

```
$ mkdir ../results
$ mv authors.txt releases.txt ../results
```

Instead of typing those commands into a file in an editor (and potentially getting them wrong) we can use `history` and redirection to save recent commands to a file. For example, we can save the last six commands to `summarize_all_books.sh`:

```
$ history 6 > summarize_all_books.sh
$ cat summarize_all_books.sh
```

```
297 for x in ../data/*.txt; do echo $x;
  bash book_summary.sh $x Author; done > authors.txt
298 for x in ../data/*.txt; do echo $x;
  bash book_summary.sh $x Release; done > releases.txt
299 ls
300 mkdir ../results
301 mv authors.txt releases.txt ../results
302 history 6 > summarize_all_books.sh
```

We can now open the file in an editor, remove the serial numbers at the start of each line, and delete the lines we don't want to create a script that captures exactly what we actually did. This is how we usually develop shell scripts: run commands interactively a few times to make sure they are doing the right thing, then save our recent history to a file and turn that into a reusable script.

4.4 Finding Things in Files

We can use `head` and `tail` to select lines from a file by position, but we also often want to select lines that contain certain values. This is called **filtering**, and we usually do it in the shell with the command `grep` that we briefly met in Section 4.1. Its name is an acronym of "global regular expression print," which was a common sequence of operations in early Unix text editors.

To show how `grep` works, we will use our sleuthing skills to explore `data/sherlock_holmes.txt`. First, let's find lines that contain the word "Sherlock." Since there are likely to be hundreds of matches, we will pipe `grep`'s output to `head` to show only the first few:

```
$ cd ~/zipf
$ grep Sherlock data/sherlock_holmes.txt | head -n 5
```

Here, `Sherlock` is our (very simple) pattern. `grep` searches the file line by line and shows those lines that contain matches, so the output is:

```
Project Gutenberg's The Adventures of Sherlock Holmes, by Arthur
Conan Doyle
Title: The Adventures of Sherlock Holmes
To Sherlock Holmes she is always THE woman. I have seldom heard
as I had pictured it from Sherlock Holmes' succinct description,
"Good-night, Mister Sherlock Holmes."
```

If we run `grep sherlock` instead, we get no output because `grep` patterns are case-sensitive. If we wanted to make the search case-insensitive, we can add the option `-i`:

```
$ grep -i sherlock data/sherlock_holmes.txt | head -n 5
```

```
Project Gutenberg's The Adventures of Sherlock Holmes, by Arthur
Conan Doyle
Title: The Adventures of Sherlock Holmes
*** START OF THIS PROJECT GUTENBERG EBOOK THE ADVENTURES OF
SHERLOCK HOLMES ***
THE ADVENTURES OF SHERLOCK HOLMES
To Sherlock Holmes she is always THE woman. I have seldom heard
```

This output is different from our previous output because of the lines containing "SHERLOCK" near the top of the file.

Next, let's search for the pattern `on`:

```
$ grep on data/sherlock_holmes.txt | head -n 5
```

```
Project Gutenberg's The Adventures of Sherlock Holmes, by Arthur
Conan Doyle
This eBook is for the use of anyone anywhere at no cost and with
almost no restrictions whatsoever.  You may copy it, give it away
or with this eBook or online at www.gutenberg.net
Author: Arthur Conan Doyle
```

In each of these lines, our pattern ("on") is part of a larger word such as "Conan." To restrict matching to lines containing `on` by itself, we can give `grep` the `-w` option (for "match words"):

```
$ grep -w on data/sherlock_holmes.txt | head -n 5
```

```
One night--it was on the twentieth of March, 1888--I was
put on seven and a half pounds since I saw you."
that I had a country walk on Thursday and came home in a dreadful
"It is simplicity itself," said he; "my eyes tell me that on the
on the right side of his top-hat to show where he has secreted
```

What if we want to search for a phrase rather than a single word?

```
$ grep on the data/sherlock_holmes.txt | head -n 5
```

```
grep: the: No such file or directory
data/sherlock_holmes.txt:Project Gutenberg's The Adventures of
  Sherlock Holmes, by Arthur Conan Doyle
data/sherlock_holmes.txt:This eBook is for the use of anyone
  anywhere at no cost and with
data/sherlock_holmes.txt:almost no restrictions whatsoever.
  You may copy it, give it away or
data/sherlock_holmes.txt:with this eBook or online at
  www.gutenberg.net
data/sherlock_holmes.txt:Author: Arthur Conan Doyle
```

In this case, `grep` uses `on` as the pattern and tries to find it in files called `the` and `data/sherlock_holmes.txt`. It then tells us that the file `the` cannot be found, but prints `data/sherlock_holmes.txt` as a prefix to each other line of output to tell us which file those lines came from. If we want to give `grep` both words as a single argument, we must wrap them in quotation marks as before:

```
$ grep "on the" data/sherlock_holmes.txt | head -n 5
```

```
One night--it was on the twentieth of March, 1888--I was
drug-created dreams and was hot upon the scent of some new
"It is simplicity itself," said he; "my eyes tell me that on the
on the right side of his top-hat to show where he has secreted
pink-tinted note-paper which had been lying open upon the table.
```

Quoting

Quotation marks aren't specific to `grep`: the shell interprets them before running commands, just as it expands wildcards to create filenames no matter what command those filenames are being passed to. This allows us to do things like `head -n 5 "My Thesis.txt"` to get lines from a file that has a space in its name. It is also why many programmers write `"$variable"` instead of just `$variable` when creating loops or shell scripts: if there's any chance at all that the variable's value will contain spaces, it's safest to put it in quotes.

One of the most useful options for `grep` is `-n`, which numbers the lines that match the search:

```
$ grep -n "on the" data/sherlock_holmes.txt | head -n 5
```

```
105:One night--it was on the twentieth of March, 1888--I was
118:drug-created dreams and was hot upon the scent of some new
155:"It is simplicity itself," said he; "my eyes tell me ...
165:on the right side of his top-hat to show where he has ...
198:pink-tinted note-paper which had been lying open upon ...
```

`grep` has many options—so many, in fact, that almost every letter of the alphabet means something to it:

```
$ man grep
```

```
GREP(1)              BSD General Commands Manual             GREP(1)

NAME
     grep, egrep, fgrep, zgrep, zegrep, zfgrep -- file pattern
      searcher

SYNOPSIS
     grep [-abcdDEFGHhIiJLlmnOopqRSsUVvwxZ] [-A num] [-B num]
          [-C[num]] [-e pattern] [-f file] [--binary-files=value]
          [--color[=when]] [--colour[=when]] [--context[=num]]
          [--label] [--line-buffered]
          [--null] [pattern] [file ...]
...more...
```

We can combine options to `grep` as we do with other Unix commands. For example, we can combine two options we've covered previously with `-v` to invert the match—i.e., to print lines that *don't* match the pattern:

```
$ grep -i -n -v the data/sherlock_holmes.txt | head -n 5
```

```
2:
4:almost no restrictions whatsoever. You may copy it, give ...
6:with this eBook or online at www.gutenberg.net
7:
8:
```

As we learned in Section 2.2, we can write this command as `grep -inv`, but probably shouldn't for the sake of readability.

If we want to search several files at once, all we have to do is give `grep` all of their names. The easiest way to do this is usually to use wildcards. For example, this command counts how many lines contain "pain" in all of our books:

```
$ grep -w pain data/*.txt | wc -l
```

```
     122
```

Alternatively, the `-r` option (for "recursive") tells `grep` to search all of the files in or below a directory:

```
$ grep -w -r pain data | wc -l
```

```
     122
```

`grep` becomes even more powerful when we start using **regular expressions**, which are sets of letters, numbers, and symbols that define complex patterns. For example, this command finds lines that start with the letter 'T':

```
$ grep -E "^T" data/sherlock_holmes.txt | head -n 5
```

```
This eBook is for the use of anyone anywhere at no cost and with
Title: The Adventures of Sherlock Holmes
THE ADVENTURES OF SHERLOCK HOLMES
To Sherlock Holmes she is always THE woman. I have seldom heard
The distinction is clear. For example, you have frequently seen
```

The `-E` option tells `grep` to interpret the pattern as a regular expression, rather than searching for an actual circumflex followed by an upper-case 'T.' The quotation marks prevent the shell from treating special characters in the pattern as wildcards, and the `^` means that a line only matches if it begins with the search term—in this case, `T`.

Many tools support regular expressions: we can use them in programming languages, database queries, online search engines, and most text editors (though not Nano—its creators wanted to keep it as small as possible). A detailed guide of regular expressions is outside the scope of this book, but a wide range of tutorials are available online, and Goyvaerts and Levithan (2012) is a useful companion if you need to go further.

4.5 Finding Files

While `grep` finds things in files, the `find` command finds files themselves. It also has a lot of options, but unlike most Unix commands they are written as full words rather than single-letter abbreviations. To show how it works, we will use the entire contents of our `zipf` directory, including files we created earlier in this chapter:

```
zipf/
├── bin
│   ├── book_summary.sh
│   ├── summarize_all_books.sh
├── data
│   ├── README.md
│   ├── dracula.txt
│   ├── frankenstein.txt
│   ├── jane_eyre.txt
│   ├── moby_dick.txt
│   ├── sense_and_sensibility.txt
│   ├── sherlock_holmes.txt
│   └── time_machine.txt
└── results
```

```
├── authors.txt
└── releases.txt
```

For our first command, let's run `find .` to find and list everything in this directory. As always, `.` on its own means the current working directory, which is where we want our search to start.

```
$ cd ~/zipf
$ find .
```

```
.
./bin
./bin/summarize_all_books.sh
./bin/book_summary.sh
./results
./results/releases.txt
./results/authors.txt
./data
./data/moby_dick.txt
./data/sense_and_sensibility.txt
./data/sherlock_holmes.txt
./data/time_machine.txt
./data/frankenstein.txt
./data/README.md
./data/dracula.txt
./data/jane_eyre.txt
```

If we only want to find directories, we can tell `find` to show us things of type `d`:

```
$ find . -type d
```

```
.
./bin
./results
./data
```

If we change `-type d` to `-type f` we get a listing of all the files instead:

```
$ find . -type f
```

```
./bin/summarize_all_books.sh
./bin/book_summary.sh
./results/releases.txt
./results/authors.txt
./data/moby_dick.txt
./data/sense_and_sensibility.txt
./data/sherlock_holmes.txt
./data/time_machine.txt
./data/frankenstein.txt
./data/README.md
./data/dracula.txt
./data/jane_eyre.txt
```

Now let's try matching by name:

```
$ find . -name "*.txt"
```

```
./results/releases.txt
./results/authors.txt
./data/moby_dick.txt
./data/sense_and_sensibility.txt
./data/sherlock_holmes.txt
./data/time_machine.txt
./data/frankenstein.txt
./data/dracula.txt
./data/jane_eyre.txt
```

Notice the quotes around `"*.txt"`. If we omit them and type:

```
$ find . -name *.txt
```

then the shell tries to expand the `*` wildcard in `*.txt` *before* running `find`. Since there aren't any text files in the current directory, the expanded list is empty, so the shell tries to run the equivalent of

```
$ find . -name
```

and gives us the error message:

```
find: -name: requires additional arguments
```

We have seen before how to combine commands using pipes. Let's use another technique to see how large our books are:

```
$ wc -l $(find . -name "*.txt")
```

```
14 ./results/releases.txt
14 ./results/authors.txt
22331 ./data/moby_dick.txt
13028 ./data/sense_and_sensibility.txt
13053 ./data/sherlock_holmes.txt
3582 ./data/time_machine.txt
7832 ./data/frankenstein.txt
15975 ./data/dracula.txt
21054 ./data/jane_eyre.txt
96883 total
```

When the shell executes our command, it runs whatever is inside `$(...)` and then replaces `$(...)` with that command's output. Since the output of `find` is the paths to our text files, the shell constructs the command:

```
$ wc -l ./results/releases.txt ... ./data/jane_eyre.txt
```

(We are using `...` in place of seven files' names in order to fit things neatly on the printed page.) This results in the output as seen above. It is exactly like expanding the wildcard in `*.txt`, but more flexible.

We will often use `find` and `grep` together. The first command finds files whose names match a pattern, while the second looks for lines inside those files that match another pattern. For example, we can look for `Authors` in all our text files:

```
$ grep "Author:" $(find . -name "*.txt")
```

```
./results/authors.txt:Author: Bram Stoker
./results/authors.txt:Author: Mary Wollstonecraft (Godwin)
Shelley
./results/authors.txt:Author: Charlotte Bronte
./results/authors.txt:Author: Herman Melville
./results/authors.txt:Author: Jane Austen
./results/authors.txt:Author: Arthur Conan Doyle
./results/authors.txt:Author: H. G. Wells
./data/moby_dick.txt:Author: Herman Melville
./data/sense_and_sensibility.txt:Author: Jane Austen
./data/sherlock_holmes.txt:Author: Arthur Conan Doyle
./data/time_machine.txt:Author: H. G. Wells
./data/frankenstein.txt:Author: Mary Wollstonecraft (Godwin)
Shelley
./data/dracula.txt:Author: Bram Stoker
./data/jane_eyre.txt:Author: Charlotte Bronte
```

We can also use `$(...)` expansion to create a list of filenames to use in a loop:

```
$ for file in $(find . -name "*.txt")
> do
> cp $file $file.bak
> done
$ find . -name "*.bak"
```

```
./results/releases.txt.bak
./results/authors.txt.bak
./data/frankenstein.txt.bak
./data/sense_and_sensibility.txt.bak
./data/dracula.txt.bak
./data/time_machine.txt.bak
./data/moby_dick.txt.bak
./data/jane_eyre.txt.bak
./data/sherlock_holmes.txt.bak
```

4.6 Configuring the Shell

As Section 3.3 explained, the shell is a program, and it has variables like any other program. Some of those variables control the shell's operations; by changing their values we can change how the shell and other programs behave.

Let's run the command `set` and look at some of the variables the shell defines:

```
$ set
```

```
COMPUTERNAME=TURING
HOME=/Users/amira
HOMEDRIVE=C:
HOSTNAME=TURING
HOSTTYPE=i686
NUMBER_OF_PROCESSORS=4
OS=Windows_NT
PATH=/Users/amira/anaconda3/bin:/usr/bin:
  /bin:/usr/sbin:/sbin:/usr/local/bin
PWD=/Users/amira
UID=1000
USERNAME=amira
...
```

There are many more than are shown here—roughly a hundred in our current shell session. And yes, using `set` to *show* things might seem a little strange, even for Unix, but if we don't give it any arguments, the command might as well show us things we *could* set.

By convention, **shell variables** that are always present have upper-case names. All shell variables' values are strings, even those (such as `UID`) that look like numbers. It's up to programs to convert these strings to other types when necessary. For example, if a program wanted to find out how many processors the computer had, it would convert the value of `NUMBER_OF_PROCESSORS` from a string to an integer.

Similarly, some variables (like `PATH`) store lists of values. In this case, the convention is to use a colon ':' as a separator. If a program wants the individual elements of such a list, it must split the variable's value into pieces.

Let's have a closer look at `PATH`. Its value defines the shell's **search path**, which is the list of directories that the shell looks in for programs when we type in a command name without specifying exactly where it is. For example, when we type a command like `analyze`, the shell needs to decide whether to run `./analyze` (in our current directory) or `/bin/analyze` (in a system directory). To do this, the shell checks each directory in the `PATH` variable in turn. As soon as it finds a program with the right name, it stops searching and runs what it has found.

To show how this works, here are the components of `PATH` listed one per line:

```
/Users/amira/anaconda3/bin
/usr/bin
/bin
/usr/sbin
/sbin
/usr/local/bin
```

Suppose that our computer has three programs called `analyze`: `/bin/analyze`, `/usr/local/bin/analyze`, and `/Users/amira/analyze`. Since the shell searches the directories in the order in which they're listed in `PATH`, it finds `/bin/analyze` first and runs that. Since `/Users/amira` is not in our path, Bash will *never* find the program `/Users/amira/analyze` unless we type the path in explicitly (for example, as `./analyze` if we are in `/Users/amira`).

If we want to see a variable's value, we can print it using the `echo` command introduced at the end of Section 3.5. Let's look at the value of the variable `HOME`, which keeps track of our home directory:

```
$ echo HOME
```

```
HOME
```

Whoops: this just prints "HOME," which isn't what we wanted. Instead, we need to run this:

```
$ echo $HOME
```

```
/Users/amira
```

As with loop variables (Section 3.3), the dollar sign before the variable names tells the shell that we want the variable's value. This works just like wildcard expansion—the shell replaces the variable's name with its value *before* running the command we've asked for. Thus, `echo $HOME` becomes `echo /Users/amira`, which displays the right thing.

Creating a variable is easy: we assign a value to a name using "=," putting quotes around the value if it contains spaces or special characters:

```
$ DEPARTMENT="Library Science"
$ echo $DEPARTMENT
```

```
Library Science
```

To change the value, we simply assign a new one:

```
$ DEPARTMENT="Information Science"
$ echo $DEPARTMENT
```

```
Information Science
```

If we want to set some variables automatically every time we run a shell, we can put commands to do this in a file called `.bashrc` in our home directory. For example, here are two lines in `/Users/amira/.bashrc`:

```
export DEPARTMENT="Library Science"
export TEMP_DIR=/tmp
export BACKUP_DIR=$TEMP_DIR/backup
```

These three lines create the variables `DEPARTMENT`, `TEMP_DIR`, and `BACKUP_DIR`, and **export** them so that any programs the shell runs can see them as well. Notice that `BACKUP_DIR`'s definition relies on the value of `TEMP_DIR`, so that if we change where we put temporary files, our backups will be relocated automatically. However, this will only happen once we restart the shell, because `.bashrc` is only executed when the shell starts up.

What's in a Name?

The '.' character at the front of the name `.bashrc` prevents `ls` from listing this file unless we specifically ask it to using `-a`. The "rc" at the end is an abbreviation for "run commands," which meant something really important decades ago, and is now just a convention everyone follows without understanding why.

While we're here, it's also common to use the `alias` command to create shortcuts for things we frequently type. For example, we can define the alias `backup` to run `/bin/zback` with a specific set of arguments:

```
alias backup=/bin/zback -v --nostir -R 20000 $HOME $BACKUP_DIR
```

Aliases can save us a lot of typing, and hence a lot of typing mistakes. The name of an alias can be the same as an existing command, so we can use them to change the behavior of a familiar command:

```
# Long list format including hidden files
alias ls='ls -la'

# Print the file paths that were copied/moved
alias mv='mv -v'
alias cp='cp -v'

# Request confirmation to remove a file and
# print the file path that is removed
alias rm='rm -iv'
```

We can find interesting suggestions for other aliases by searching online for "sample bashrc."

While searching for additional aliases, you're likely to encounter references to other common shell features to customize, such as the color of your shell's background and text. As mentioned in Chapter 2, another important feature to consider customizing is your shell prompt. In addition to a standard symbol (like `$`), your computer may include other information as well, such as the working directory, username, and/or date/time. If your shell does not include that information and you would like to see it, or if your current prompt is too long and you'd like to shorten it, you can include a line in your `.bashrc` file that defines `$PS1`:

```
PS1="\u \w $ "
```

This changes the prompt to include your username and current working directory:

```
amira ~/Desktop $
```

4.7 Summary

As powerful as the Unix shell is, it does have its shortcomings: dollar signs, quotes, and other punctuation can make a complex shell script look as though it was created by a cat dancing on a keyboard. However, it is the glue that holds data science together: shell scripts are used to create pipelines from miscellaneous sets of programs, while shell variables are used to do everything from specifying package installation directories to managing database login credentials. And while `grep` and `find` may take some getting used to, they and their cousins can handle enormous datasets very efficiently. If you would like to go further, Ray and Ray (2014) is an excellent general introduction, while Janssens (2014) looks specifically at how to process data on the command line.

4.8 Exercises

As with the previous chapter, extra files and directories created during these exercises may need to be removed when you are done.

4.8.1 Cleaning up

As we have gone through this chapter, we have created several files that we won't need again. We can clean them up with the following commands; briefly explain what each line does.

```
$ cd ~/zipf
$ for file in $(find . -name "*.bak")
> do
>   rm $file
> done
$ rm bin/summarize_all_books.sh
$ rm -r results
```

4.8.2 Variables in shell scripts

Imagine you have a shell script called `script.sh` that contains:

```
head -n $2 $1
tail -n $3 $1
```

With this script in your `data` directory, you type the following command:

```
$ bash script.sh '*.txt' 1 1
```

Which of the following outputs would you expect to see?

1. All of the lines between the first and the last lines of each file ending in `.txt` in the `data` directory
2. The first and the last line of each file ending in `.txt` in the `data` directory
3. The first and the last line of each file in the `data` directory
4. An error because of the quotes around `*.txt`

4.8.3 Find the longest file with a given extension

Write a shell script called `longest.sh` that takes the name of a directory and a filename extension as its arguments, and prints out the name of the file with the most lines in that directory with that extension. For example:

```
$ bash longest.sh data/ txt
```

would print the name of the `.txt` file in `data` that has the most lines.

4.8.4 Script reading comprehension

For this question, consider your `data` directory once again. Explain what each of the following three scripts would do when run as `bash script1.sh *.txt`, `bash script2.sh *.txt`, and `bash script3.sh *.txt` respectively.

```
# script1.sh
echo *.*
```

```
# script2.sh
for filename in $1 $2 $3
  do
    cat $filename
  done
```

```
# script3.sh
echo $@.txt
```

(You may need to search online to find the meaning of `$@`.)

4.8.5 Using grep

Assume the following text from *The Adventures of Sherlock Holmes* is contained in a file called `excerpt.txt`:

```
To Sherlock Holmes she is always THE woman. I have seldom heard
him mention her under any other name. In his eyes she eclipses
and predominates the whole of her sex. It was not that he felt
any emotion akin to love for Irene Adler.
```

Which of the following commands would provide the following output:

```
and predominates the whole of her sex. It was not that he felt
```

1. `grep "he" excerpt.txt`
2. `grep -E "he" excerpt.txt`
3. `grep -w "he" excerpt.txt`
4. `grep -i "he" excerpt.txt`

4.8.6 Tracking publication years

In Exercise 3.8.6 you examined code that extracted the publication year from a list of book titles. Write a shell script called `year.sh` that takes any number of filenames as command-line arguments, and uses a variation of the code you used earlier to print a list of the unique publication years appearing in each of those files separately.

4.8.7 Counting names

You and your friend have just finished reading *Sense and Sensibility* and are now having an argument. Your friend thinks that the elder of the two Dashwood sisters, Elinor, was mentioned more frequently in the book, but you are certain it was the younger sister, Marianne. Luckily, `sense_and_sensibility.txt` contains the full text of the novel. Using a `for` loop, how would you tabulate the number of times each of the sisters is mentioned?

Hint: one solution might employ the commands `grep` and `wc` and a `|`, while another might utilize `grep` options. There is often more than one way to solve a problem with the shell; people choose solutions based on readability, speed, and what commands they are most familiar with.

4.8.8 Matching and subtracting

Assume you are in the root directory of the `zipf` project. Which of the following commands will find all files in `data` whose names end in `e.txt`, but do *not* contain the word `machine`?

1. `find data -name '*e.txt' | grep -v machine`
2. `find data -name *e.txt | grep -v machine`
3. `grep -v "machine" $(find data -name '*e.txt')`
4. None of the above.

4.8.9 `find` pipeline reading comprehension

Write a short explanatory comment for the following shell script:

```
wc -l $(find . -name '*.dat') | sort -n
```

4.8.10 Finding files with different properties

The `find` command can be given criteria called "tests" to locate files with specific attributes, such as creation time, size, or ownership. Use `man find` to explore these, then write a single command using `-type`, `-mtime`, and `-user` to find all files in or below your Desktop directory that are owned by you and were modified in the last 24 hours. Explain why the value for `-mtime` needs to be negative.

4.9 Key Points

- Save commands in files (usually called **shell scripts**) for re-use.
- `bash filename` runs the commands saved in a file.
- `$@` refers to all of a shell script's command-line arguments.
- `$1`, `$2`, etc., refer to the first command-line argument, the second command-line argument, etc.
- Place variables in quotes if the values might have spaces or other special characters in them.
- `find` prints a list of files with specific properties or whose names match patterns.
- `$(command)` inserts a command's output in place.
- `grep` selects lines in files that match patterns.
- Use the `.bashrc` file in your home directory to set shell variables each time the shell runs.
- Use `alias` to create shortcuts for things you type frequently.

5

Building Command-Line Tools with Python

Multiple exclamation marks are a sure sign of a diseased mind.

— Terry Pratchett

The Jupyter Notebook[1], PyCharm, and other graphical interfaces are great for prototyping code and exploring data, but eventually we may need to apply our code to thousands of data files, run it with many different parameters, or combine it with other programs as part of a data analysis pipeline. The easiest way to do this is often to turn our code into a standalone program that can be run in the Unix shell just like other command-line tools (Taschuk and Wilson 2017).

In this chapter we will develop some command-line Python programs that handle input and output in the same way as other shell commands, can be controlled by several option flags, and provide useful information when things go wrong. The result will have more scaffolding than useful application code, but that scaffolding stays more or less the same as programs get larger.

After the previous chapters, our Zipf's Law project should have the following files and directories:

```
zipf/
├── bin
│   └── book_summary.sh
└── data
    ├── README.md
    ├── dracula.txt
    ├── frankenstein.txt
    ├── jane_eyre.txt
    ├── moby_dick.txt
    ├── sense_and_sensibility.txt
    ├── sherlock_holmes.txt
    └── time_machine.txt
```

[1] https://jupyter.org/

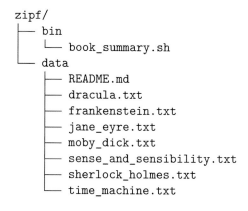

Python Style

When writing Python code there are many style choices to make. How many spaces should I put between functions? Should I use capital letters in variable names? How should I order all the different elements of a Python script? Fortunately, there are well established conventions and guidelines for good Python style. We follow those guidelines throughout this book and discuss them in detail in Appendix F.

5.1 Programs and Modules

To create a Python program that can run from the command line, the first thing we do is to add the following to the bottom of the file:

```
if __name__ == '__main__':
```

This strange-looking check tells us whether the file is running as a standalone program or whether it is being imported as a module by some other program. When we import a Python file as a module in another program, the `__name__` variable is automatically set to the name of the file. When we run a Python file as a standalone program, on the other hand, `__name__` is always set to the special string `"__main__"`. To illustrate this, let's consider a script named `print_name.py` that prints the value of the `__name__` variable:

```
print(__name__)
```

When we run this file directly, it will print `__main__`:

```
$ python print_name.py
```

```
__main__
```

But if we import `print_name.py` from another file or from the Python interpreter, it will print the name of the file, i.e., `print_name`.

```
$ python
```

```
Python 3.7.6 (default, Jan  8 2020, 13:42:34)
[Clang 4.0.1 (tags/RELEASE_401/final)] ::
Anaconda, Inc. on darwin
Type "help", "copyright", "credits" or "license"
for more information.
```

```
>>> import print_name
```

```
print_name
```

Checking the value of the variable `__name__` therefore tells us whether our file is the top-level program or not. If it is, we can handle command-line options, print help, or whatever else is appropriate; if it isn't, we should assume that some other code is doing this.

We could put the main program code directly under the `if` statement like this:

```
if __name__ == "__main__":
    # code goes here
```

but that is considered poor practice, since it makes testing harder (Chapter 11). Instead, we put the high-level logic in a function, then call that function if our file is being run directly:

```
def main():
    # code goes here

if __name__ == "__main__":
    main()
```

This top-level function is usually called `main`, but we can use whatever name we want.

5.2 Handling Command-Line Options

The main function in a program usually starts by parsing any options the user gave on the command line. The most commonly used library for doing this in Python is `argparse`[2], which can handle options with or without arguments, convert arguments from strings to numbers or other types, display help, and many other things.

The simplest way to explain how `argparse` works is by example. Let's create a short Python program called `script_template.py`:

```
import argparse


def main(args):
    print('Input file:', args.infile)
    print('Output file:', args.outfile)


if __name__ == '__main__':
    USAGE = 'Brief description of what the script does.'
    parser = argparse.ArgumentParser(description=USAGE)
    parser.add_argument('infile', type=str,
                        help='Input file name')
    parser.add_argument('outfile', type=str,
                        help='Output file name')
    args = parser.parse_args()
    main(args)
```

Empty Lines, Again

As we discussed in the last chapter for shell scripts, remember to end your Python scripts in a newline character (which we view as an empty line).

If `script_template.py` is run as a standalone program at the command line, then `__name__ == '__main__'` is true, so the program uses `argparse` to create an argument parser. It then specifies that it expects two command-line arguments: an input filename (`infile`) and an output filename (`outfile`).

[2] `https://docs.python.org/3/library/argparse.html`

The program uses `parser.parse_args()` to parse the actual command-line arguments given by the user and stores the result in a variable called `args`, which it passes to `main`. That function can then get the values using the names specified in the `parser.add_argument` calls.

Specifying Types

We have passed `type=str` to `add_argument` to tell `argparse` that we want `infile` and `outfile` to be treated as strings. `str` is not quoted because it is not a string itself: instead, it is the built-in Python function that converts things to strings. As we will see below, we can pass in other functions like `int` if we want arguments converted to numbers.

If we run `script_template.py` at the command line, the output shows us that `argparse` has successfully handled the arguments:

```
$ cd ~/zipf
$ python script_template.py in.csv out.png
```

```
Input file: in.csv
Output file: out.png
```

It also displays an error message if we give the program invalid arguments:

```
$ python script_template.py in.csv
```

```
usage: script_template.py [-h] infile outfile
script_template.py: error: the following arguments are
  required: outfile
```

Finally, it automatically generates help information (which we can get using the `-h` option):

```
$ python script_template.py -h
```

```
usage: script_template.py [-h] infile outfile

Brief description of what the script does.
```

```
positional arguments:
  infile      Input file name
  outfile     Output file name

optional arguments:
  -h, --help  show this help message and exit
```

5.3 Documentation

Our program template is a good starting point, but we improve it right away by adding a bit of documentation. To demonstrate, let's write a function that doubles a number:

```
def double(num):
    'Double the input.'
    return 2 * num
```

The first line of this function is a string that isn't assigned to a variable. Such a string is called a documentation string, or **docstring** for short. If we call our function it does what we expect:

```
double(3)
```

```
6
```

However, we can also ask for the function's documentation, which is stored in `double.__doc__`:

```
double.__doc__
```

```
'Double the input.'
```

Python creates the variable `__doc__` automatically for every function, just as it creates the variable `__name__` for every file. If we don't write a docstring for a function, `__doc__`'s value is an empty string. We can put whatever text we want into a function's docstring, but it is usually used to provide online documentation.

We can also put a docstring at the start of a file, in which case it is assigned to a variable called `__doc__` that is visible inside the file. If we add documentation to our template, it becomes:

```
"""Brief description of what the script does."""

import argparse


def main(args):
    """Run the program."""
    print('Input file:', args.infile)
    print('Output file:', args.outfile)


if __name__ == '__main__':
    parser = argparse.ArgumentParser(description=__doc__)
    parser.add_argument('infile', type=str,
                        help='Input file name')
    parser.add_argument('outfile', type=str,
                        help='Output file name')
    args = parser.parse_args()
    main(args)
```

Note that docstrings are usually written using triple-quoted strings, since these can span multiple lines. Note also how we pass `description=__doc__` to `argparse.ArgumentParser`. This saves us from typing the same information twice, but more importantly ensures that the help message provided in response to the `-h` option will be the same as the interactive help.

Let's try this out in an interactive Python session. (Remember, do not type the `>>>` prompt: Python provides this for us.)

```
$ python
```

```
Python 3.7.6 (default, Jan  8 2020, 13:42:34)
[Clang 4.0.1 (tags/RELEASE_401/final)] ::
Anaconda, Inc. on darwin
Type "help", "copyright", "credits" or "license"
for more information.
```

```
>>> import script_template
>>> script_template.__doc__
```

```
'Brief description of what the script does.'
```

```
>>> help(script_template)
```

```
Help on module script_template:

NAME
    script_template - Brief description of what the script does.

FUNCTIONS
    main(args)
        Run the program.

FILE
    /Users/amira/script_template.py
```

As this example shows, if we ask for help on the module, Python formats and displays all of the docstrings for everything in the file. We talk more about what to put in a docstring in Appendix G.

5.4 Counting Words

Now that we have a template for command-line Python programs, we can use it to check Zipf's Law for our collection of classic novels. We start by moving the template into the directory where we store our runnable programs (Section 1.1.2):

```
$ mv script_template.py bin
```

Next, let's write a function that counts how often words appear in a file. Our function splits the text on **whitespace** characters (which is the default behavior of the string object's `split` method), then strips leading and trailing punctuation. This isn't completely correct—if two words are joined by a long dash like "correct" and "if" in this sentence, for example, they will be treated as one word—but given that long dashes are used relatively infrequently, it's close enough to correct for our purposes. (We will submit a bug report about the long dash issue in Section 8.6). We also use the `Counter` class from the `collections` library to count how many times each word occurs. If we give `Counter` a list of words, the result is an object that contains the number of times each one appears in the list:

```
import string
from collections import Counter


def count_words(reader):
    """Count the occurrence of each word in a string."""
    text = reader.read()
    chunks = text.split()
    npunc = [word.strip(string.punctuation) for word in chunks]
    word_list = [word.lower() for word in npunc if word]
    word_counts = Counter(word_list)
    return word_counts
```

Let's try our function on *Dracula*:

```
with open('data/dracula.txt', 'r') as reader:
    word_counts = count_words(reader)
print(word_counts)
```

```
Counter({'the': 8036, 'and': 5896, 'i': 4712, 'to': 4540,
         'of': 3738, 'a': 2961, 'in': 2558, 'he': 2543,
         'that': 2455, 'it': 2141, 'was': 1877, 'as': 1581,
         'we': 1535, 'for': 1534, ...})
```

If we want the word counts in a format like CSV for easier processing, we can write another small function that takes our `Counter` object, orders its contents from most to least frequent, and then writes it to **standard output** as CSV:

```
import sys
import csv


def collection_to_csv(collection):
    """Write collection of items and counts in csv format."""
    collection = collection.most_common()
    writer = csv.writer(sys.stdout)
    writer.writerows(collection)
```

Running this would print all the distinct words in the book along with their counts. This list could well be several thousand lines long, so to make the output a little easier to view on our screen, we can add an option to limit the output to the most frequent words. We set its default value to `None` so that we can easily tell if the caller *hasn't* specified a cutoff, in which case we display the whole collection:

```
def collection_to_csv(collection, num=None):
    """Write collection of items and counts in csv format."""
    collection = collection.most_common()
    if num is None:
        num = len(collection)
    writer = csv.writer(sys.stdout)
    writer.writerows(collection[0:num])
```

```
collection_to_csv(word_counts, num=10)
```

```
the,8036
and,5896
i,4712
to,4540
of,3738
a,2961
in,2558
he,2543
that,2455
it,2141
```

To make our `count_words` and `collection_to_csv` functions available at the command line, we need to insert them into our script template and call them from within the `main` function. Let's call our program `countwords.py` and put it in the `bin` subdirectory of the `zipf` project:

```
"""
Count the occurrences of all words in a text
and output them in CSV format.
"""

import sys
import argparse
import string
import csv
from collections import Counter


def collection_to_csv(collection, num=None):
    """Write collection of items and counts in csv format."""
    collection = collection.most_common()
    if num is None:
        num = len(collection)
    writer = csv.writer(sys.stdout)
    writer.writerows(collection[0:num])


def count_words(reader):
    """Count the occurrence of each word in a string."""
    text = reader.read()
    chunks = text.split()
    npunc = [word.strip(string.punctuation) for word in chunks]
    word_list = [word.lower() for word in npunc if word]
    word_counts = Counter(word_list)
    return word_counts


def main(args):
    """Run the command line program."""
    with open(args.infile, 'r') as reader:
        word_counts = count_words(reader)
    collection_to_csv(word_counts, num=args.num)


if __name__ == '__main__':
```

```
    parser = argparse.ArgumentParser(description=__doc__)
    parser.add_argument('infile', type=str,
                        help='Input file name')
    parser.add_argument('-n', '--num',
                        type=int, default=None,
                        help='Output n most frequent words')
    args = parser.parse_args()
    main(args)
```

Note that we have replaced the `'outfile'` argument from our template script with an optional `-n` (or `--num`) flag to control how much output is printed and modified `collection_to_csv` so that it always prints to standard output. If we want that output in a file, we can redirect with `>`.

Let's take our program for a test drive:

```
$ python bin/countwords.py data/dracula.txt -n 10
```

```
the,8036
and,5896
i,4712
to,4540
of,3738
a,2961
in,2558
he,2543
that,2455
it,2141
```

5.5 Pipelining

As discussed in Section 3.2, most Unix commands follow a useful convention: if the user doesn't specify the names of any input files, they read from **standard input**. Similarly, if no output file is specified, the command sends its results to **standard output**. This makes it easy to use the command in a pipeline.

Our program always sends its output to standard output; as noted above, we can always redirect it to a file with `>`. If we want `countwords.py` to read from standard input, we only need to change the handling of `infile` in the argument parser and simplify `main` to match:

```
def main(args):
    """Run the command line program."""
    word_counts = count_words(args.infile)
    collection_to_csv(word_counts, num=args.num)


if __name__ == '__main__':
    parser = argparse.ArgumentParser(description=__doc__)
    parser.add_argument('infile', type=argparse.FileType('r'),
                        nargs='?', default='-',
                        help='Input file name')
    parser.add_argument('-n', '--num',
                        type=int, default=None,
                        help='Output n most frequent words')
    args = parser.parse_args()
    main(args)
```

There are two changes to how `add_argument` handles `infile`:

1. Setting `type=argparse.FileType('r')` tells `argparse` to treat the argument as a filename and open that file for reading. This is why we no longer need to call `open` ourselves, and why `main` can pass `args.infile` directly to `count_words`.
2. The number of expected arguments (`nargs`) is set to `?`. This means that if an argument is given it will be used, but if none is provided, a default of `'-'` will be used instead. `argparse.FileType('r')` understands `'-'` to mean "read from standard input"; this is another Unix convention that many programs follow.

After these changes, we can create a pipeline like this to count the words in the first 500 lines of a book:

```
$ head -n 500 data/dracula.txt | python bin/countwords.py --n 10
```

```
the,227
and,121
of,116
i,98
```

```
to,80
in,58
a,49
it,45
was,42
that,41
```

5.6 Positional and Optional Arguments

We have met two kinds of command-line arguments while writing `countwords.py`. **Optional arguments** are defined using a leading `-` or `--` (or both), which means that all three of the following definitions are valid:

```
parser.add_argument('-n', type=int, help='Limit output')
parser.add_argument('--num', type=int, help='Limit output')
parser.add_argument('-n', '--num',
                    type=int, help='Limit output')
```

The convention is for `-` to precede a **short** (single letter) option and `--` a **long** (multi-letter) option. The user can provide optional arguments at the command line in any order they like.

Positional arguments have no leading dashes and are not optional: the user must provide them at the command line in the order in which they are specified to `add_argument` (unless `nargs='?'` is provided to say that the value is optional).

5.7 Collating Results

Ultimately, we want to save the word counts to a CSV file for further analysis and plotting. Let's create a subdirectory to hold our results (following the structure described in Section 1.1):

```
$ mkdir results
```

and then save the counts for various files:

```
$ python bin/countwords.py data/dracula.txt > results/dracula.csv
```

```
$ python bin/countwords.py data/moby_dick.txt >
  results/moby_dick.csv
```

```
$ python bin/countwords.py data/jane_eyre.txt >
  results/jane_eyre.csv
```

As in the previous chapter, we've split long lines of code onto separate lines for formatting purposes; each of the three code chunks above should be run as a single line of code.

Now that we can get word counts for individual books we can collate the counts for several books. This can be done using a loop that adds up the counts of a word from each of the CSV files created by `countwords.py`. Using the same template as before, we can write a program called `collate.py`:

```
"""
Combine multiple word count CSV-files
into a single cumulative count.
"""

import sys
import csv
import argparse
from collections import Counter


def collection_to_csv(collection, num=None):
    """Write collection of items and counts in csv format."""
    collection = collection.most_common()
    if num is None:
        num = len(collection)
    writer = csv.writer(sys.stdout)
    writer.writerows(collection[0:num])
```

```
def update_counts(reader, word_counts):
    """Update word counts with data from another reader/file."""
    for word, count in csv.reader(reader):
        word_counts[word] += int(count)


def main(args):
    """Run the command line program."""
    word_counts = Counter()
    for fname in args.infiles:
        with open(fname, 'r') as reader:
            update_counts(reader, word_counts)
    collection_to_csv(word_counts, num=args.num)


if __name__ == '__main__':
    parser = argparse.ArgumentParser(description=__doc__)
    parser.add_argument('infiles', type=str, nargs='*',
                        help='Input file names')
    parser.add_argument('-n', '--num',
                        type=int, default=None,
                        help='Output n most frequent words')
    args = parser.parse_args()
    main(args)
```

The loop in the `main` function iterates over each filename in `infiles`, opens the CSV file, and calls `update_counts` with the input stream as one parameter and the counter as the other. `update_counts` then iterates through all the words in the CSV files and increments the counts using the `+=` operator.

Note that we have not used `type=argparse.FileType('r')` here. Instead, we have called the option `infiles` (plural) and specified `nargs='*'` to tell `argparse` that we will accept zero or more filenames. We must then open the files ourselves.

Let's give `collate.py` a try (using `-n 10` to limit the number of lines of output):

```
$ python bin/collate.py results/dracula.csv
  results/moby_dick.csv results/jane_eyre.csv -n 10
```

```
the,30505
and,18916
of,14908
to,14369
i,13572
a,12059
in,9547
that,6984
it,6821
he,6142
```

5.8 Writing Our Own Modules

`countwords.py` and `collate.py` both now contain the function `collection_to_csv`. Having the same function in two or more places is a bad idea: if we want to improve it or fix a bug, we have to find and change every single script that contains a copy.

The solution is to put the shared functions in a separate file and load that file as a module. Let's create a file called `utilities.py` in the `bin` directory that looks like this:

```python
"""Collection of commonly used functions."""

import sys
import csv


def collection_to_csv(collection, num=None):
    """
    Write out collection of items and counts in csv format.

    Parameters
    ----------
    collection : collections.Counter
        Collection of items and counts
    num : int
        Limit output to N most frequent items
    """
    collection = collection.most_common()
```

```
    if num is None:
        num = len(collection)
    writer = csv.writer(sys.stdout)
    writer.writerows(collection[0:num])
```

Note that we have written a much more detailed docstring for `collection_to_csv`: as a rule, the more widely used code is, the more it's worth explaining exactly what it does.

We can now import our utilities into our programs just as we would import any other Python module using either `import utilities` (to get the whole thing) or something like `from utilities import collection_to_csv` (to get a single function). After making this change, `countwords.py` looks like this:

```
"""
Count the occurrences of all words in a text
and write them to a CSV-file.
"""

import argparse
import string
from collections import Counter

import utilities as util


def count_words(reader):
    """Count the occurrence of each word in a string."""
    text = reader.read()
    chunks = text.split()
    npunc = [word.strip(string.punctuation) for word in chunks]
    word_list = [word.lower() for word in npunc if word]
    word_counts = Counter(word_list)
    return word_counts


def main(args):
    """Run the command line program."""
    word_counts = count_words(args.infile)
    util.collection_to_csv(word_counts, num=args.num)
```

```
if __name__ == '__main__':
    parser = argparse.ArgumentParser(description=__doc__)
    parser.add_argument('infile', type=argparse.FileType('r'),
                        nargs='?', default='-',
                        help='Input file name')
    parser.add_argument('-n', '--num',
                        type=int, default=None,
                        help='Output n most frequent words')
    args = parser.parse_args()
    main(args)
```

`collate.py` is now:

```
"""
Combine multiple word count CSV-files
into a single cumulative count.
"""

import csv
import argparse
from collections import Counter

import utilities as util


def update_counts(reader, word_counts):
    """Update word counts with data from another reader/file."""
    for word, count in csv.reader(reader):
        word_counts[word] += int(count)


def main(args):
    """Run the command line program."""
    word_counts = Counter()
    for fname in args.infiles:
        with open(fname, 'r') as reader:
            update_counts(reader, word_counts)
    util.collection_to_csv(word_counts, num=args.num)


if __name__ == '__main__':
    parser = argparse.ArgumentParser(description=__doc__)
```

```
parser.add_argument('infiles', type=str, nargs='*',
                    help='Input file names')
parser.add_argument('-n', '--num',
                    type=int, default=None,
                    help='Output n most frequent words')
args = parser.parse_args()
main(args)
```

Any Python source file can be imported by any other. This is why Python files should be named using **snake case** (e.g., `some_thing`) instead of **kebab case** (e.g., `some-thing`): an expression like `import some-thing` isn't allowed because `some-thing` isn't a legal variable name. When a file is imported, the statements in it are executed as it loads. Variables, functions, and items defined in the file are then available as `module.thing`, where `module` is the filename (without the `.py` extension) and `thing` is the name of the item.

The `__pycache__` Directory

When we import a file, Python translates the source code into instructions called **byte codes** that it can execute efficiently. Since the byte codes only change when the source changes, Python saves the byte code in a separate file, and reloads that file instead of re-translating the source code the next time it's asked to import the file (unless the file has changed, in which case Python starts from the beginning).

Python creates a subdirectory called `__pycache__` that holds the byte code for the files imported from that directory. We typically don't want to put the files in `__pycache__` in version control, so we normally tell Git to ignore it as discussed in Section 6.9.

5.9 Plotting

The last thing for us to do is to plot the word count distribution. Recall that Zipf's Law[3] states the second most common word in a body of text appears half as often as the most common, the third most common appears a third as often, and so on. Mathematically, this might be written as "word frequency is proportional to 1/rank."

[3] `https://en.wikipedia.org/wiki/Zipf%27s_law`

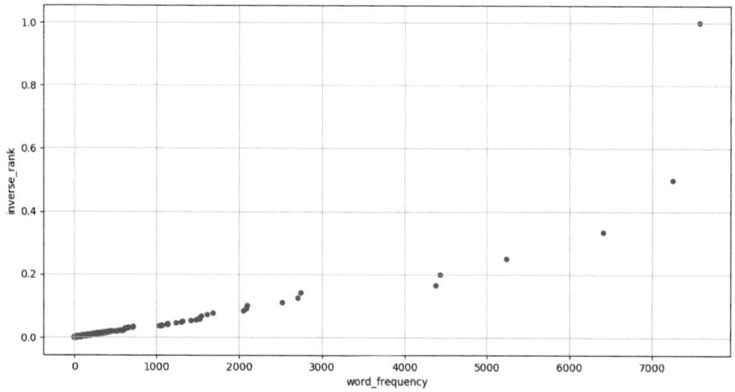

FIGURE 5.1: Word frequency distribution for Jane Eyre.

The following code plots the word frequency against the inverse rank using the pandas[4] library:

```
import pandas as pd


input_csv = 'results/jane_eyre.csv'
df = pd.read_csv(input_csv, header=None,
                 names=('word', 'word_frequency'))
df['rank'] = df['word_frequency'].rank(ascending=False,
                                       method='max')
df['inverse_rank'] = 1 / df['rank']
scatplot = df.plot.scatter(x='word_frequency',
                           y='inverse_rank',
                           figsize=[12, 6],
                           grid=True)
fig = scatplot.get_figure()
fig.savefig('results/jane_eyre.png')
```

You'll build on this code to create a plotting script for your project in Exercise 5.11.4.

[4]https://pandas.pydata.org/

5.10 Summary

Why is building a simple command-line tool so complex? One answer is that the conventions for command-line programs have evolved over several decades, so libraries like `argparse` must now support several different generations of option handling. Another is that the things we want to do genuinely *are* complex: read from either standard input or a list of files, display help when asked to, respect parameters that might not be there, and so on. As with many other things in programming (and life), everyone wishes it was simpler, but no one can agree on what to throw away.

The good news is that this complexity is a fixed cost: our template for command-line tools can be re-used for programs that are much larger than the examples shown in this chapter. Making tools that behave in ways people expect greatly increases the chances that others will find them useful.

5.11 Exercises

5.11.1 Running Python statements from the command line

We don't need to open the interactive interpreter to run Python code. Instead, we can invoke Python with the command flag `-c` and the statement we want to run:

```
$ python -c "print(2+3)"
```

```
5
```

When and why is this useful?

5.11.2 Listing files

A Python library called glob[5] can be used to create a list of files matching a pattern, much like the `ls` shell command.

[5] `https://docs.python.org/3/library/glob.html`

```
$ python
```

```
Python 3.7.6 (default, Jan  8 2020, 13:42:34)
[Clang 4.0.1 (tags/RELEASE_401/final)] ::
Anaconda, Inc. on darwin
Type "help", "copyright", "credits" or "license"
for more information.
```

```
>>> import glob
>>> glob.glob('data/*.txt')
```

```
['data/moby_dick.txt', 'data/sense_and_sensibility.txt',
'data/sherlock_holmes.txt', 'data/time_machine.txt',
'data/frankenstein.txt', 'data/dracula.txt',
'data/jane_eyre.txt']
```

Using `script_template.py` as a guide, write a new script called `my_ls.py` that takes as input a directory and a suffix (e.g., py, txt, md, sh) and outputs a list of the files (sorted alphabetically) in that directory ending in that suffix.

The help information for the new script should read as follows:

```
$ python bin/my_ls.py -h
```

```
usage: my_ls.py [-h] dir suffix

List the files in a given directory with a given suffix.

positional arguments:
  dir         Directory
  suffix      File suffix (e.g. py, sh)

optional arguments:
  -h, --help  show this help message and exit
```

and an example of the output would be:

```
$ python bin/my_ls.py data/ txt
```

```
data/dracula.txt
data/frankenstein.txt
data/jane_eyre.txt
data/moby_dick.txt
data/sense_and_sensibility.txt
data/sherlock_holmes.txt
data/time_machine.txt
```

Note: we will not be including this script in subsequent chapters.

5.11.3 Sentence ending punctuation

Our `countwords.py` script strips the punctuation from a text, which means it provides no information on sentence endings. Using `script_template.py` and `countwords.py` as a guide, write a new script called `sentence_endings.py` that counts the occurrence of full stops, question marks and exclamation points and prints that information to the screen.

Hint: String objects have a `count` method:

```
$ python
```

```
Python 3.7.6 (default, Jan  8 2020, 13:42:34)
[Clang 4.0.1 (tags/RELEASE_401/final)] ::
Anaconda, Inc. on darwin
Type "help", "copyright", "credits" or "license"
for more information.
```

```
>>> "Hello! Are you ok?".count('!')
```

```
1
```

When you're done, the script should be able to accept an input file:

```
$ python bin/sentence_endings.py data/dracula.txt
```

```
Number of . is 8505
Number of ? is 492
Number of ! is 752
```

or standard input:

```
$ head -n 500 data/dracula.txt | python bin/sentence_endings.py
```

```
Number of . is 148
Number of ? is 8
Number of ! is 8
```

Note: we will not be including this script in subsequent chapters.

5.11.4 A better plotting program

Using `script_template.py` as a guide, take the plotting code from Section 5.9 and write a new Python program called `plotcounts.py`. The script should do the following:

1. Use the `type=argparse.FileType('r')`, `nargs='?'` and `default='-'` options for the input file argument (i.e., similar to the `countwords.py` script) so that `plotcounts.py` uses standard input if no CSV file is given.
2. Include an optional `--outfile` argument for the name of the output image file. The default value should be `plotcounts.png`.
3. Include an optional `--xlim` argument so that the user can change the x-axis bounds.

When you are done, generate a plot for *Jane Eyre* by passing the word counts to `plotcounts.py` via a CSV file:

```
$ python bin/plotcounts.py results/jane_eyre.csv
  --outfile results/jane_eyre.png
```

and by standard input:

```
$ python bin/countwords.py data/jane_eyre.txt | python
  bin/plotcounts.py --outfile results/jane_eyre.png
```

Note: the solution to this exercise is used in following chapters.

5.12 Key Points

- Write command-line Python programs that can be run in the **Unix shell** like other command-line tools.
- If the user does not specify any input files, read from **standard input**.
- If the user does not specify any output files, write to **standard output**.
- Place all `import` statements at the start of a module.
- Use the value of `__name__` to determine if a file is being run directly or being loaded as a module.
- Use `argparse`[6] to handle command-line arguments in standard ways.
- Use **short options** for common controls and **long options** for less common or more complicated ones.
- Use **docstrings** to document functions and scripts.
- Place functions that are used across multiple scripts in a separate file that those scripts can import.

[6] `https://docs.python.org/3/library/argparse.html`

6

Using Git at the Command Line

> +++ Divide By Cucumber Error. Please Reinstall Universe And Reboot +++
>
> — Terry Pratchett

A **version control system** tracks changes to files and helps people share those changes with each other. These things can be done by emailing files to colleagues or by using Microsoft Word and Google Docs, but version control does both more accurately and efficiently. Originally developed to support software development, over the past fifteen years it has become the cornerstone of **reproducible research**.

Version control works by storing a master copy of your code in a repository, which you can't edit directly. Instead, you check out a working copy of the code, edit that code, then commit changes back to the repository. In this way, version control records a complete revision history (i.e., of every commit), so that you can retrieve and compare previous versions at any time. This is useful from an individual viewpoint, because you don't need to store multiple (but slightly different) copies of the same script (Figure 6.1). It's also useful from a collaboration viewpoint, because the system keeps a record of who made what changes and when.

There are many different version control systems, such as CVS, Subversion, and Mercurial, but the most widely used version control system today is **Git**. Many people first encounter it through a GUI like GitKraken[1] or the RStudio IDE[2]. However, these tools are actually wrappers around Git's original command-line interface, which gives us access to all of Git's features. This lesson describes how to perform fundamental operations using that interface; Chapter 7 then introduces more advanced operations that can be used to implement a smoother research workflow.

To show how Git works, we will apply it to the Zipf's Law project. Our project directory should currently include:

[1] `https://www.gitkraken.com/`
[2] `https://www.rstudio.com/products/rstudio/`

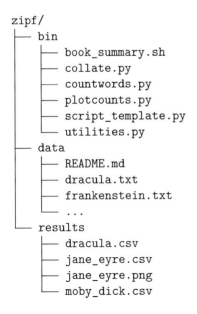

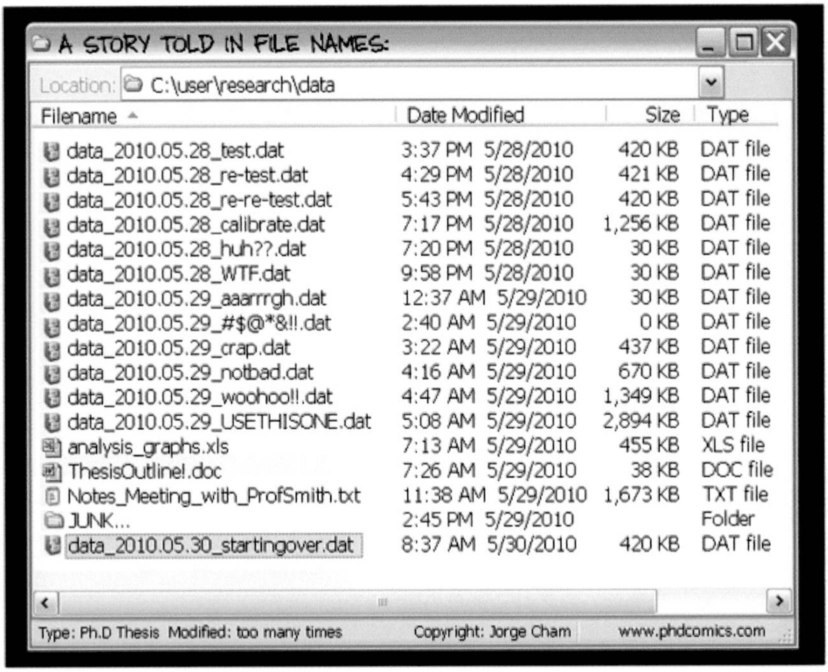

FIGURE 6.1: Managing different versions of a file without version control.

```
zipf/
├── bin
│   ├── book_summary.sh
│   ├── collate.py
│   ├── countwords.py
│   ├── plotcounts.py
│   ├── script_template.py
│   └── utilities.py
├── data
│   ├── README.md
│   ├── dracula.txt
│   ├── frankenstein.txt
│   └── ...
└── results
    ├── dracula.csv
    ├── jane_eyre.csv
    ├── jane_eyre.png
    └── moby_dick.csv
```

bin/plotcounts.py is the solution to Exercise 5.11.4; over the course of this chapter we will edit it to produce more informative plots. Initially, it looks like this:

```python
"""Plot word counts."""

import argparse

import pandas as pd


def main(args):
    """Run the command line program."""
    df = pd.read_csv(args.infile, header=None,
                     names=('word', 'word_frequency'))
    df['rank'] = df['word_frequency'].rank(ascending=False,
                                           method='max')
    df['inverse_rank'] = 1 / df['rank']
    ax = df.plot.scatter(x='word_frequency',
                         y='inverse_rank',
                         figsize=[12, 6],
                         grid=True,
                         xlim=args.xlim)
    ax.figure.savefig(args.outfile)


if __name__ == '__main__':
    parser = argparse.ArgumentParser(description=__doc__)
    parser.add_argument('infile', type=argparse.FileType('r'),
                        nargs='?', default='-',
                        help='Word count csv file name')
    parser.add_argument('--outfile', type=str,
                        default='plotcounts.png',
                        help='Output image file name')
    parser.add_argument('--xlim', type=float, nargs=2,
                        metavar=('XMIN', 'XMAX'),
                        default=None, help='X-axis limits')
    args = parser.parse_args()
    main(args)
```

6.1 Setting Up

We write Git commands as `git verb options`, where the **subcommand** `verb` tells Git what we want to do and `options` provide whatever additional information that subcommand needs. Using this syntax, the first thing we need to do is configure Git.

```
$ git config --global user.name "Amira Khan"
$ git config --global user.email "amira@zipf.org"
```

(Please use your own name and email address instead of the one shown.) Here, `config` is the verb and the rest of the command are options. We put the name in quotation marks because it contains a space; we don't actually need to quote the email address, but do so for consistency. Since we are going to be using GitHub, the email address should be the same as you have or intend to use when setting up your GitHub account.

The `--global` option tells Git to use the settings for all of our projects on this computer, so these two commands only need to be run once. However, we can re-run them any time if we want to change our details. We can also check our settings using the `--list` option:

```
$ git config --list
```

```
user.name=Amira Khan
user.email=amira@zipf.org
core.autocrlf=input
core.editor=nano
core.repositoryformatversion=0
core.filemode=true
core.bare=false
core.ignorecase=true
...
```

Depending on your operating system and version of Git, your configuration list may look a bit different. Most of these differences shouldn't matter right now, as long as your username and email are accurate.

Git Help and Manual

If we forget a Git command, we can list which ones are available using `--help`:

```
$ git --help
```

This option also gives us more information about specific commands:

```
$ git config --help
```

6.2 Creating a New Repository

Once Git is configured, we can use it to track work on our Zipf's Law project. Let's make sure we are in the top-level directory of our project:

```
$ cd ~/zipf
$ ls
```

```
bin        data       results
```

We want to make this directory a **repository**, i.e., a place where Git can store versions of our files. We do this using the `init` command with `.` to mean "the current directory":

```
$ git init .
```

```
Initialized empty Git repository in /Users/amira/zipf/.git/
```

`ls` seems to show that nothing has changed:

```
$ ls
```

```
bin     data     results
```

but if we add the `-a` flag to show everything, we can see that Git has created a hidden directory within `zipf` called `.git`:

```
$ ls -a
```

```
.       ..       .git    bin      data     results
```

Git stores information about the project in this special subdirectory. If we ever delete it, we will lose that history.

We can check that everything is set up correctly by asking Git to tell us the status of our project:

```
$ git status
```

```
On branch master

No commits yet

Untracked files:
  (use "git add <file>..." to include in what will be
  committed)

        bin/
        data/
        results/

nothing added to commit but untracked files
present (use "git add" to track)
```

"No commits yet" means that Git hasn't recorded any history yet, while "Untracked files" means Git has noticed that there are things in `bin/`, `data/` and `results/` that it is not yet keeping track of.

6.3 Adding Existing Work

Now that our project is a repository, we can tell Git to start recording its history. To do this, we add things to the list of things Git is tracking using `git add`. We can do this for single files:

```
$ git add bin/countwords.py
```

or entire directories:

```
$ git add bin
```

The easiest thing to do with an existing project is to tell Git to add everything in the current directory using `.`:

```
$ git add .
```

We can then check the repository's status to see what files have been added:

```
$ git status
```

```
On branch master

No commits yet

Changes to be committed:
  (use "git rm --cached <file>..." to unstage)
      new file:   bin/book_summary.sh
      new file:   bin/collate.py
      new file:   bin/countwords.py
      new file:   bin/plotcounts.py
      new file:   bin/script_template.py
      new file:   bin/utilities.py
      new file:   data/README.md
      new file:   data/dracula.txt
```

```
        new file:   data/frankenstein.txt
        new file:   data/jane_eyre.txt
        new file:   data/moby_dick.txt
        new file:   data/sense_and_sensibility.txt
        new file:   data/sherlock_holmes.txt
        new file:   data/time_machine.txt
        new file:   results/dracula.csv
        new file:   results/jane_eyre.csv
        new file:   results/jane_eyre.png
        new file:   results/moby_dick.csv
```

Adding all of our existing files this way is easy, but we can accidentally add things that should never be in version control, such as files containing passwords or other sensitive information. The output of `git status` tells us that we can remove such files from the list of things to be saved using `git rm --cached`; we will practice this in Exercise 6.11.2.

What to Save

We always want to save programs, manuscripts, and everything else we have created by hand in version control. In this project, we have also chosen to save our data files and the results we have generated (including our plots). This is a project-specific decision: if these files are very large, for example, we may decide to save them elsewhere, while if they are easy to re-create, we may not save them at all. We will explore this issue further in Chapter 13.

We no longer have any untracked files, but the tracked files haven't been **committed** (i.e., saved permanently in our project's history). We can do this using `git commit`:

```
$ git commit -m "Add scripts, novels, word counts, and plots"
```

```
[master (root-commit) 173222b] Add scripts, novels, word
    counts, and plots
 18 files changed, 145296 insertions(+)
 create mode 100644 bin/book_summary.sh
 create mode 100644 bin/collate.py
 create mode 100644 bin/countwords.py
 create mode 100644 bin/plotcounts.py
 create mode 100644 bin/script_template.py
 create mode 100644 bin/utilities.py
```

```
 create mode 100644 data/README.md
 create mode 100644 data/dracula.txt
 create mode 100644 data/frankenstein.txt
 create mode 100644 data/jane_eyre.txt
 create mode 100644 data/moby_dick.txt
 create mode 100644 data/sense_and_sensibility.txt
 create mode 100644 data/sherlock_holmes.txt
 create mode 100644 data/time_machine.txt
 create mode 100644 results/dracula.csv
 create mode 100644 results/jane_eyre.csv
 create mode 100644 results/jane_eyre.png
 create mode 100644 results/moby_dick.csv
```

`git commit` takes everything we have told Git to save using `git add` and stores a copy permanently inside the repository's `.git` directory. This permanent copy is called a **commit** or a **revision**. Git gives is a unique identifier, and the first line of output from `git commit` displays its **short identifier** `2dc78f0`, which is the first few characters of that unique label.

We use the `-m` option (short for **m**essage) to record a short comment with the commit to remind us later what we did and why. (Once again, we put it in double quotes because it contains spaces.) If we run `git status` now:

```
$ git status
```

the output tells us that all of our existing work is tracked and up to date:

```
On branch master
nothing to commit, working tree clean
```

This first commit becomes the starting point of our project's history: we won't be able to see changes made before this point. This implies that we should make our project a Git repository as soon as we create it rather than after we have done some work.

6.4 Describing Commits

If we run `git commit` *without* the `-m` option, Git opens a text editor so that we can write a longer **commit message**. In this message, the first line is referred to as the "subject" and the rest as the "body," just as in an email.

When we use `-m`, we are only writing the subject line; this makes things easier in the short run, but if our project's history fills up with one-liners like "Fixed problem" or "Updated," our future self will wish that we had taken a few extra seconds to explain things in a little more detail. Following these guidelines[3] will help:

1. Separate the subject from the body with a blank line so that it is easy to spot.
2. Limit subject lines to 50 characters so that they are easy to scan.
3. Write the subject line in Title Case (like a section heading).
4. Do not end the subject line with a period.
5. Write as if giving a command (e.g., "Make each plot half the width of the page").
6. Wrap the body (i.e., insert line breaks to format text as paragraphs rather than relying on editors to wrap lines automatically).
7. Use the body to explain what and why rather than how.

Which Editor?

The default editor in the Unix shell is called Vim. It has many useful features, but no one has ever claimed that its interface is intuitive. ("How do I exit the Vim editor?" is one of the most frequently read questions on Stack Overflow.)

To configure Git to use the `nano` editor introduced in Chapter 2 instead, execute the following command:

```
$ git config --global core.editor "nano -w"
```

6.5 Saving and Tracking Changes

Our initial commit gave us a starting point. The process to build on top of it is similar: first add the file, then commit changes. Let's check that we're in the right directory:

[3]`https://chris.beams.io/posts/git-commit/`

```
$ pwd
```

```
/Users/amira/zipf
```

Let's use `plotcounts.py` to plot the word counts in `results/dracula.csv`:

```
$ python bin/plotcounts.py results/dracula.csv --outfile
  results/dracula.png
```

If we check the status of our repository again, Git tells us that we have a new file:

```
$ git status
```

```
On branch master
Untracked files:
  (use "git add <file>..." to include in what will be
  committed)
    results/dracula.png

nothing added to commit but untracked files
present (use "git add" to track)
```

Git isn't tracking this file yet because we haven't told it to. Let's do that with `git add` and then commit our change:

```
$ git add results/dracula.png
$ git commit -m "Add plot of word counts for 'Dracula'"
```

```
[master 851d590] Add plot of word counts for 'Dracula'
 1 file changed, 0 insertions(+), 0 deletions(-)
 create mode 100644 results/dracula.png
```

If we want to know what we've done recently, we can display the project's history using `git log`:

```
$ git log
```

```
commit 851d590a214c7859eafa0998c6c951f8e0eb359b (HEAD -> master)
Author: Amira Khan <amira@zipf.org>
Date:   Sat Dec 19 09:32:41 2020 -0800

    Add plot of word counts for 'Dracula'

commit 173222bf90216b408c8997f4e143572b99637750
Author: Amira Khan <amira@zipf.org>
Date:   Sat Dec 19 09:30:23 2020 -0800

    Add scripts, novels, word counts, and plots
```

`git log` lists all commits made to a repository in reverse chronological order. The listing for each commit includes the commit's **full identifier** (which starts with the same characters as the short identifier printed by `git commit`), the commit's author, when it was created, and the commit message that we wrote.

Scrolling through Logs

Our log this time isn't very long, so you were likely able to see it printed to your screen without needing to scroll. When you begin working with longer logs (like later in this chapter), you'll notice that the commits are shown in a pager program, as you saw in Section 2.8 with manual pages. You can apply the same keystrokes to scroll through the log and exit the paging program.

The plot we have made is shown in Figure 6.2. It could be better: most of the visual space is devoted to a few very common words, which makes it hard to see what is happening with the other ten thousand or so words.

An alternative way to visually evaluate Zipf's Law is to plot the word frequency against rank on log-log axes. Let's change the section:

```
ax = df.plot.scatter(x='word_frequency',
                     y='inverse_rank',
                     figsize=[12, 6],
                     grid=True,
                     xlim=args.xlim)
```

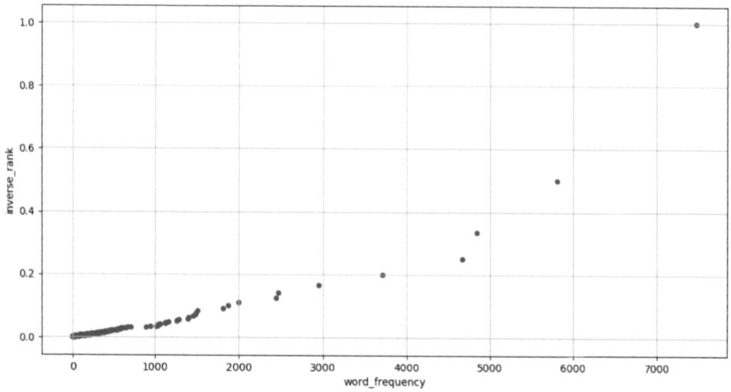

FIGURE 6.2: Inverse rank versus word frequency for *Dracula*.

to put `'rank'` on the y-axis and add `loglog=True`:

```
ax = df.plot.scatter(x='word_frequency',
                     y='rank', loglog=True,
                     figsize=[12, 6],
                     grid=True,
                     xlim=args.xlim)
```

When we check our status again, it prints:

```
$ git status
```

```
On branch master
Changes not staged for commit:
  (use "git add <file>..." to update what will be
  committed)
  (use "git restore <file>..." to discard changes in
  working directory)
      modified:   bin/plotcounts.py

no changes added to commit (use "git add" or "git commit -a")
```

The last line tells us that a file Git already knows about has been modified.

Hints from Git

After executing Git commands, you may see output that differs slightly from what is shown here. For example, you may see a suggestion for `git checkout` in place of `git restore` after executing the code above, which means you're running an different version of Git. As with most tasks in coding, there are often multiple commands to accomplish the same action with Git. This chapter will show output from Git version 2.29. If you see something different in your Git output, you can try the commands we present here, or follow the suggestions included in the output you see. When in doubt, check the documentation (e.g., `git checkout --help`) if you get stuck.

To save those changes in the repository's history, we must `git add` and then `git commit`. Before we do, though, let's review the changes using `git diff`. This command shows us the differences between the current state of our repository and the most recently saved version:

```
$ git diff
```

```
diff --git a/bin/plotcounts.py b/bin/plotcounts.py
index f274473..c4c5b5a 100644
--- a/bin/plotcounts.py
+++ b/bin/plotcounts.py
@@ -13,7 +13,7 @@ def main(args):
                                      method='max')
     df['inverse_rank'] = 1 / df['rank']
     ax = df.plot.scatter(x='word_frequency',
-                         y='inverse_rank',
+                         y='rank', loglog=True,
                          figsize=[12, 6],
                          grid=True,
                          xlim=args.xlim)
```

The output is cryptic, even by the standards of the Unix command line, because it is actually a series of commands telling editors and other tools how to turn the file we *had* into the file we *have*. If we break it down into pieces:

1. The first line tells us that Git is producing output in the format of the Unix `diff` command.

2. The second line tells exactly which versions of the file Git is comparing: `f274473` and `c4c5b5a` are the short identifiers for those versions.
3. The third and fourth lines once again show the name of the file being changed; the name appears twice in case we are renaming a file as well as modifying it.
4. The remaining lines show us the changes and the lines on which they occur. A minus sign `-` in the first column indicates a line that is being removed, while a plus sign `+` shows a line that is being added. Lines without either plus or minus signs have not been changed, but are provided around the lines that have been changed to add context.

Git's default is to compare line by line, but it can be instructive to instead compare word by word using the `--word-diff` or `--color-words` options. These are particularly useful when running `git diff` on prose rather than code.

After reviewing our change we can commit it just as we did before:

```
$ git commit -m "Plot frequency against rank on log-log axes"
```

```
On branch master
Changes not staged for commit:
  (use "git add <file>..." to update what will be
  committed)
  (use "git restore <file>..." to discard changes in
  working directory)
      modified:   bin/plotcounts.py

no changes added to commit (use "git add" or "git commit -a")
```

Whoops: we forgot to add the file to the set of things we want to commit. Let's do that and then try the commit again:

```
$ git add bin/plotcounts.py
$ git status
```

```
On branch master
Changes to be committed:
  (use "git restore --staged <file>..." to unstage)
      modified:   bin/plotcounts.py
```

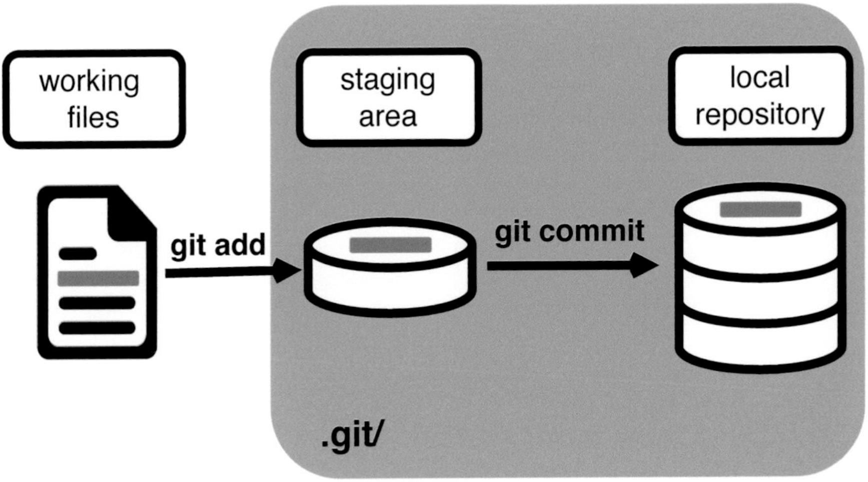

FIGURE 6.3: The staging area in Git.

```
$ git commit -m "Plot frequency against rank on log-log axes"
```

```
[master 582f7f6] Plot frequency against rank on log-log axes
1 file changed, 1 insertion(+), 1 deletion(-)
```

The Staging Area

Git insists that we add files to the set we want to commit before actually committing anything. This allows us to commit our changes in stages and capture changes in logical portions rather than only large batches. For example, suppose we add a few citations to the introduction of our thesis, which is in the file `introduction.tex`. We might want to commit those additions but not commit the changes to `conclusion.tex` (which we haven't finished writing yet). To allow for this, Git has a special **staging area** where it keeps track of things that have been added to the current changeset but not yet committed (Figure 6.3).

Let's take a look at our new plot (Figure 6.4):

```
$ python bin/plotcounts.py results/dracula.csv --outfile
  results/dracula.png
```

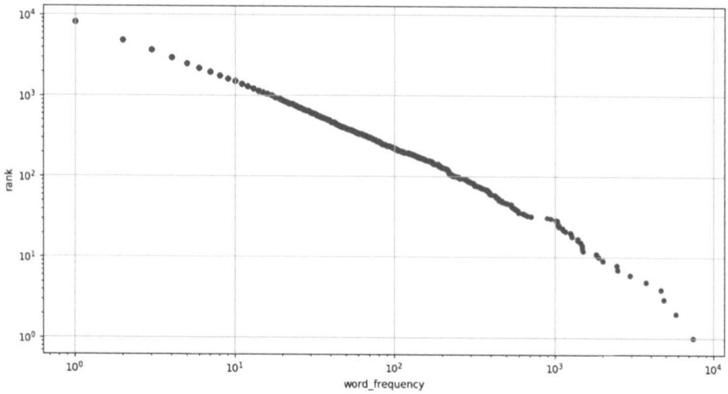

FIGURE 6.4: Rank versus word frequency on log-log axes for 'Dracula'.

Interpreting Our Plot

If Zipf's Law holds, we should still see a linear relationship, although now it will be negative, rather than positive (since we're plotting the rank instead of the reverse rank). The low-frequency words (below about 120 instances) seem to follow a straight line very closely, but we currently have to make this evaluation by eye. In the next chapter, we'll write code to fit and add a line to our plot.

Running `git status` again shows that our plot has been modified:

```
On branch master
Changes not staged for commit:
  (use "git add <file>..." to update what will be
  committed)
  (use "git restore <file>..." to discard changes in
  working directory)
      modified:   results/dracula.png

no changes added to commit (use "git add" or "git commit -a")
```

Since `results/dracula.png` is a binary file rather than text, `git diff` can't show what has changed. It therefore simply tells us that the new file is different from the old one:

```
diff --git a/results/dracula.png b/results/dracula.png
index c1f62fd..57a7b70 100644
Binary files a/results/dracula.png and
b/results/dracula.png differ
```

This is one of the biggest weaknesses of Git (and other version control systems): they are built to handle text. They can track changes to images, PDFs, and other formats, but they cannot do as much to show or merge differences. In a better world than ours, programmers fixed this years ago.

If we are sure we want to save all of our changes, we can add and commit in a single command by giving `git commit` the `-a` option:

```
$ git commit -a -m "Update dracula plot"
```

```
[master ee8684c] Update dracula plot
 1 file changed, 0 insertions(+), 0 deletions(-)
 rewrite results/dracula.png (99%)
```

The Git commands we've covered so far (`git add`, `git commit`, `git diff`) represent the tasks you perform in a basic Git workflow in a local repository (Figure 6.5a).

6.6 Synchronizing with Other Repositories

Sooner or later our computer will experience a hardware failure, be stolen, or be thrown in the lake by someone who thinks that we shouldn't spend the entire vacation working on our thesis. Even before that happens we will probably want to collaborate with others, which we can do by linking our local repository to one stored on a hosting service such as GitHub[4].

[4]`https://github.com`

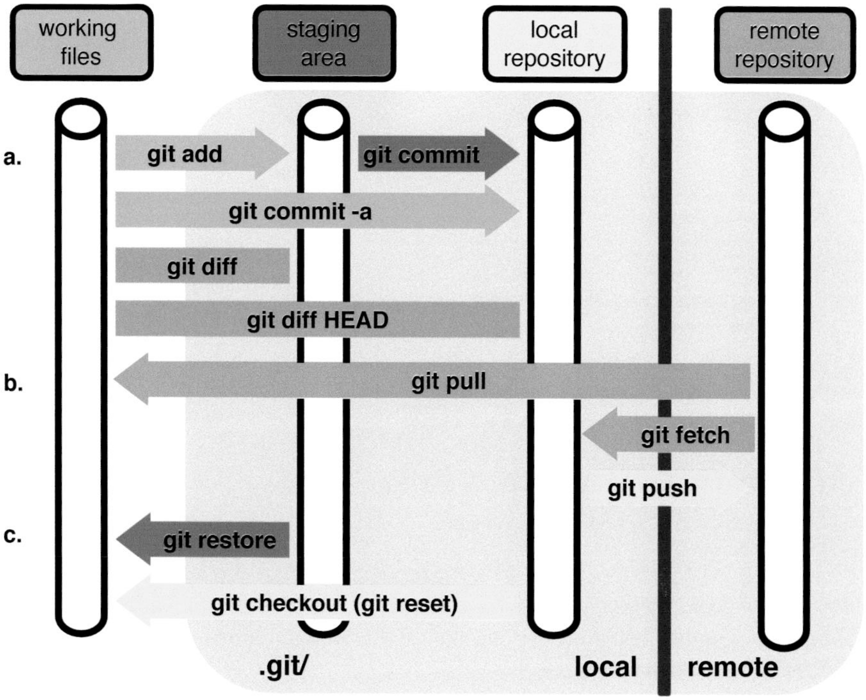

FIGURE 6.5: Commands included in basic Git workflows.

Where's My Repository?

So far we've worked with repositories located on your own computer, which we'll also refer to as local or desktop repositories. The alternative is hosting repositories on GitHub or another server, which we'll refer to as a remote or GitHub repository.

The first steps are to create an account on GitHub, and then select the option there to create a new remote repository to synchronize with our local repository. Select the option on GitHub to create a new repository, then add the requested information for your Zipf's Law project. The remote repository doesn't have to have the same name as the local one, but we will probably get confused if they are different, so the repository we create on GitHub will also be called `zipf`. The other default options are likely appropriate for your remote repository. Because we are synchronizing with an existing repository, do not add a README, `.gitignore`, or license; we'll discuss these additions in other chapters.

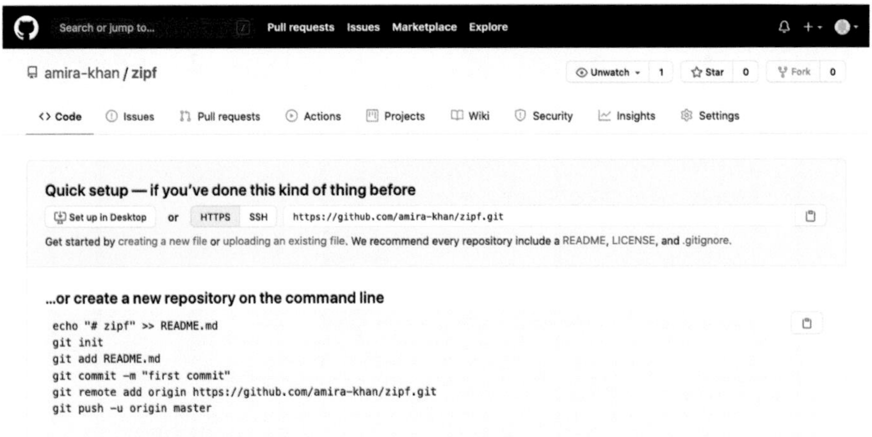

FIGURE 6.6: Find the link for a GitHub repository.

Next, we need to connect our desktop repository with the one on GitHub. We do this by making the GitHub repository a **remote** of the local repository. The home page of our new repository on GitHub includes the string we need to identify it (Figure 6.6).

We can click on "HTTPS" to change the URL from SSH to HTTPS and then copy that URL.

HTTPS vs. SSH

We use HTTPS here because it does not require additional configuration. If we want to set up SSH access so that we do not have to type in our password as often, the tutorials from GitHub[5], BitBucket[6], or GitLab[7] explain the steps required.

Next, let's go into the local `zipf` repository and run this command:

```
$ cd ~/zipf
$ git remote add origin https://github.com/amira-khan/zipf.git
```

[5] https://help.github.com/articles/generating-ssh-keys
[6] https://confluence.atlassian.com/bitbucket/set-up-ssh-for-git-728138079.html
[7] https://about.gitlab.com/2014/03/04/add-ssh-key-screencast/

Make sure to use the URL for your repository instead of the one shown: the only difference should be that it includes your username instead of `amira-khan`.

A Git remote is like a bookmark: it gives a short name to a URL. In this case the remote's name is `origin`; we could use anything we want, but `origin` is Git's default, so we will stick with it. We can check that the command has worked by running `git remote -v` (where the `-v` option is short for **v**erbose):

```
$ git remote -v
```

```
origin  https://github.com/amira-khan/zipf.git (fetch)
origin  https://github.com/amira-khan/zipf.git (push)
```

Git displays two lines because it's actually possible to set up a remote to download from one URL but upload to another. Sensible people don't do this, so we won't explore this possibility any further.

Now that we have configured a remote, we can **push** the work we have done so far to the repository on GitHub:

```
$ git push origin master
```

This may prompt us to enter our username and password; once we do that, Git prints a few lines of administrative information:

```
Enumerating objects: 35, done.
Counting objects: 100% (35/35), done.
Delta compression using up to 4 threads
Compressing objects: 100% (35/35), done.
Writing objects: 100% (35/35), 2.17 MiB | 602.00 KiB/s, done.
Total 35 (delta 7), reused 0 (delta 0), pack-reused 0
remote: Resolving deltas: 100% (7/7), done.
To https://github.com/amira-khan/zipf.git
 * [new branch]      master -> master
```

If we view our GitHub repository in the browser, it now includes all of our project files, along with all of the commits we have made so far (Figure 6.7).

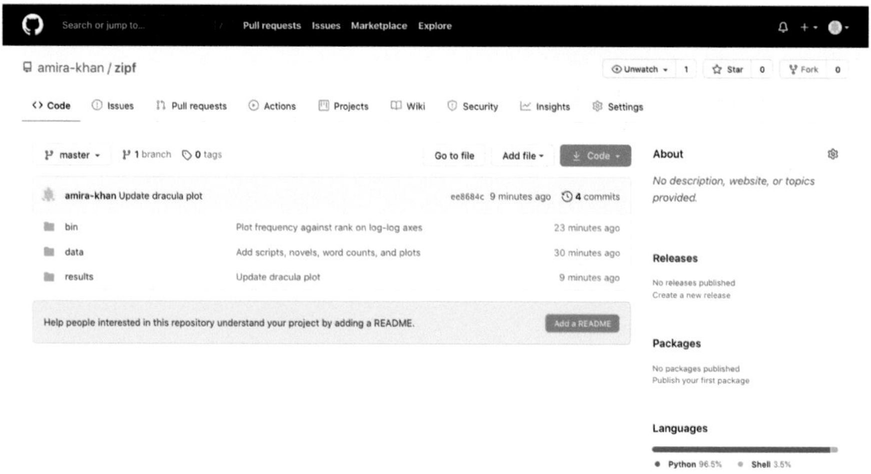

FIGURE 6.7: Viewing the repository's history on GitHub.

We can also **pull** changes from the remote repository to the local one:

```
$ git pull origin master
```

```
From https://github.com/amira-khan/zipf
 * branch            master     -> FETCH_HEAD
Already up-to-date.
```

Pulling has no effect in this case because the two repositories are already synchronized.

Fetching

The second line in the remote configuration we viewed earlier is labeled `push`, which makes sense given the command we used (`git push`) to upload changes from our local to remote repositories. Why is the other line labeled `fetch` instead of `pull`? Fetching and pulling both download new data from a remote repository, but only pulling integrates those changes into your local repository's version history. Because `git fetch` doesn't alter your local files, it's used to view changes between local and remote versions.

The Git commands we've covered in this section (`git pull`, `git push`) are the main tasks associated with incorporating remote repositories into your workflow (Figure 6.5b).

Amira's Repository

Amira's repository referenced in this section exists on GitHub at amira-khan/zipf[8]; you may find it a useful reference point when proceeding through the rest of the book.

6.7 Exploring History

Git lets us look at previous versions of files and restore specific files to earlier states if we want to. In order to do these things, we need to identify the versions we want.

The two ways to do this are analogous to **absolute** and **relative** paths. The "absolute" version is the unique identifier that Git gives to each commit. These identifiers are 40 characters long, but in most situations Git will let us use just the first half dozen characters or so. For example, if we run `git log` right now, it shows us something like this:

```
commit ee8684ca123e1e829fc995d672e3d7e4b00f2610
(HEAD -> master, origin/master)
Author: Amira Khan <amira@zipf.org>
Date:   Sat Dec 19 09:52:04 2020 -0800

    Update dracula plot

commit 582f7f6f536d520b1328c04c9d41e24b54170656
Author: Amira Khan <amira@zipf.org>
Date:   Sat Dec 19 09:37:25 2020 -0800

    Plot frequency against rank on log-log axes

commit 851d590a214c7859eafa0998c6c951f8e0eb359b
Author: Amira Khan <amira@zipf.org>
Date:   Sat Dec 19 09:32:41 2020 -0800

    Add plot of word counts for 'Dracula'

commit 173222bf90216b408c8997f4e143572b99637750
```

[8] `https://github.com/amira-khan/zipf`

```
Author: Amira Khan <amira@zipf.org>
Date:   Sat Dec 19 09:30:23 2020 -0800

    Add scripts, novels, word counts, and plots
```

The commit in which we changed `plotcounts.py` has the absolute identifier `582f7f6f536d520b1328c04c9d41e24b54170656`, but we can use `582f7f6` to reference it in almost all situations.

While `git log` includes the commit message, it doesn't tell us exactly what changes were made in each commit. If we add the `-p` option (short for **p**atch), we get the same kind of details `git diff` provides to describe the changes in each commit:

```
$ git log -p
```

The first part of the output is shown below; we have truncated the rest, since it is very long:

```
commit ee8684ca123e1e829fc995d672e3d7e4b00f2610
(HEAD -> master, origin/master)
Author: Amira Khan <amira@zipf.org>
Date:   Sat Dec 19 09:52:04 2020 -0800

    Update dracula plot

diff --git a/results/dracula.png b/results/dracula.png
index c1f62fd..57a7b70 100644
Binary files a/results/dracula.png and
b/results/dracula.png differ
...
```

Alternatively, we can use `git diff` directly to examine the differences between files at any stage in the repository's history. Let's explore this with the `plotcounts.py` file. We no longer need the line of code in `plotcounts.py` that calculates the inverse rank:

```
df['inverse_rank'] = 1 / df['rank']
```

If we delete that line from `bin/plotcounts.py`, `git diff` on its own will show the difference between the file as it is now and the most recent version:

```
diff --git a/bin/plotcounts.py b/bin/plotcounts.py
index c4c5b5a..c511da1 100644
--- a/bin/plotcounts.py
+++ b/bin/plotcounts.py
@@ -11,7 +11,6 @@ def main(args):
                       names=('word', 'word_frequency'))
     df['rank'] = df['word_frequency'].rank(ascending=False,
                                            method='max')
-    df['inverse_rank'] = 1 / df['rank']
     ax = df.plot.scatter(x='word_frequency',
                          y='rank', loglog=True,
                          figsize=[12, 6],
```

`git diff 582f7f6`, on the other hand, shows the difference between the current state and the commit referenced by the short identifier:

```
diff --git a/bin/plotcounts.py b/bin/plotcounts.py
index c4c5b5a..c511da1 100644
--- a/bin/plotcounts.py
+++ b/bin/plotcounts.py
@@ -11,7 +11,6 @@ def main(args):
                       names=('word', 'word_frequency'))
     df['rank'] = df['word_frequency'].rank(ascending=False,
                                            method='max')
-    df['inverse_rank'] = 1 / df['rank']
     ax = df.plot.scatter(x='word_frequency',
                          y='rank', loglog=True,
                          figsize=[12, 6],
diff --git a/results/dracula.png b/results/dracula.png
index c1f62fd..57a7b70 100644
Binary files a/results/dracula.png and
b/results/dracula.png differ
```

Note that you will need to reference your `git log` to replace `582f7f6` in the code above, since Git assigned your commit a different unique identifier. Note also that we have *not* committed the last change to `plotcounts.py`; we will look at ways of undoing it in the next section.

The "relative" version of history relies on a special identifier called `HEAD`, which always refers to the most recent version in the repository. `git diff HEAD` therefore shows the same thing as `git diff`, but instead of typing in a version identifier to back up one commit, we can use `HEAD~1` (where `~` is the tilde symbol). This shorthand is read "HEAD minus one," and gives us the difference to the previous saved version. `git diff HEAD~2` goes back two revisions and so on. We can also look at the differences between two saved versions by separating their identifiers with two dots `..` like this:

```
$ git diff HEAD~1..HEAD~2
```

```
diff --git a/bin/plotcounts.py b/bin/plotcounts.py
index c4c5b5a..f274473 100644
--- a/bin/plotcounts.py
+++ b/bin/plotcounts.py
@@ -13,7 +13,7 @@ def main(args):
                                         method='max')
     df['inverse_rank'] = 1 / df['rank']
     ax = df.plot.scatter(x='word_frequency',
-                         y='rank', loglog=True,
+                         y='inverse_rank',
                          figsize=[12, 6],
                          grid=True,
                          xlim=args.xlim)
```

If we want to see the changes made in a particular commit, we can use `git show` with an identifier and a filename:

```
$ git show HEAD~1 bin/plotcounts.py
```

```
commit 582f7f6f536d520b1328c04c9d41e24b54170656
Author: Amira Khan <amira@zipf.org>
Date:   Sat Dec 19 09:37:25 2020 -0800

    Plot frequency against rank on log-log axes

diff --git a/bin/plotcounts.py b/bin/plotcounts.py
index f274473..c4c5b5a 100644
--- a/bin/plotcounts.py
+++ b/bin/plotcounts.py
@@ -13,7 +13,7 @@ def main(args):
```

```
                                                     method='max')
     df['inverse_rank'] = 1 / df['rank']
     ax = df.plot.scatter(x='word_frequency',
-                           y='inverse_rank',
+                           y='rank', loglog=True,
                            figsize=[12, 6],
                            grid=True,
                            xlim=args.xlim)
```

If we wanted to view the contents of a file at a given point in the version history, we could use the same command, but separating the identifier and file with a colon:

```
$ git show HEAD~1:bin/plotcounts.py
```

This allows us to look through the file using a paging program.

6.8 Restoring Old Versions of Files

We can see what we changed, but how can we restore it? Suppose we change our mind about the last update to `bin/plotcounts.py` (removing `df['inverse_rank'] = 1 / df['rank']`) before we add or commit it. `git status` tells us that the file has been changed, but those changes haven't been **staged**:

```
$ git status
```

```
On branch master
Changes not staged for commit:
  (use "git add <file>..." to update what will be
  committed)
  (use "git restore <file>..." to discard changes in
  working directory)
    modified:   bin/plotcounts.py

no changes added to commit (use "git add" or "git commit -a")
```

We can put things back the way they were in the last saved revision using `git restore`, as the screen output suggests:

```
$ git restore bin/plotcounts.py
$ git status
```

```
On branch master
nothing to commit, working tree clean
```

As its name suggests, `git restore` restores an earlier version of a file. In this case, we used it to recover the version of the file in the most recent commit.

Checking Out with Git

If you're running a different version of Git, you may see a suggestion for `git checkout` instead of `git restore`. As of Git version 2.29, `git restore` is still an experimental command, and operates as a specialized form of `git checkout`. `git checkout HEAD bin/plotcounts.py` is equivalent to the last command run.

We can confirm the file has been restored by printing the relevant lines of the file:

```
$ head -n 19 bin/plotcounts.py | tail -n 8
```

```
    df['rank'] = df['word_frequency'].rank(ascending=False,
                                           method='max')
    df['inverse_rank'] = 1 / df['rank']
    ax = df.plot.scatter(x='word_frequency',
                         y='rank', loglog=True,
                         figsize=[12, 6],
                         grid=True,
                         xlim=args.xlim)
```

Because `git restore` is designed to restore working files, we'll need to use `git checkout` to revert to earlier versions of files. We can use a specific commit identifier rather than `HEAD` to go back as far as we want:

```
$ git checkout 851d590 bin/plotcounts.py
```

```
Updated 1 path from c8d6a33
```

Doing this does not change the history: `git log` still shows our four commits. Instead, it replaces the content of the file with the old content:

```
$ git status
```

```
On branch master
Changes to be committed:
  (use "git restore --staged <file>..." to unstage)
      modified:   bin/plotcounts.py
```

```
$ head -n 19 bin/plotcounts.py | tail -n 8
```

```
    df['rank'] = df['word_frequency'].rank(ascending=False,
                                            method='max')
    df['inverse_rank'] = 1 / df['rank']
    ax = df.plot.scatter(x='word_frequency',
                         y='inverse_rank',
                         figsize=[12, 6],
                         grid=True,
                         xlim=args.xlim)
```

If we change our mind again, we can use the suggestion in the output to restore the earlier version. Because checking out the changes added them to the staging area, we need to first remove them from the staging area:

```
$ git restore --staged bin/plotcounts.py
```

However, the changes have been unstaged but still exist in the file. We can return the file to the state of the most recent commit:

```
$ git restore bin/plotcounts.py
$ git status
```

```
On branch master
nothing to commit, working tree clean
```

```
$ head -n 19 bin/plotcounts.py | tail -n 8
```

```
    df['rank'] = df['word_frequency'].rank(ascending=False,
                                            method='max')
    df['inverse_rank'] = 1 / df['rank']
    ax = df.plot.scatter(x='word_frequency',
                         y='rank', loglog=True,
                         figsize=[12, 6],
                         grid=True,
                         xlim=args.xlim)
```

We have restored the most recent commit. Since we didn't commit the change that removed the line that calculates the inverse rank, that work is now lost: Git can only go back and forth between committed versions of files.

This section has demonstrated a few different ways to view differences among versions, and to work with those changes (Figure 6.5c). These commands can operate on either individual files or entire commits, and the behavior of them can sometimes differ based on your version of Git. Remember to reference documentation, and use `git status` and `git log` frequently to understand your workflow.

6.9 Ignoring Files

We don't always want Git to track every file's history. For example, we might want to track text files with names ending in `.txt` but not data files with names ending in `.dat`.

To stop Git from telling us about these files every time we call `git status`, we can create a file in the root directory of our project called `.gitignore`. This file can contain filenames like `thesis.pdf` or **wildcard** patterns like `*.dat`. Each must be on a line of its own, and Git will ignore anything that matches any of these lines. For now we only need one entry in our `.gitignore` file:

```
__pycache__
```

which tells Git to ignore any `__pycache__` directory created by Python (Section 5.8).

Remember to Ignore

Don't forget to commit `.gitignore` to your repository so that Git knows to use it.

6.10 Summary

The biggest benefit of version control for individual research is that we can always go back to the precise set of files that we used to produce a particular result. While Git is complex (Perez De Rosso and Jackson 2013), being able to back up our changes on sites like GitHub with just a few keystrokes can save us a lot of pain, and some of Git's advanced features make it even more powerful. We will explore these in the next chapter.

6.11 Exercises

6.11.1 Places to create Git repositories

Along with information about the Zipf's Law project, Amira would also like to keep some notes on Heaps' Law[9]. Despite her colleagues' concerns, Amira creates a `heaps-law` project inside her `zipf` project as follows:

```
$ cd ~/zipf
$ mkdir heaps-law
$ cd heaps-law
$ git init heaps-law
```

Is the `git init` command that she runs inside the `heaps-law` subdirectory required for tracking files stored there?

[9] https://en.wikipedia.org/wiki/Heaps%27_law

6.11.2 Removing before saving

If you're working on an older version of Git, you may see an output from `git status` suggesting you can take files out of the staging area using `git rm --cached`. Try this out:

1. Create a new file in an initialized Git repository called `example.txt`.
2. Use `git add example.txt` to add this file.
3. Use `git status` to check that Git has noticed it.
4. Use `git rm --cached example.txt` to remove it from the list of things to be saved.

What does `git status` now show? What (if anything) has happened to the file?

6.11.3 Viewing changes

Make a few changes to a file in a Git repository, then view those differences using both `git diff` and `git diff --word-diff`. Which output do you find easiest to understand?

Note: If you performed this exercise in your Zipf's Law project, we recommend discarding (not committing) your changes made to a file.

6.11.4 Committing changes

Which command(s) below would save changes to `myfile.txt` to a local Git repository?

```
# Option 1
$ git commit -m "Add recent changes"
```

```
# Option 2
$ git init myfile.txt
$ git commit -m "Add recent changes"
```

```
# Option 3
$ git add myfile.txt
$ git commit -m "Add recent changes"


# Option 4
$ git commit -m myfile.txt "Add recent changes"
```

6.11.5 Write your biography

1. Create a new Git repository on your computer called `bio`. Make sure the directory containing this repository is outside your `zipf` project directory!
2. Write a three-line biography for yourself in a file called `me.txt` and commit your changes.
3. Modify one line and add a fourth line.
4. Display the differences between the file's original state and its updated state.

6.11.6 Committing multiple files

The staging area can hold changes from any number of files that you want to commit as a single snapshot. From your new `bio` directory, following after the previous exercise (wich leaves you with uncommitted changes to `me.txt`):

1. Create another new file `employment.txt` and add the details of your most recent job.
2. Add the new changes to both `me.txt` and `employment.txt` to the staging area and commit those changes.

6.11.7 GitHub timestamps

1. Create a remote repository on GitHub for your new `bio` repository.
2. Push the contents of your local repository to the remote.
3. Make a change to your local repository (e.g., edit `me.txt` or `employment.txt`) and push these changes as well.
4. Go to the repo you just created on GitHub and check the timestamps of the files.

How does GitHub record times, and why?

6.11.8 Workflow and history

Assume you made the following changes in your `bio` repository. What is the output of the last command in the sequence below?

```
$ echo "Sharing information about myself." > motivation.txt
$ git add motivation.txt
$ echo "Documenting major milestones." > motivation.txt
$ git commit -m "Motivate project"
$ git restore motivation.txt
$ cat motivation.txt
```

1. `Sharing information about myself.`
2. `Documenting major milestones.`
3. `Sharing information about myself.`
   `Documenting major milestones.`
4. An error message because we have changed `motivation.txt` without committing first.

6.11.9 Ignoring nested files

Suppose our project has a directory `results` with two subdirectories called `data` and `plots`. How would we ignore all of the files in `results/plots` but not ignore files in `results/data`?

6.11.10 Including specific files

How would you ignore all `.dat` files in your root directory except for `final.dat`? (Hint: find out what the exclamation mark `!` means in a `.gitignore` file.)

6.11.11 Exploring the GitHub interface

Browse to your `zipf` repository on GitHub. Under the `Code` tab, find and click on the text that says "NN commits" (where "NN" is some number). Hover over and click on the three buttons to the right of each commit. What information can you gather/explore from these buttons? How would you get that same information in the shell?

6.11.12 Push versus commit

Explain in one or two sentences how `git push` is different from `git commit`.

6.11.13 License and README files

When we initialized our remote `zipf` GitHub repo, we didn't add a `README.md` or license file. If we had, what would have happened when we tried to link our local and remote repositories?

6.11.14 Recovering older versions of a file

Amira made changes this morning to a shell script called `data_cruncher.sh` that she has been working on for weeks. Her changes broke the script, and she has now spent an hour trying to get it back in working order. Luckily, she has been keeping track of her project's versions using Git. Which of the commands below can she use to recover the last committed version of her script?

1. `$ git checkout HEAD`
2. `$ git checkout HEAD data_cruncher.sh`
3. `$ git checkout HEAD~1 data_cruncher.sh`
4. `$ git checkout <unique ID of last commit> data_cruncher.sh`
5. `$ git restore data_cruncher.sh`
6. `$ git restore HEAD`

6.11.15 Understanding `git diff`

Using your `zipf` project directory:

1. What would the command `git diff HEAD~9 bin/plotcounts.py` do if we run it?
2. What does it actually do?
3. What does `git diff HEAD bin/plotcounts.py` do?

6.11.16 Getting rid of staged changes

`git checkout` can be used to restore a previous commit when unstaged changes have been made, but will it also work for changes that have been staged but not committed? To find out, use your `zipf` project directory to:

1. Change `bin/plotcounts.py`.
2. Use `git add` on those changes to `bin/plotcounts.py`.
3. Use `git checkout` to see if you can remove your change.

Does it work?

6.11.17 Figuring out who did what

We never committed the last edit that removes the calculation of inverse rank. Remove this line from `plotcounts.py`, then commit the change:

```
df['inverse_rank'] = 1 / df['rank']
```

Run the command `git blame bin/plotcounts.py`. What does each line of the output show?

6.12 Key Points

- Use `git config` with the `--global` option to configure your username, email address, and other preferences once per machine.
- `git init` initializes a **repository**.
- Git stores all repository management data in the `.git` subdirectory of the repository's root directory.
- `git status` shows the status of a repository.
- `git add` puts files in the repository's staging area.
- `git commit` saves the staged content as a new commit in the local repository.
- `git log` lists previous commits.
- `git diff` shows the difference between two versions of the repository.
- Synchronize your local repository with a **remote repository** on a **forge** such as GitHub[10].
- `git remote` manages bookmarks pointing at remote repositories.
- `git push` copies changes from a local repository to a remote repository.
- `git pull` copies changes from a remote repository to a local repository.
- `git restore` and `git checkout` recover old versions of files.
- The `.gitignore` file tells Git what files to ignore.

[10] `https://github.com`

7

Going Further with Git

> It's got three keyboards and a hundred extra knobs, including twelve with '?' on them.
>
> — Terry Pratchett

Two of Git's advanced features let us to do much more than just track our work. **Branches** let us work on multiple things simultaneously in a single repository; **pull requests** (PRs) let us submit our work for review, get feedback, and make updates. Used together, they allow us to go through the write-review-revise cycle familiar to anyone who has ever written a journal paper in hours rather than weeks.

Your `zipf` project directory should now include:

```
zipf/
├── .gitignore
├── bin
│   ├── book_summary.sh
│   ├── collate.py
│   ├── countwords.py
│   ├── plotcounts.py
│   ├── script_template.py
│   └── utilities.py
├── data
│   ├── README.md
│   ├── dracula.txt
│   ├── frankenstein.txt
│   └── ...
└── results
    ├── dracula.csv
    ├── dracula.png
    ├── jane_eyre.csv
    ├── jane_eyre.png
    └── moby_dick.csv
```

All of these files should also be tracked in your version history. We'll use them and some additional analyses to explore Zipf's Law using Git's advanced features.

7.1 What's a Branch?

So far we have only used a sequential timeline with Git: each change builds on the one before, and *only* on the one before. However, there are times when we want to try things out without disrupting our main work. To do this, we can use **branches** to work on separate tasks in parallel. Each branch is a parallel timeline; changes made on the branch only affect that branch unless and until we explicitly combine them with work done in another branch.

We can see what branches exist in a repository using this command:

```
$ git branch
```

```
* master
```

When we initialize a repository, Git automatically creates a branch called `master`. It is often considered the "official" version of the repository. The asterisk `*` indicates that it is currently active, i.e., that all changes we make will take place in this branch by default. (The active branch is like the **current working directory** in the shell.)

> **Default Branches**
>
> In mid-2020, GitHub changed the name of the default branch (the first branch created when a repository is initialized) from "master" to "main." Owners of repositories may also change the name of the default branch. This means that the name of the default branch may be different among repositories based on when and where it was created, as well as who manages it.

In the previous chapter, we foreshadowed some experimental changes that we could try and make to `plotcounts.py`.

Making sure our project directory is our working directory, we can inspect our current `plotcounts.py`:

```
$ cd ~/zipf
$ cat bin/plotcounts.py
```

```python
"""Plot word counts."""

import argparse

import pandas as pd


def main(args):
    """Run the command line program."""
    df = pd.read_csv(args.infile, header=None,
                     names=('word', 'word_frequency'))
    df['rank'] = df['word_frequency'].rank(ascending=False,
                                            method='max')
    ax = df.plot.scatter(x='word_frequency',
                         y='rank', loglog=True,
                         figsize=[12, 6],
                         grid=True,
                         xlim=args.xlim)
    ax.figure.savefig(args.outfile)


if __name__ == '__main__':
    parser = argparse.ArgumentParser(description=__doc__)
    parser.add_argument('infile', type=argparse.FileType('r'),
                        nargs='?', default='-',
                        help='Word count csv file name')
    parser.add_argument('--outfile', type=str,
                        default='plotcounts.png',
                        help='Output image file name')
    parser.add_argument('--xlim', type=float, nargs=2,
                        metavar=('XMIN', 'XMAX'),
                        default=None, help='X-axis limits')
    args = parser.parse_args()
    main(args)
```

We used this version of `plotcounts.py` to display the word counts for *Dracula* on a log-log plot (Figure 6.4). The relationship between word count and rank looked linear, but since the eye is easily fooled, we should fit a curve to the data. Doing this will require more than just a trivial change to the script, so to ensure that this version of `plotcounts.py` keeps working while we try to build a new one, we will do our work in a separate branch. Once we have successfully added curve fitting to `plotcounts.py`, we can decide if we want to merge our changes back into the `master` branch.

7.2 Creating a Branch

To create a new branch called `fit`, we run:

```
$ git branch fit
```

We can check that the branch exists by running `git branch` again:

```
$ git branch
```

```
* master
  fit
```

Our branch is there, but the asterisk `*` shows that we are still in the `master` branch. (By analogy, creating a new directory doesn't automatically move us into that directory.) As a further check, let's see what our repository's status is:

```
$ git status
```

```
On branch master
nothing to commit, working directory clean
```

To switch to our new branch we can use the `checkout` command that we first saw in Chapter 6:

```
$ git checkout fit
$ git branch
```

```
  master
* fit
```

In this case, we're using `git checkout` to check out a whole repository, i.e., switch it from one saved state to another.

We should choose the name to signal the purpose of the branch, just as we choose the names of files and variables to indicate what they are for. We haven't made any changes since switching to the `fit` branch, so at this point `master` and `fit` are at the same point in the repository's history. Commands like `ls` and `git log` therefore show that the files and history haven't changed.

Where Are Branches Saved?

Git saves every version of every file in the `.git` directory that it creates in the project's root directory. When we switch from one branch to another, it replaces the files we see with their counterparts from the branch we're switching to. It also rearranges directories as needed so that those files are in the right places.

7.3 What Curve Should We Fit?

Before we make any changes to our new branch, we need to figure out how to fit a line to the word count data. Zipf's Law says:

> The second most common word in a body of text appears half as often as the most common, the third most common appears a third as often, and so on.

In other words the frequency of a word f is proportional to its inverse rank r:

$$f \propto \frac{1}{r^\alpha}$$

with a value of α close to one. The reason α must be close to one for Zipf's Law to hold becomes clear if we include it in a modified version of the earlier definition:

> The most frequent word will occur approximately 2^α times as often as the second most frequent word, 3^α times as often as the third most frequent word, and so on.

This mathematical expression for Zipf's Law is an example of a **power law**. In general, when two variables x and y are related through a power law, so that

$$y = ax^b$$

taking logarithms of both sides yields a linear relationship:

$$\log(y) = \log(a) + b\log(x)$$

Hence, plotting the variables on a log-log scale reveals this linear relationship. If Zipf's Law holds, we should have

$$r = cf^{\frac{-1}{\alpha}}$$

where c is a constant of proportionality. The linear relationship between the log word frequency and log rank is then

$$\log(r) = \log(c) - \frac{1}{\alpha}\log(f)$$

This suggests that the points on our log-log plot should fall on a straight line with a slope of $-\frac{1}{\alpha}$ and intercept $\log(c)$. To fit a line to our word count data we therefore need to estimate the value of α; we'll see later that c is completely defined.

In order to determine the best method for estimating α, we turn to Moreno-Sánchez, Font-Clos, and Corral (2016), which suggests using a method called **maximum likelihood estimation**. The likelihood function is the probability of our observed data as a function of the parameters in the statistical model that we assume generated it. We estimate the parameters in the model by choosing them to maximize this likelihood; computationally, it is often easier to minimize the negative log likelihood function. Moreno-Sánchez, Font-Clos, and Corral (2016) define the likelihood using a parameter β, which is related to the α parameter in our definition of Zipf's Law through $\alpha = \frac{1}{\beta-1}$. Under their model, the value of c is the total number of unique words, or equivalently the largest value of the rank.

Expressed as a Python function, the negative log likelihood function is:

```
import numpy as np


def nlog_likelihood(beta, counts):
    """Log-likelihood function."""
    likelihood = - np.sum(np.log((1/counts)**(beta - 1)
                          - (1/(counts + 1))**(beta - 1)))
    return likelihood
```

Obtaining an estimate of β (and thus α) then becomes a numerical optimization problem, for which we can use the `scipy.optimize`[1] library. Again following Moreno-Sánchez, Font-Clos, and Corral (2016), we use Brent's Method with $1 < \beta \leq 4$.

```
from scipy.optimize import minimize_scalar


def get_power_law_params(word_counts):
    """Get the power law parameters."""
    mle = minimize_scalar(nlog_likelihood,
                          bracket=(1 + 1e-10, 4),
                          args=word_counts,
                          method='brent')
    beta = mle.x
    alpha = 1 / (beta - 1)
    return alpha
```

We can then plot the fitted curve on the plot axes (`ax`) defined in the `plotcounts.py` script:

```
def plot_fit(curve_xmin, curve_xmax, max_rank, alpha, ax):
    """
    Plot the power law curve that was fitted to the data.

    Parameters
    ----------
    curve_xmin : float
```

[1] https://docs.scipy.org/doc/scipy/reference/optimize.html

```
        Minimum x-bound for fitted curve
    curve_xmax : float
        Maximum x-bound for fitted curve
    max_rank : int
        Maximum word frequency rank.
    alpha : float
        Estimated alpha parameter for the power law.
    ax : matplotlib axes
        Scatter plot to which the power curve will be added.
    """
    xvals = np.arange(curve_xmin, curve_xmax)
    yvals = max_rank * (xvals**(-1 / alpha))
    ax.loglog(xvals, yvals, color='grey')
```

where the maximum word frequency rank corresponds to c, and $-1/\alpha$ the exponent in the power law. We have followed the `numpydoc`[2] format for the detailed docstring in `plot_fit`—see Appendix G for more information about docstring formats.

7.4 Verifying Zipf's Law

Now that we can fit a curve to our word count plots, we can update `plotcounts.py` so the entire script reads as follows:

```
"""Plot word counts."""

import argparse

import numpy as np
import pandas as pd
from scipy.optimize import minimize_scalar


def nlog_likelihood(beta, counts):
    """Log-likelihood function."""
    likelihood = - np.sum(np.log((1/counts)**(beta - 1)
```

[2]https://numpydoc.readthedocs.io/en/latest/

```python
                                  - (1/(counts + 1))**(beta - 1)))
    return likelihood


def get_power_law_params(word_counts):
    """Get the power law parameters."""
    mle = minimize_scalar(nlog_likelihood,
                          bracket=(1 + 1e-10, 4),
                          args=word_counts,
                          method='brent')
    beta = mle.x
    alpha = 1 / (beta - 1)
    return alpha


def plot_fit(curve_xmin, curve_xmax, max_rank, alpha, ax):
    """
    Plot the power law curve that was fitted to the data.

    Parameters
    ----------
    curve_xmin : float
        Minimum x-bound for fitted curve
    curve_xmax : float
        Maximum x-bound for fitted curve
    max_rank : int
        Maximum word frequency rank.
    alpha : float
        Estimated alpha parameter for the power law.
    ax : matplotlib axes
        Scatter plot to which the power curve will be added.
    """
    xvals = np.arange(curve_xmin, curve_xmax)
    yvals = max_rank * (xvals**(-1 / alpha))
    ax.loglog(xvals, yvals, color='grey')


def main(args):
    """Run the command line program."""
    df = pd.read_csv(args.infile, header=None,
                     names=('word', 'word_frequency'))
    df['rank'] = df['word_frequency'].rank(ascending=False,
                                           method='max')
    ax = df.plot.scatter(x='word_frequency',
```

```
                     y='rank', loglog=True,
                     figsize=[12, 6],
                     grid=True,
                     xlim=args.xlim)

    word_counts = df['word_frequency'].to_numpy()
    alpha = get_power_law_params(word_counts)
    print('alpha:', alpha)

    # Since the ranks are already sorted, we can take the last
    # one instead of computing which row has the highest rank
    max_rank = df['rank'].to_numpy()[-1]

    # Use the range of the data as the boundaries
    # when drawing the power law curve
    curve_xmin = df['word_frequency'].min()
    curve_xmax = df['word_frequency'].max()

    plot_fit(curve_xmin, curve_xmax, max_rank, alpha, ax)
    ax.figure.savefig(args.outfile)


if __name__ == '__main__':
    parser = argparse.ArgumentParser(description=__doc__)
    parser.add_argument('infile', type=argparse.FileType('r'),
                        nargs='?', default='-',
                        help='Word count csv file name')
    parser.add_argument('--outfile', type=str,
                        default='plotcounts.png',
                        help='Output image file name')
    parser.add_argument('--xlim', type=float, nargs=2,
                        metavar=('XMIN', 'XMAX'),
                        default=None, help='X-axis limits')
    args = parser.parse_args()
    main(args)
```

We can then run the script to obtain the α value for *Dracula* and a new plot with a line fitted.

```
$ python bin/plotcounts.py results/dracula.csv --outfile
  results/dracula.png
```

```
alpha: 1.0866646252515038
```

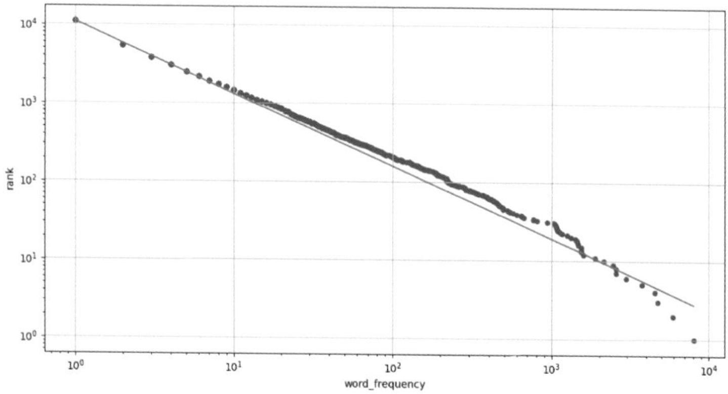

FIGURE 7.1: Word frequency distribution for Dracula.

So according to our fit, the most frequent word will occur approximately $2^{1.1} = 2.1$ times as often as the second most frequent word, $3^{1.1} = 3.3$ times as often as the third most frequent word, and so on. Figure 7.1 shows the plot.

The script appears to be working as we'd like, so we can go ahead and commit our changes to the `fit` development branch:

```
$ git add bin/plotcounts.py results/dracula.png
$ git commit -m "Added fit to word count data"
```

```
[fit 38c209b] Added fit to word count data
 2 files changed, 57 insertions(+)
 rewrite results/dracula.png (99%)
```

If we look at the last couple of commits using `git log`, we see our most recent change:

```
$ git log --oneline -n 2
```

```
38c209b (HEAD -> fit) Added fit to word count data
ddb00fb (origin/master, master) removing inverse rank
        calculation
```

(We use `--oneline` and `-n 2` to shorten the log display.) But if we switch back to the `master` branch:

```
$ git checkout master
$ git branch
```

```
  fit
* master
```

and look at the log, our change is not there:

```
$ git log --oneline -n 2
```

```
ddb00fb (HEAD -> master, origin/master) removing inverse rank
        calculation
7de9877 ignoring __pycache__
```

We have not lost our work: it just isn't included in this branch. We can prove this by switching back to the `fit` branch and checking the log again:

```
$ git checkout fit
$ git log --oneline -n 2
```

```
38c209b (HEAD -> fit) Added fit to word count data
ddb00fb (origin/master, master) removing inverse rank
        calculation
```

We can also look inside `plotcounts.py` and see our changes. If we make another change and commit it, that change will also go into the `fit` branch. For instance, we could add some additional information to one of our docstrings to make it clear what equations were used in estimating α.

```
def get_power_law_params(word_counts):
    """
    Get the power law parameters.

    References
    ----------
    Moreno-Sanchez et al (2016) define alpha (Eq. 1),
      beta (Eq. 2) and the maximum likelihood estimation (mle)
```

```
      of beta (Eq. 6).

    Moreno-Sanchez I, Font-Clos F, Corral A (2016)
      Large-Scale Analysis of Zipf's Law in English Texts.
      PLoS ONE 11(1): e0147073.
      https://doi.org/10.1371/journal.pone.0147073
    """
    mle = minimize_scalar(nlog_likelihood,
                          bracket=(1 + 1e-10, 4),
                          args=word_counts,
                          method='brent')
    beta = mle.x
    alpha = 1 / (beta - 1)
    return alpha
```

```
$ git add bin/plotcounts.py
$ git commit -m "Adding Moreno-Sanchez et al (2016) reference"
```

```
[fit 1577404] Adding Moreno-Sanchez et al (2016) reference
 1 file changed, 14 insertions(+), 1 deletion(-)
```

Finally, if we want to see the differences between two branches, we can use `git diff` with the same double-dot `..` syntax used to view differences between two revisions:

```
$ git diff master..fit
```

```
diff --git a/bin/plotcounts.py b/bin/plotcounts.py
index c511da1..6905b6e 100644
--- a/bin/plotcounts.py
+++ b/bin/plotcounts.py
@@ -2,7 +2,62 @@

 import argparse

+import numpy as np
 import pandas as pd
```

```
+from scipy.optimize import minimize_scalar
+
+
+def nlog_likelihood(beta, counts):
+    """Log-likelihood function."""
+    likelihood = - np.sum(np.log((1/counts)**(beta - 1)
+                          - (1/(counts + 1))**(beta - 1)))
+    return likelihood
+
+
+def get_power_law_params(word_counts):
+    """
+    Get the power law parameters.
+
+    References
+    ----------
+    Moreno-Sanchez et al (2016) define alpha (Eq. 1),
+      beta (Eq. 2) and the maximum likelihood estimation (mle)
+      of beta (Eq. 6).
+
+    Moreno-Sanchez I, Font-Clos F, Corral A (2016)
+      Large-Scale Analysis of Zipf's Law in English Texts.
+      PLoS ONE 11(1): e0147073.
+      https://doi.org/10.1371/journal.pone.0147073
+    """
+    mle = minimize_scalar(nlog_likelihood,
+                          bracket=(1 + 1e-10, 4),
+                          args=word_counts,
+                          method='brent')
+    beta = mle.x
+    alpha = 1 / (beta - 1)
+    return alpha
+
+
+def plot_fit(curve_xmin, curve_xmax, max_rank, alpha, ax):
+    """
+    Plot the power law curve that was fitted to the data.
+
+    Parameters
+    ----------
+    curve_xmin : float
+        Minimum x-bound for fitted curve
+    curve_xmax : float
+        Maximum x-bound for fitted curve
```

```
+     max_rank : int
+         Maximum word frequency rank.
+     alpha : float
+         Estimated alpha parameter for the power law.
+     ax : matplotlib axes
+         Scatter plot to which the power curve will be added.
+     """
+     xvals = np.arange(curve_xmin, curve_xmax)
+     yvals = max_rank * (xvals**(-1 / alpha))
+     ax.loglog(xvals, yvals, color='grey')


 def main(args):
@@ -16,6 +71,21 @@ def main(args):
                               figsize=[12, 6],
                               grid=True,
                               xlim=args.xlim)
+
+     word_counts = df['word_frequency'].to_numpy()
+     alpha = get_power_law_params(word_counts)
+     print('alpha:', alpha)
+
+     # Since the ranks are already sorted, we can take the last
+     # one instead of computing which row has the highest rank
+     max_rank = df['rank'].to_numpy()[-1]
+
+     # Use the range of the data as the boundaries
+     # when drawing the power law curve
+     curve_xmin = df['word_frequency'].min()
+     curve_xmax = df['word_frequency'].max()
+
+     plot_fit(curve_xmin, curve_xmax, max_rank, alpha, ax)
      ax.figure.savefig(args.outfile)


diff --git a/results/dracula.png b/results/dracula.png
index 57a7b70..5f10271 100644
Binary files a/results/dracula.png and b/results/dracula.png
differ
```

Why Branch?

Why go to all this trouble? Imagine we are in the middle of debugging a change like this when we are asked to make final revisions to a paper that was created using the old code. If we revert `plotcount.py` to its previous state we might lose our changes. If instead we have been doing the work on a branch, we can switch branches, create the plot, and switch back in complete safety.

7.5 Merging

We could proceed in three ways at this point:

1. Add our changes to `plotcounts.py` once again in the `master` branch.
2. Stop working in `master` and start using the `fit` branch for future development.
3. **Merge** the `fit` and `master` branches.

The first option is tedious and error-prone; the second will lead to a bewildering proliferation of branches, but the third option is simple, fast, and reliable. To start, let's make sure we're in the `master` branch:

```
$ git checkout master
$ git branch
```

```
  fit
* master
```

We can now merge the changes in the `fit` branch into our current branch with a single command:

```
$ git merge fit
```

```
Updating ddb00fb..1577404
Fast-forward
```

```
 bin/plotcounts.py   |  70 ++++++++++++++++++++++++++++++++++++
                            +++++++++++++++++++++++++++++++
 results/dracula.png | Bin 23291 -> 38757 bytes
 2 files changed, 70 insertions(+)
```

Merging doesn't change the source branch `fit`, but once the merge is done, all of the changes made in `fit` are also in the history of `master`:

```
$ git log --oneline -n 4
```

```
1577404 (HEAD -> master, fit) Adding Moreno-Sanchez et al
        (2016) reference
38c209b Added fit to word count data
ddb00fb (origin/master) removing inverse rank calculation
7de9877 ignoring __pycache__
```

Note that Git automatically creates a new commit (in this case, `1577404`) to represent the merge. If we now run `git diff master..fit`, Git doesn't print anything because there aren't any differences to show.

Now that we have merged all of the changes from `fit` into `master` there is no need to keep the `fit` branch, so we can delete it:

```
$ git branch -d fit
```

```
Deleted branch fit (was 1577404).
```

Not Just the Command Line

We have been creating, merging, and deleting branches on the command line, but we can do all of these things using GitKraken[3], the RStudio IDE[4], and other GUIs. The operations stay the same; all that changes is how we tell the computer what we want to do.

[3] https://www.gitkraken.com/
[4] https://www.rstudio.com/products/rstudio/

7.6 Handling Conflicts

A **conflict** occurs when a line has been changed in different ways in two separate branches or when a file has been deleted in one branch but edited in the other. Merging `fit` into `master` went smoothly because there were no conflicts between the two branches, but if we are going to use branches, we must learn how to merge conflicts.

To start, use `nano` to add the project's title to a new file called `README.md` in the `master` branch, which we can then view:

```
$ cat README.md
```

```
# Zipf's Law
```

```
$ git add README.md
$ git commit -m "Initial commit of README file"
```

```
[master 232b564] Initial commit of README file
 1 file changed, 1 insertion(+)
 create mode 100644 README.md
```

Now let's create a new development branch called `docs` to work on improving the documentation for our code. We will use `git checkout -b` to create a new branch and switch to it in a single step:

```
$ git checkout -b docs
```

```
Switched to a new branch 'docs'
```

```
$ git branch
```

```
* docs
  master
```

On this new branch, let's add some information to the README file:

```
# Zipf's Law

These Zipf's Law scripts tally the occurrences of words in text
files and plot each word's rank versus its frequency.
```

```
$ git add README.md
$ git commit -m "Added repository overview"
```

```
[docs a0b88e5] Added repository overview
 1 file changed, 3 insertions(+)
```

In order to create a conflict, let's switch back to the `master` branch. The changes we made in the `docs` branch are not present:

```
$ git checkout master
```

```
Switched to branch 'master'
```

```
$ cat README.md
```

```
# Zipf's Law
```

Let's add some information about the contributors to our work:

```
# Zipf's Law

## Contributors

- Amira Khan <amira@zipf.org>
```

```
$ git add README.md
$ git commit -m "Added contributor list"
```

```
[master 45a576b] Added contributor list
 1 file changed, 4 insertions(+)
```

We now have two branches, master and docs, in which we have changed README.md in different ways:

```
$ git diff docs..master
```

```
diff --git a/README.md b/README.md
index f40e895..71f67db 100644
--- a/README.md
+++ b/README.md
@@ -1,4 +1,5 @@
 # Zipf's Law

-These Zipf's Law scripts tally occurrences of words in text
-files and plot each word's rank versus its frequency.
+## Contributors
+
+- Amira Khan <amira@zipf.org>
```

When we try to merge docs into master, Git doesn't know which of these changes to keep:

```
$ git merge docs master
```

```
Auto-merging README.md
CONFLICT (content): Merge conflict in README.md
Automatic merge failed; fix conflicts and then commit the result.
```

If we look in README.md, we see that Git has kept both sets of changes, but has marked which came from where:

```
$ cat README.md
```

```
# Zipf's Law

<<<<<<< HEAD
## Contributors
```

```
- Amira Khan <amira@zipf.org>
=======
These Zipf's Law scripts tally the occurrences of words in text
files and plot each word's rank versus its frequency.
>>>>>>> docs
```

The lines from `<<<<<<< HEAD` to `=======` are what was in `master`, while the lines from there to `>>>>>>> docs` show what was in `docs`. If there were several conflicting regions in the same file, Git would mark each one this way.

We have to decide what to do next: keep the `master` changes, keep those from `docs`, edit this part of the file to combine them, or write something new. Whatever we do, we must remove the `>>>`, `===`, and `<<<` markers. Let's combine the two sets of changes so the resulting file reads:

```
# Zipf's Law

These Zipf's Law scripts tally the occurrences of words in text
files and plot each word's rank versus its frequency.

## Contributors

- Amira Khan <amira@zipf.org>
```

We can now add the file and commit the change, just as we would after any other edit:

```
$ git add README.md
$ git commit -m "Merging README additions"
```

```
[master 55c63d0] Merging README additions
```

Our branch's history now shows a single sequence of commits, with the `master` changes on top of the earlier `docs` changes:

```
$ git log --oneline -n 4
```

```
55c63d0 (HEAD -> master) Merging README additions
45a576b Added contributor list
a0b88e5 (docs) Added repository overview
232b564 Initial commit of README file
```

If we want to see what really happened, we can add the `--graph` option to `git log`:

```
$ git log --oneline --graph -n 4
```

```
*   55c63d0 (HEAD -> master) Merging README additions
|\
| * a0b88e5 (docs) Added repository overview
* | 45a576b Added contributor list
|/
* 232b564 Initial commit of README file
```

At this point we can delete the `docs` branch:

```
$ git branch -d docs
```

```
Deleted branch docs (was a0b88e5).
```

Alternatively, we can keep using `docs` for documentation updates. Each time we switch to it, we merge changes *from* `master` *into* `docs`, do our editing (while switching back to `master` or other branches as needed to work on the code), and then merge *from* `docs` *to* `master` once the documentation is updated.

Remember to Push

If you are using a remote repository, don't forget to use `git push` to keep your version on GitHub up to date with your local version.

7.7 A Branch-Based Workflow

What is the best way to incorporate branching into our regular coding practice? If we are working on our own computer, this workflow will help us keep track of what we are doing:

1. `git checkout master` to make sure we are in the `master` branch.
2. `git checkout -b name-of-feature` to create a new branch. We *always* create a branch when making changes, since we never know what else might come up. The branch name should be as descriptive as a variable name or filename would be.
3. Make our changes. If something occurs to us along the way—for example, if we are writing a new function and realize that the documentation for some other function should be updated—we do *not* do that work in this branch just because we happen to be there. Instead, we commit our changes, switch back to `master`, and create a new branch for the other work.
4. When the new feature is complete, we `git merge master name-of-feature` to get any changes we merged into `master` after creating `name-of-feature` and resolve any conflicts. This is an important step: we want to do the merge and test that everything still works in our feature branch, not in `master`.
5. Finally, we switch back to `master` and `git merge name-of-feature master` to merge our changes into `master`. We should not have any conflicts, and all of our tests should pass.

Most experienced developers use this **branch-per-feature workflow**, but what exactly is a "feature?" These rules make sense for small projects:

1. Anything cosmetic that is only one or two lines long can be done in `master` and committed right away. Here, "cosmetic" means changes to comments or documentation: nothing that affects how code runs, not even a simple variable renaming.
2. A pure addition that doesn't change anything else is a feature and goes into a branch. For example, if we run a new analysis and save the results, that should be done on its own branch because it might take several tries to get the analysis to run, and we might interrupt ourselves to fix things that we discover aren't working.
3. Every change to code that someone might want to undo later in one step is a feature. For example, if a new parameter is added to a function, then every call to the function has to be updated. Since neither alteration makes sense without the other, those changes are considered a single feature and should be done in one branch.

The hardest thing about using a branch-per-feature workflow is sticking to it for small changes. As the first point in the list above suggests, most people are pragmatic about this on small projects; on large ones, where dozens of people might be committing, even the smallest and most innocuous change needs to be in its own branch so that it can be reviewed (which we discuss below).

7.8 Using Other People's Work

So far we have used Git to manage individual work, but it really comes into its own when we are working with other people. We can do this in two ways:

1. Everyone has read and write access to a single shared repository.
2. Everyone can read from the project's main repository, but only a few people can commit changes to it. The project's other contributors **fork** the main repository to create one that they own, do their work in that, and then submit their changes to the main repository.

The first approach works well for teams of up to half a dozen people who are all comfortable using Git, but if the project is larger, or if contributors are worried that they might make a mess in the `master` branch, the second approach is safer.

Git itself doesn't have any notion of a "main repository," but **forges** like GitHub[5], GitLab[6], and BitBucket[7] all encourage people to use Git in ways that effectively create one. Suppose, for example, that Sami wants to contribute to the Zipf's Law code that Amira is hosting on GitHub at `https://github.com/amira-khan/zipf`. Sami can go to that URL and click on the "Fork" button in the upper right corner (Figure 7.2). GitHub immediately creates a copy of Amira's repository within Sami's account on GitHub's own servers.

When the command completes, the setup on GitHub now looks like Figure 7.3. Nothing has happened yet on Sami's own machine: the new repository exists only on GitHub. When Sami explores its history, they see that it contains all of the changes Amira made.

A copy of a repository is called a **clone**. In order to start working on the project, Sami needs a clone of *their* repository (not Amira's) on their own computer. We will modify Sami's prompt to include their desktop user ID (`sami`) and working directory (initially `~`) to make it easier to follow what's happening:

[5]`https://github.com`
[6]`https://gitlab.com/`
[7]`https://bitbucket.org/`

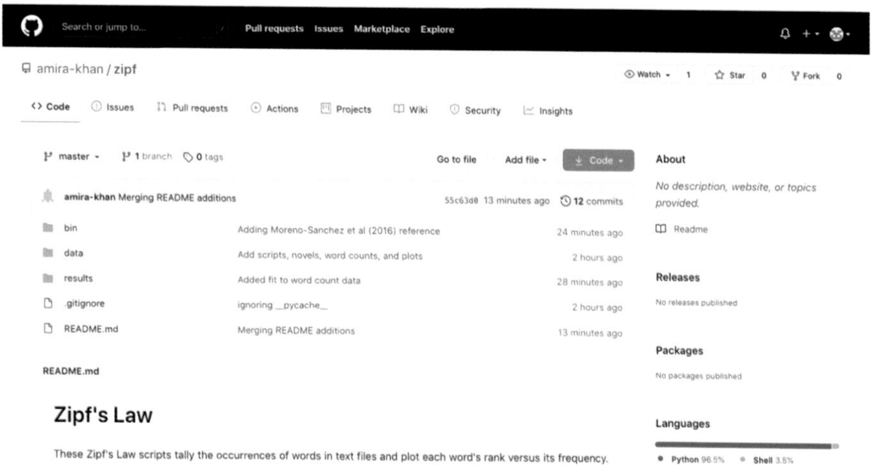

FIGURE 7.2: Forking a repository.

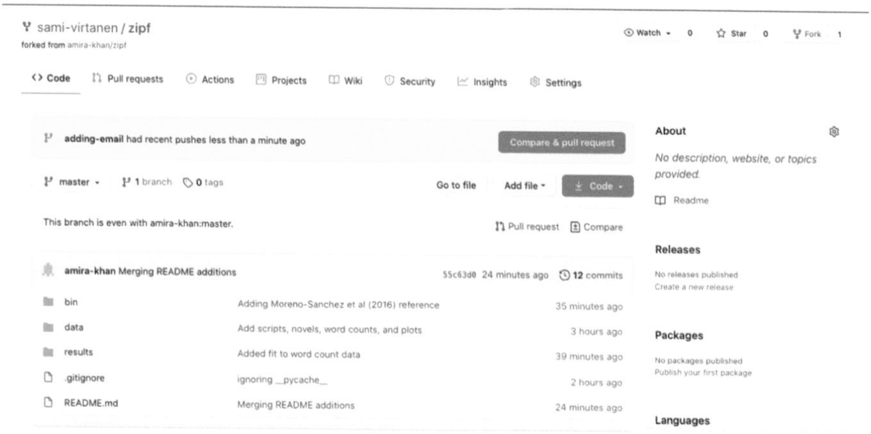

FIGURE 7.3: Repositories on GitHub after forking.

```
sami:~ $ git clone https://github.com/sami-virtanen/zipf.git
```

```
Cloning into 'zipf'...
remote: Enumerating objects: 64, done.
remote: Counting objects: 100% (64/64), done.
remote: Compressing objects: 100% (43/43), done.
remote: Total 64 (delta 20), reused 63 (delta 19), pack-reused 0
Receiving objects: 100% (64/64), 2.20 MiB | 2.66 MiB/s, done.
Resolving deltas: 100% (20/20), done.
```

This command creates a new directory with the same name as the project, i.e., `zipf`. When Sami goes into this directory and runs `ls` and `git log`, they see that all of the project's files and history are there:

```
sami:~ $ cd zipf
sami:~/zipf $ ls
```

```
README.md       bin             data            results
```

```
sami:~/zipf $ git log --oneline -n 4
```

```
55c63d0 (HEAD -> master, origin/master, origin/HEAD)
        Merging README additions
45a576b Added contributor list
a0b88e5 Added repository overview
232b564 Initial commit of README file
```

Sami also sees that Git has automatically created a **remote** for their repository that points back at their repository on GitHub:

```
sami:~/zipf $ git remote -v
```

```
origin  https://github.com/sami-virtanen/zipf.git (fetch)
origin  https://github.com/sami-virtanen/zipf.git (push)
```

Sami can pull changes from their fork and push work back there, but needs to do one more thing before getting the changes from Amira's repository:

```
sami:~/zipf $ git remote add upstream
              https://github.com/amira-khan/zipf.git
sami:~/zipf $ git remote -v
```

```
origin      https://github.com/sami-virtanen/zipf.git (fetch)
origin      https://github.com/sami-virtanen/zipf.git (push)
upstream    https://github.com/amira-khan/zipf.git (fetch)
upstream    https://github.com/amira-khan/zipf.git (push)
```

Sami has called their new remote `upstream` because it points at the repository from which theirs is derived. They could use any name, but `upstream` is a nearly universal convention.

With this remote in place, Sami is finally set up. Suppose, for example, that Amira has modified the project's `README.md` file to add Sami as a contributor. (Again, we show Amira's user ID and working directory in her prompt to make it clear who's doing what):

```
# Zipf's Law

These Zipf's Law scripts tally the occurrences of words in text
files and plot each word's rank versus its frequency.

## Contributors

- Amira Khan <amira@zipf.org>
- Sami Virtanen
```

Amira commits her changes and pushes them to *her* repository on GitHub:

```
amira:~/zipf $ git commit -a -m "Adding Sami as a contributor"
```

```
[master 35fca86] Adding Sami as a contributor
 1 file changed, 1 insertion(+)
```

```
amira:~/zipf $ git push origin master
```

```
Enumerating objects: 5, done.
Counting objects: 100% (5/5), done.
Delta compression using up to 4 threads
Compressing objects: 100% (3/3), done.
Writing objects: 100% (3/3), 315 bytes | 315.00 KiB/s, done.
Total 3 (delta 2), reused 0 (delta 0), pack-reused 0
remote: Resolving deltas: 100% (2/2), completed with 2 local
        objects.
To https://github.com/amira-khan/zipf.git
   55c63d0..35fca86  master -> master
```

Amira's changes are now on her desktop and in her GitHub repository but not in either of Sami's repositories (local or remote). Since Sami has created a remote that points at Amira's GitHub repository, though, they can easily pull those changes to their desktop:

```
sami:~/zipf $ git pull upstream master
```

```
From https://github.com/amira-khan/zipf
 * branch              master     -> FETCH_HEAD
 * [new branch]        master     -> upstream/master
Updating 55c63d0..35fca86
Fast-forward
 README.md | 1 +
 1 file changed, 1 insertion(+)
```

Pulling from a repository owned by someone else is no different than pulling from a repository we own. In either case, Git merges the changes and asks us to resolve any conflicts that arise. The only significant difference is that, as with `git push` and `git pull`, we have to specify both a remote and a branch: in this case, `upstream` and `master`.

7.9 Pull Requests

Sami can now get Amira's work, but how can Amira get Sami's? She could create a remote that pointed at Sami's repository on GitHub and periodically pull in Sami's changes, but that would lead to chaos, since we could never be sure that everyone's work was in any one place at the same time. Instead, almost everyone uses **pull requests**. They aren't part of Git itself, but are supported by all major online **forges**.

A pull request is essentially a note saying, "Someone would like to merge branch A of repository B into branch X of repository Y." The pull request does not contain the changes, but instead points at two particular branches. That way, the difference displayed is always up to date if either branch changes.

But a pull request can store more than just the source and destination branches: it can also store comments people have made about the proposed merge. Users can comment on the pull request as a whole, or on particular lines, and mark comments as out of date if the author of the pull request updates the code that the comment is attached to. Complex changes can go through several rounds of review and revision before being merged, which makes pull requests the review system we all wish journals actually had.

To see this in action, suppose Sami wants to add their email address to `README.md`. They create a new branch and switch to it:

```
sami:~/zipf $ git checkout -b adding-email
```

```
Switched to a new branch 'adding-email'
```

then make a change and commit it:

```
sami:~/zipf $ git commit -a -m "Adding my email address"
```

```
[adding-email 3e73dc0] Adding my email address
 1 file changed, 1 insertion(+), 1 deletion(-)
```

```
sami:~/zipf $ git diff HEAD~1
```

```
diff --git a/README.md b/README.md
index e8281ee..e1bf630 100644
--- a/README.md
+++ b/README.md
@@ -6,4 +6,4 @@ and plot each word's rank versus its frequency.
 ## Contributors

 - Amira Khan <amira@zipf.org>
-- Sami Virtanen
+- Sami Virtanen <sami@zipf.org>
```

Sami's changes are only in their local repository. They cannot create a pull request until those changes are on GitHub, so they push their new branch to their repository on GitHub:

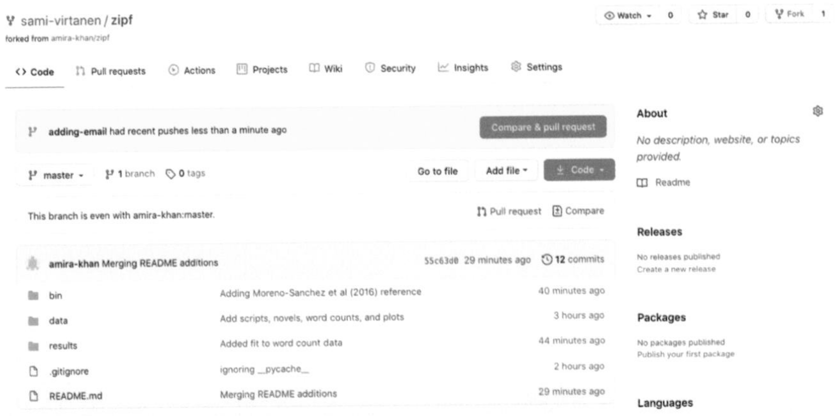

FIGURE 7.4: Repository state after Sami pushes.

```
sami:~/zipf $ git push origin adding-email
```

```
Enumerating objects: 5, done.
Counting objects: 100% (5/5), done.
Delta compression using up to 4 threads
Compressing objects: 100% (3/3), done.
Writing objects: 100% (3/3), 315 bytes | 315.00 KiB/s, done.
Total 3 (delta 2), reused 0 (delta 0), pack-reused 0
remote: Resolving deltas: 100% (2/2), completed with 2 local
  objects.
remote:
remote: Create a pull request for 'adding-email' on GitHub by
  visiting:
  https://github.com/sami-virtanen/zipf/pull/new/adding-email
remote:
To https://github.com/sami-virtanen/zipf.git
 * [new branch]      adding-email -> adding-email
```

When Sami goes to their GitHub repository in the browser, GitHub notices that they have just pushed a new branch and asks them if they want to create a pull request (Figure 7.4).

When Sami clicks on the button, GitHub displays a page showing the default source and destination of the pull request and a pair of editable boxes for the pull request's title and a longer comment (Figure 7.5).

If they scroll down, Sami can see a summary of the changes that will be in the pull request (Figure 7.6).

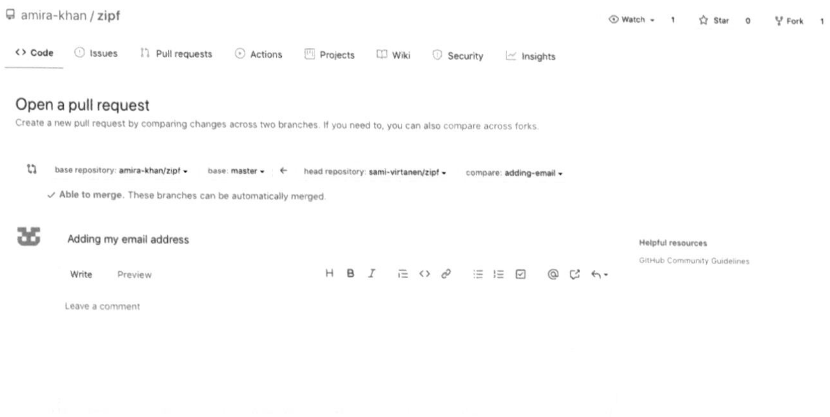

FIGURE 7.5: Starting a pull request.

FIGURE 7.6: Summary of a pull request.

The top (title) box is autofilled with the previous commit message, so Sami adds an extended explanation to provide additional context before clicking on "Create Pull Request" (Figure 7.7). When they do, GitHub displays a page showing the new pull request, which has a unique serial number (Figure 7.8). Note that this pull request is displayed in Amira's repository rather than Sami's, since it is Amira's repository that will be affected if the pull request is merged.

Amira's repository now shows a new pull request (Figure 7.9). Clicking on the "Pull requests" tab brings up a list of PRs (Figure 7.10) and clicking on the pull request link itself displays its details (Figure 7.11). Sami and Amira can both see and interact with these pages, though only Amira has permission to merge.

Since there are no conflicts, GitHub will let Amira merge the PR immediately using the "Merge pull request" button. She could also discard or reject it

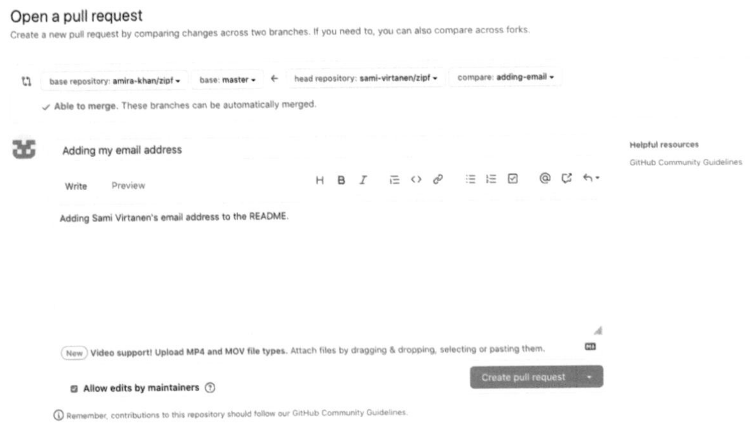

FIGURE 7.7: Filling in a pull request.

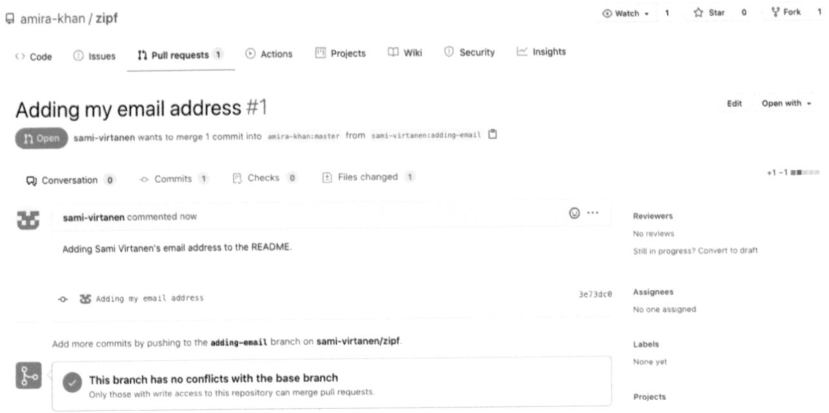

FIGURE 7.8: Creating a new pull request.

without merging using the "Close pull request" button. Instead, she clicks on the "Files changed" tab to see what Sami has changed (Figure 7.12).

If she moves her mouse over particular lines, a white-on-blue cross appears near the numbers to indicate that she can add comments (Figure 7.13). She clicks on the marker beside her own name and writes a comment: She only wants to make one comment rather than write a lengthier multi-comment review, so she chooses "Add single comment" (Figure 7.14). GitHub redisplays the page with her remarks inserted (Figure 7.15).

While Amira is working, GitHub has been emailing notifications to both Sami and Amira. When Sami clicks on the link in their email notification, it takes them to the PR and shows Amira's comment. Sami changes `README.md`, com-

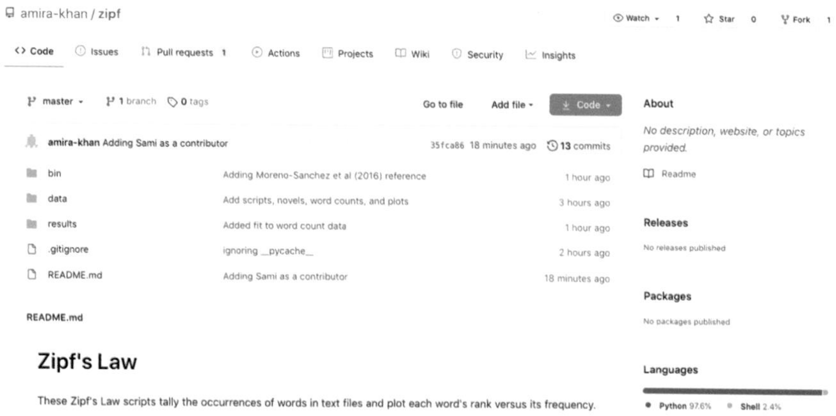

FIGURE 7.9: Viewing a pull request.

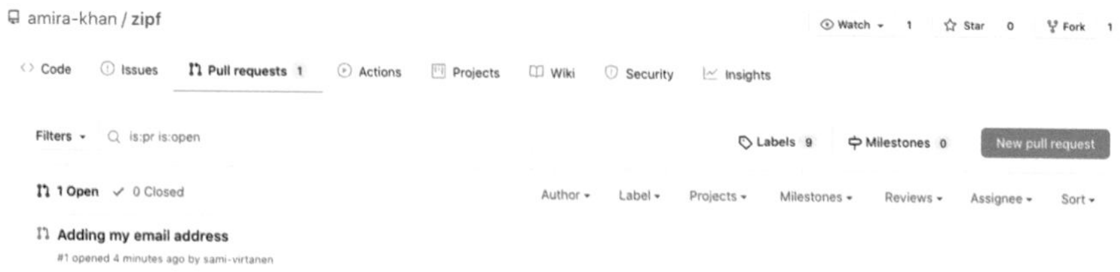

FIGURE 7.10: Listing pull requests.

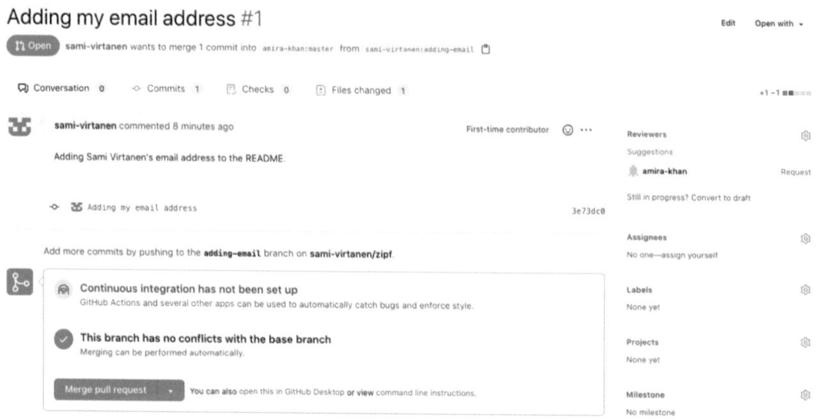

FIGURE 7.11: Details of pull requests.

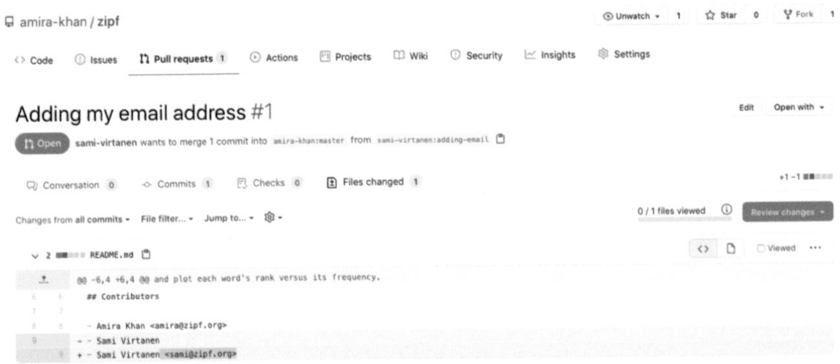

FIGURE 7.12: Viewing changes to files.

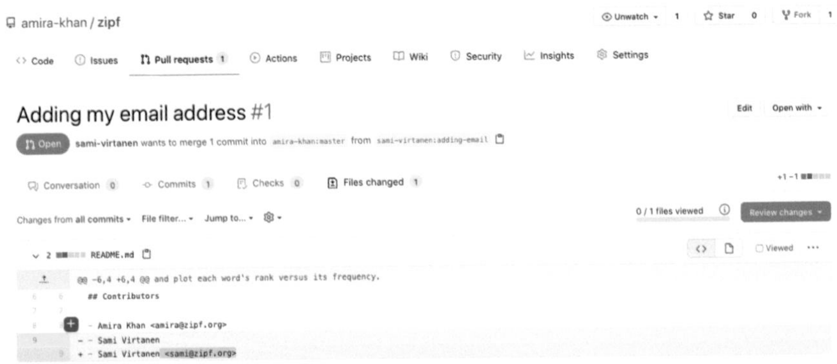

FIGURE 7.13: A GitHub comment marker.

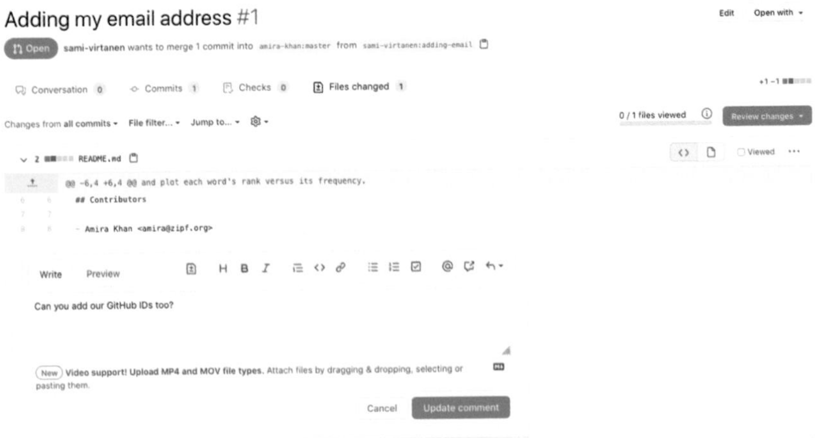

FIGURE 7.14: Writing a comment on a pull request.

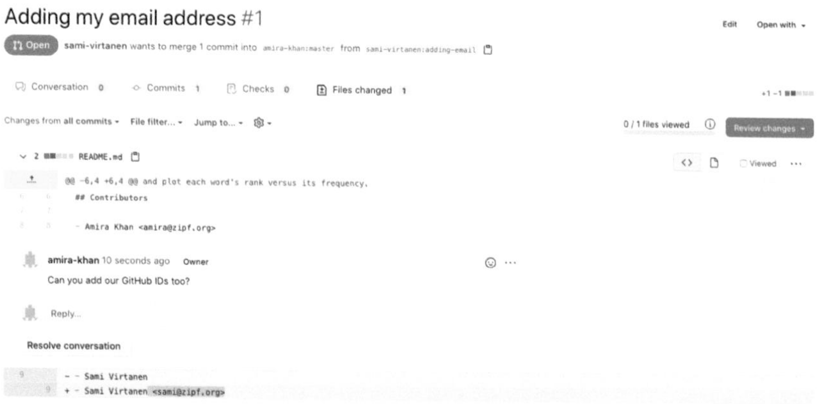

FIGURE 7.15: Viewing a comment on a pull request.

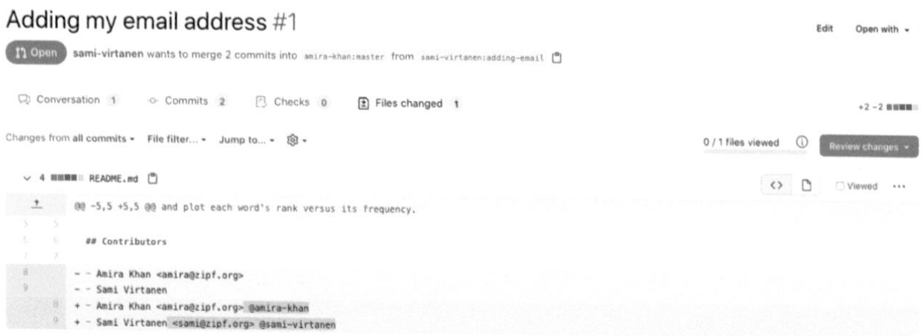

FIGURE 7.16: A pull request with a fix.

mits, and pushes, but does *not* create a new pull request or do anything to the existing one. As explained above, a PR is a note asking that two branches be merged, so if either end of the merge changes, the PR updates automatically.

Sure enough, when Amira looks at the PR again a few moments later she sees Sami's changes (Figure 7.16). Satisfied, she goes back to the "Conversation" tab and clicks on "Merge." The icon at the top of the PR's page changes text and color to show that the merge was successful (Figure 7.17).

To get those changes from GitHub to her desktop repository, Amira uses `git pull`:

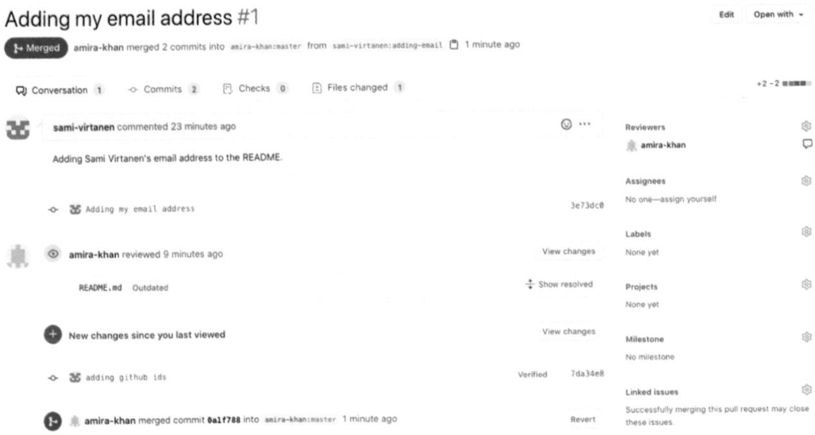

FIGURE 7.17: After a successful merge.

```
amira:~/zipf $ git pull origin master
```

```
From https://github.com/amira-khan/zipf
 * branch            master     -> FETCH_HEAD
Updating 35fca86..a04e3b9
Fast-forward
 README.md | 4 ++--
 1 file changed, 2 insertions(+), 2 deletions(-)
```

To get the change they just made from their `adding-email` branch into their `master` branch, Sami could use `git merge` on the command line. It's a little clearer, though, if they also use `git pull` from their `upstream` repository (i.e., Amira's repository) so that they're sure to get any other changes that Amira may have merged:

```
sami:~/zipf $ git checkout master
```

```
Switched to branch 'master'
Your branch is up to date with 'origin/master'.
```

```
sami:~/zipf $ git pull upstream master
```

```
From https://github.com/amira-khan/zipf
 * branch            master     -> FETCH_HEAD
Updating 35fca86..a04e3b9
Fast-forward
 README.md | 4 ++--
 1 file changed, 2 insertions(+), 2 deletions(-)
```

Finally, Sami can push their changes back to the `master` branch in their own remote repository:

```
sami:~/zipf $ git push origin master
```

```
Total 0 (delta 0), reused 0 (delta 0), pack-reused 0
To https://github.com/sami-virtanen/zipf.git
   35fca86..a04e3b9  master -> master
```

All four repositories are now synchronized.

7.10 Handling Conflicts in Pull Requests

Finally, suppose that Amira and Sami have decided to collaborate more extensively on this project. Amira has added Sami as a collaborator to the GitHub repository. Now Sami can make contributions directly to the repository, rather than via a pull request from a forked repository.

Sami makes a change to `README.md` in the `master` branch on GitHub. Meanwhile, Amira is making a conflicting change to the same file in a different branch. When Amira creates her pull request, GitHub will detect the conflict and report that the PR cannot be merged automatically (Figure 7.18).

Amira can solve this problem with the tools she already has. If she has made her changes in a branch called `editing-readme`, the steps are:

1. Pull Sami's changes from the `master` branch of the GitHub repository into the `master` branch of her desktop repository.
2. Merge *from* the `master` branch of her desktop repository *to* the `editing-readme` branch in the same repository.
3. Push her updated `editing-readme` branch to her repository on GitHub. The pull request from there back to the `master` branch of the main repository will update automatically.

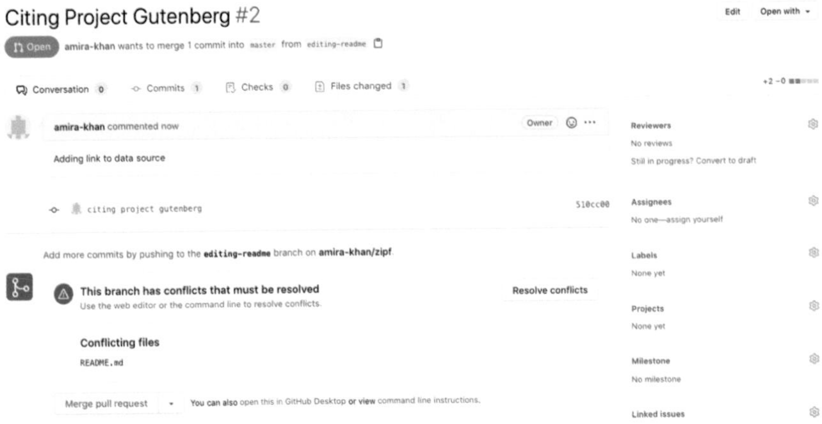

FIGURE 7.18: Showing a conflict in a pull request.

GitHub and other forges do allow people to merge conflicts through their browser-based interfaces, but doing it on our desktop means we can use our favorite editor to resolve the conflict. It also means that if the change affects the project's code, we can run everything to make sure it still works.

But what if Sami or someone else merges another change while Amira is resolving this one, so that by the time she pushes to her repository there is another, different, conflict? In theory this cycle could go on forever; in practice, it reveals a communication problem that Amira (or someone) needs to address. If two or more people are constantly making incompatible changes to the same files, they should discuss who's supposed to be doing what, or rearrange the project's contents so that they aren't stepping on each other's toes.

7.11 Summary

Branches and pull requests seem complicated at first, but they quickly become second nature. Everyone involved in the project can work at their own pace on what they want to, picking up others' changes and submitting their own whenever they want. More importantly, this workflow gives everyone has a chance to review each other's work. As we discuss in Section F.5, doing reviews doesn't just prevent errors from creeping in: it is also an effective way to spread understanding and skills.

7.12 Exercises

7.12.1 Explaining options

1. What do the `--oneline` and `-n` options for `git log` do?
2. What other options does `git log` have that you would find useful?

7.12.2 Modifying prompt

Modify your shell prompt so that it shows the branch you are on when you are in a repository.

7.12.3 Ignoring files

GitHub maintains a collection of `.gitignore` files[8] for projects of various kinds. Look at the sample `.gitignore` file for Python: how many of the ignored files do you recognize? Where could you look for more information about them?

7.12.4 Creating the same file twice

Create a branch called `same`. In it, create a file called `same.txt` that contains your name and the date.

Switch back to `master`. Check that `same.txt` does not exist, then create the same file with exactly the same contents.

1. What will `git diff master..same` show? (Try to answer the question *before* running the command.)
2. What will `git merge same master` do? (Try to answer the question *before* running the command.)

7.12.5 Deleting a branch without merging

Create a branch called `experiment`. In it, create a file called `experiment.txt` that contains your name and the date, then switch back to `master`.

[8] `https://github.com/github/gitignore`

1. What happens when you try to delete the `experiment` branch using `git branch -d experiment`? Why?
2. What option can you give Git to delete the `experiment` branch? Why should you be very careful using it?
3. What do you think will happen if you try to delete the branch you are currently on using this flag?

7.12.6 Tracing changes

Chartreuse and Fuchsia are collaborating on a project. Describe what is in each of the four repositories involved after each of the steps below.

1. Chartreuse creates a repository containing a `README.md` file on GitHub and clones it to their desktop.
2. Fuchsia forks that repository on GitHub and clones their copy to their desktop.
3. Fuchsia adds a file `fuchsia.txt` to the `master` branch of their desktop repository and pushes that change to their repository on GitHub.
4. Fuchsia creates a pull request from the `master` branch of their repository on GitHub to the `master` branch of Chartreuse's repository on GitHub.
5. Chartreuse does *not* merge Fuchsia's PR. Instead, they add a file `chartreuse.txt` to the `master` branch of their desktop repository and push that change to their repository on GitHub.
6. Fuchsia adds a remote to their desktop repository called `upstream` that points at Chartreuse's repository on GitHub and runs `git pull upstream master`, then merges any changes or conflicts.
7. Fuchsia pushes from the `master` branch of their desktop repository to the `master` branch of their GitHub repository.
8. Chartreuse merges Fuchsia's pull request.
9. Chartreuse runs `git pull origin master` on the desktop.

7.13 Key Points

- Use a **branch-per-feature** workflow to develop new features while leaving the master branch in working order.
- `git branch` creates a new branch.
- `git checkout` switches between branches.

- `git merge` **merges** changes from another branch into the current branch.
- **Conflicts** occur when files or parts of files are changed in different ways on different branches.
- Version control systems do not allow people to overwrite changes silently; instead, they highlight conflicts that need to be resolved.
- **Forking** a repository makes a copy of it on a server.
- **Cloning** a repository with `git clone` creates a local copy of a remote repository.
- Create a remote called `upstream` to point to the repository a fork was derived from.
- Create **pull requests** to submit changes from your fork to the upstream repository.

8

Working in Teams

> Evil begins when you begin to treat people as things.
>
> — Terry Pratchett

Projects can run for years with poorly written code, but none will survive for long if people are confused, pulling in different directions, or hostile to each other. This chapter therefore looks at how to create a culture of collaboration that will help people who want to contribute to your project, and introduce a few ways to manage projects and teams as they develop. Our recommendations draw on Fogel (2005), which describes how good open source software projects are run, and on Bollier (2014), which explains what a **commons** is and when it's the right model to use.

At this point, the Zipf's Law project should include:

```
zipf/
├── .gitignore
├── README.md
├── bin
│   ├── book_summary.sh
│   ├── collate.py
│   ├── countwords.py
│   ├── plotcounts.py
│   ├── script_template.py
│   └── utilities.py
├── data
│   ├── README.md
│   ├── dracula.txt
│   └── ...
└── results
    ├── dracula.csv
    ├── dracula.png
    └── ...
```

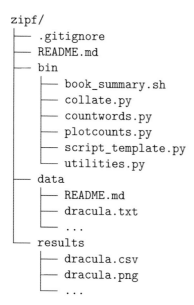

8.1 What Is a Project?

The first decision we have to make is what exactly constitutes a "project" (Wilson et al. 2017). Some examples are:

- A dataset that is being used by several research projects. The project includes the raw data, the programs used to tidy that data, the tidied data, the extra files needed to make the dataset a package, and a few text files describing the data's authors, license, and **provenance**.
- A set of annual reports written for an **NGO**. The project includes several Jupyter notebooks, some supporting Python libraries used by those notebooks, copies of the HTML and PDF versions of the reports, a text file containing links to the datasets used in the report (which can't be stored on GitHub since they contain personal identifying information), and a text file explaining details of the analysis that the authors didn't include in the reports themselves.
- A software library that provides an interactive glossary of data science terms in both Python and R. The project contains the files needed to create a package in both languages, a Markdown file full of terms and definitions, and a Makefile with targets to check cross-references, compile packages, and so on.

Some common criteria for creating projects are one per publication, one per deliverable piece of software, or one per team. The first tends to be too small: a good dataset will result in several reports, and the goal of some projects is to produce a steady stream of reports (such as monthly forecasts). The second is a good fit for software engineering projects whose primary aim is to produce tools rather than results, but can be an awkward fit for data analysis work. The third tends to be too large: a team of half a dozen people may work on many different things at once, and a repository that holds them all quickly looks like someone's basement.

One way to decide what makes up a project is to ask what people have meetings about. If the same group needs to get together on a regular basis to talk about something, that "something" probably deserves its own repository. And if the list of people changes slowly over time but the meetings continue, that's an even stronger sign.

8.2 Include Everyone

Most research software projects begin as the work of one person, who may continue to do the bulk of the coding and data analysis throughout its existence (Majumder et al. 2019). As projects become larger, though, they eventually need more contributors to sustain them. Involving more people also improves the functionality and robustness of the code, since newcomers bring their own expertise or see old problems in new ways.

In order to leverage a group's expertise, a project must do more than *allow* people to contribute: its leaders must communicate that the project *wants* contributions, and that newcomers are welcome and valued (Sholler et al. 2019). Saying "the door is open" is not enough: many potential contributors have painful personal experience of being less welcome than others. In order to create a truly welcoming environment for everyone, the project must explicitly acknowledge that some people are treated unfairly and actively take steps to remedy this. Doing this increases diversity within the team, which makes it more productive (Zhang 2020). More importantly, it is the right thing to do.

Terminology

Privilege is an unearned advantage given to some people but not all, while **oppression** is systemic inequality that benefits the privileged and harms those without privilege (Aurora and Gardiner 2018). In Europe, the Americas, Australia, and New Zealand, a straight, white, affluent, physically able male is less likely to be interrupted when speaking, more likely to be called on in class, and more likely to get a job interview based on an identical CV than someone who is outside these categories. People who are privileged are often not aware of it, as they've lived in a system that provides unearned advantages their entire lives. In John Scalzi's memorable phrase, they've been playing on the lowest difficulty setting there is their whole lives, and as a result don't realize how much harder things are for others (Scalzi 2012).

The targets of oppression are often called "members of a marginalized group," but targets don't choose to be marginalized: people with privilege marginalize them. Finally, an **ally** is a member of a privileged group who is working to understand their own privilege and end oppression.

Encouraging inclusivity is a shared responsibility. If we are privileged, we should educate ourselves and call out peers who are marginalizing others, even if (or especially if) they aren't conscious of doing it. As project leaders, part of our job is to teach contributors how to be allies and to ensure an inclusive culture (Lee 1962).

8.3 Establish a Code of Conduct

The starting point for making a project more inclusive is to establish a Code of Conduct. This does four things:

1. It promotes fairness within a group.
2. It reassures members of marginalized groups who have experienced harassment or unwelcoming behavior before that this project takes inclusion seriously.
3. It ensures that everyone knows what the rules are. (This is particularly important when people come from different cultural backgrounds.)
4. It prevents anyone who misbehaves from pretending that they didn't know what they did was unacceptable.

More generally, a Code of Conduct makes it easier for people to contribute by reducing uncertainty about what behaviors are acceptable. Some people may push back claiming that it's unnecessary, or that it infringes freedom of speech, but what they usually mean is that thinking about how they might have benefited from past inequity makes them feel uncomfortable. If having a Code of Conduct leads to them going elsewhere, that will probably make the project run more smoothly.

By convention, we add a Code of Conduct to our project by creating a file called `CONDUCT.md` in the project's root directory. Writing a Code of Conduct that is both comprehensive and readable is hard. We therefore recommend using one that other groups have drafted, refined, and tested. The Contributor Covenant[1] is relevant for projects being developed online, such as those based on GitHub:

```
# Contributor Covenant Code of Conduct

## Our Pledge

We as members, contributors, and leaders pledge to make
participation in our community a harassment-free experience for
everyone, regardless of age, body size, visible or invisible
```

[1] `https://www.contributor-covenant.org`

```
disability, ethnicity, sex characteristics, gender identity and
expression, level of experience, education, socio-economic
status, nationality, personal appearance, race, religion, or
sexual identity and orientation.

We pledge to act and interact in ways that contribute to an
open, welcoming, diverse, inclusive, and healthy community.

## Our Standards

Examples of behavior that contributes to a positive environment
for our community include:

* Demonstrating empathy and kindness toward other people
* Being respectful of differing opinions, viewpoints, and
  experiences
* Giving and gracefully accepting constructive feedback
* Accepting responsibility and apologizing to those affected by
  our mistakes, and learning from the experience
* Focusing on what is best not just for us as individuals, but
  for the overall community

Examples of unacceptable behavior include:

* The use of sexualized language or imagery, and sexual
  attention or advances of any kind
* Trolling, insulting or derogatory comments, and personal or
  political attacks
* Public or private harassment
* Publishing others' private information, such as a physical or
  email address, without their explicit permission
* Other conduct which could reasonably be considered
  inappropriate in a professional setting

## Enforcement Responsibilities

Community leaders are responsible for clarifying and enforcing
our standards of acceptable behavior and will take appropriate
and fair corrective action in response to any behavior that
they deem inappropriate, threatening, offensive, or harmful.

Community leaders have the right and responsibility to remove,
edit, or reject comments, commits, code, wiki edits, issues,
and other contributions that are not aligned to this
Code of Conduct, and will communicate reasons for moderation
```

```
decisions when appropriate.

## Scope

This Code of Conduct applies within all community spaces, and
also applies when an individual is officially representing the
community in public spaces. Examples of representing our
community include using an official e-mail address, posting via
an official social media account, or acting as an appointed
representative at an online or offline event.

## Enforcement

Instances of abusive, harassing, or otherwise unacceptable
behavior may be reported to the community leaders responsible
for enforcement at [INSERT CONTACT METHOD]. All complaints will
be reviewed and investigated promptly and fairly.

All community leaders are obligated to respect the privacy and
security of the reporter of any incident.

## Enforcement Guidelines

Community leaders will follow these Community Impact Guidelines
in determining the consequences for any action they deem in
violation of this Code of Conduct:

### 1. Correction

**Community Impact**: Use of inappropriate language or other
behavior deemed unprofessional or unwelcome in the community.

**Consequence**: A private, written warning from community
leaders, providing clarity around the nature of the violation
and an explanation of why the behavior was inappropriate.
A public apology may be requested.

### 2. Warning

**Community Impact**: A violation through a single incident or
series of actions.

**Consequence**: A warning with consequences for continued
behavior. No interaction with the people involved, including
unsolicited interaction with those enforcing the Code of
```

```
Conduct, for a specified period of time. This includes avoiding
interactions in community spaces as well as external channels
like social media. Violating these terms may lead to a
temporary or permanent ban.

### 3. Temporary Ban

**Community Impact**: A serious violation of community
standards, including sustained inappropriate behavior.

**Consequence**: A temporary ban from any sort of interaction
or public communication with the community for a specified
period of time. No public or private interaction with the
people involved, including unsolicited interaction with those
enforcing the Code of Conduct, is allowed during this period.
Violating these terms may lead to a permanent ban.

### 4. Permanent Ban

**Community Impact**: Demonstrating a pattern of violation of
community standards, including sustained inappropriate
behavior, harassment of an individual, or aggression toward or
disparagement of classes of individuals.

**Consequence**: A permanent ban from any sort of public
interaction within the community.

## Attribution

This Code of Conduct is adapted from the
[Contributor Covenant][covenant],
version 2.0, available at
www.contributor-covenant.org/version/2/0/code_of_conduct.html

Community Impact Guidelines were inspired by
[Mozilla's code of conduct enforcement
ladder](https://github.com/mozilla/diversity).

[homepage]: https://www.contributor-covenant.org

For answers to common questions about this code of conduct,
see the FAQ at https://www.contributor-covenant.org/faq.
Translations are available at
https://www.contributor-covenant.org/translations.
```

As you can see, the Contributor Covenant defines expectations for behavior, the consequences of non-compliance, and the mechanics of reporting and handling violations. The third part is as important as the first two, since rules are meaningless without a method to enforce them; Aurora and Gardiner (2018) is a short, practical guide that every project lead should read.

In-Person Events

The Contributor Covenant works well for interactions that are largely online, which is the case for many research software projects. The best option for in-person events is the model Code of Conduct[2] from the Geek Feminism Wiki[3], which is used by many open source organizations and conferences. If your project is sited at a university or within a company, it may already have Code of Conduct: the Human Resources department is usually the most helpful place to ask.

8.4 Include a License

While a Code of Conduct describes how contributors should interact with each other, a license dictates how project materials can be used and redistributed. If the license or a publication agreement makes it difficult for people to contribute, the project is less likely to attract new members, so the choice of license is crucial to the project's long-term sustainability.

Open Except...

Projects that are only developing software may not have any problem making everything open. Teams working with sensitive data, on the other hand, must be careful to ensure that what should be private isn't inadvertently shared. In particular, people who are new to Git (and even people who aren't) occasionally add raw data files containing personal identifying information to repositories. It's possible to rewrite the project's history to remove things when this happens, but that doesn't automatically erase copies people may have in forked repositories.

Every creative work has some sort of license; the only question is whether authors and users know what it is and choose to enforce it. Choosing a license

[2] `https://geekfeminism.wikia.com/wiki/Conference_anti-harassment/Policy`
[3] `https://geekfeminism.wikia.com/`

for a project can be complex, not least because the law hasn't kept up with everyday practice. Morin, Urban, and Sliz (2012) and VanderPlas (2014) are good starting points to understand licensing and intellectual property from a researcher's point of view, while Lindberg (2008) is a deeper dive for those who want details. Depending on country, institution, and job role, most creative works are automatically eligible for intellectual property protection. However, members of the team may have different levels of copyright protection. For example, students and faculty may have a copyright on the research work they produce, but university staff members may not, since their employment agreement may state that what they create on the job belongs to their employer.

To avoid legal messiness, every project should include an explicit license. This license should be chosen early, since changing a license can be complicated. For example, each collaborator may hold copyright on their work and therefore need to be asked for approval when a license is changed. Similarly, changing a license does not change it retroactively, so different users may wind up operating under different licensing structures.

Leave It to the Professionals

Don't write your own license. Legalese is a highly technical language, and words don't mean what you think they do.

To make license selection for code as easy as possible, GitHub allows us to select one of several common software licenses when creating a repository. The Open Source Initiative maintains a list[4] of **open licenses**, and choosealicense.com[5] will help us find a license that suits our needs. Some of the things we need to think about are:

1. Do we want to license the work at all?
2. Is the content we are licensing source code?
3. Do we require people distributing derivative works to also distribute their code?
4. Do we want to address patent rights?
5. Is our license compatible with the licenses of the software we depend on?
6. Do our institutions have any policies that may overrule our choices?
7. Are there any copyright experts within our institution who can assist us?

Unfortunately, GitHub's list does not include common licenses for data or written works like papers and reports. Those can be added in manually, but it's

[4]`https://opensource.org/licenses`
[5]`https://choosealicense.com/`

often hard to understand the interactions among multiple licenses on different kinds of material (Almeida et al. 2017).

Just as the project's Code of Conduct is usually placed in a root-level file called `CONDUCT.md`, its license is usually put in a file called `LICENSE.md` that is also in the project's root directory.

8.4.1 Software

In order to choose the right license for our software, we need to understand the difference between two kinds of license. The **MIT License** (and its close sibling the BSD License) say that people can do whatever they want to with the software as long as they cite the original source, and that the authors accept no responsibility if things go wrong. The **GNU Public License** (GPL) gives people similar rights, but requires them to share their own work on the same terms:

> You may copy, distribute and modify the software as long as you track changes/dates in source files. Any modifications to or software including (via compiler) GPL-licensed code must also be made available under the GPL along with build and install instructions.
>
> — tl;dr[6]

In other words, if someone modifies GPL-licensed software or incorporates it into their own project, and then distributes what they have created, they have to distribute the source code for their own work as well.

The GPL was created to prevent companies from taking advantage of open software without contributing anything back. The last thirty years have shown that this restriction isn't necessary: many projects have survived and thrived without this safeguard. We therefore recommend that projects choose the MIT license, as it places the fewest restrictions on future action.

```
MIT License

Copyright (c) 2020 Amira Khan

Permission is hereby granted, free of charge, to any person
obtaining a copy of this software and associated documentation
files (the "Software"), to deal in the Software without
restriction, including without limitation the rights to use,
copy, modify, merge, publish, distribute, sublicense, and/or
```

[6]`https://tldrlegal.com/license/gnu-general-public-license-v3-(gpl-3)`

```
sell copies of the Software, and to permit persons to whom the
Software is furnished to do so, subject to the following
conditions:

The above copyright notice and this permission notice shall be
included in all copies or substantial portions of the Software.

THE SOFTWARE IS PROVIDED "AS IS", WITHOUT WARRANTY OF ANY KIND,
EXPRESS OR IMPLIED, INCLUDING BUT NOT LIMITED TO THE WARRANTIES
OF MERCHANTABILITY, FITNESS FOR A PARTICULAR PURPOSE AND
NONINFRINGEMENT. IN NO EVENT SHALL THE AUTHORS OR COPYRIGHT
HOLDERS BE LIABLE FOR ANY CLAIM, DAMAGES OR OTHER LIABILITY,
WHETHER IN AN ACTION OF CONTRACT, TORT OR OTHERWISE, ARISING
FROM, OUT OF OR IN CONNECTION WITH THE SOFTWARE OR THE USE OR
OTHER DEALINGS IN THE SOFTWARE.
```

First, Do No Harm

The Hippocratic License[7] is a newer license that is quickly becoming popular. Where the GPL requires people to share their work, the Hippocratic License requires them to do no harm. More precisely, it forbids people from using the software in ways that violate the Universal Declaration of Human Rights[8]. We have learned the hard way that software and science can be mis-used; adopting the Hippocratic License is a small step toward preventing this.

8.4.2 Data and reports

The MIT license, the GPL, and the Hippocratic License are intended for use with software. When it comes to data and reports, the most widely used family of licenses are those produced by Creative Commons[9]. These have been written and checked by lawyers and are well understood by the community.

The most liberal option is referred to as **CC-0**, where the "0" stands for "zero restrictions." This puts work in the public domain, i.e., allows anyone who wants to use it to do so, however they want, with no restrictions. CC-0 is usually the best choice for data, since it simplifies aggregate analysis involving datasets from different sources. It does not negate the scholarly tradition and requirement of citing sources; it just doesn't make it a legal requirement.

[7] `https://firstdonoharm.dev/`
[8] `https://en.wikipedia.org/wiki/Universal_Declaration_of_Human_Rights`
[9] `https://creativecommons.org/`

The next step up from CC-0 is the Creative Commons–Attribution license, usually referred to as **CC-BY**. This allows people to do whatever they want to with the work as long as they cite the original source. This is the best license to use for manuscripts: we want people to share them widely but also want to get credit for our work.

Other Creative Commons licenses incorporate various restrictions, and are usually referred to using two-letter abbreviations:

- ND (no derivative works) prevents people from creating modified versions of our work. Unfortunately, this also inhibits translation and reformatting.
- SA (share-alike) requires people to share work that incorporates ours on the same terms that we used. Again, it is fine in principle, but in practice it makes aggregation and recombination difficult.
- NC (no commercial use) does *not* mean that people cannot charge money for something that includes our work, though some publishers still try to imply that in order to scare people away from open licensing. Instead, the NC clause means that people cannot charge for something that uses our work without our explicit permission, which we can give under whatever terms we want.

To apply these concepts to our Zipf's Law project, we need to consider both our data (which other people created) and our results (which we create). We can view the license for the novels by looking in `data/README.md`, which tells us that the Gutenberg Project books are in the public domain (i.e., CC-0). This is a good choice for our results as well, but after reflection, we decide to choose CC-BY for our papers so that everyone can read them (and cite them).

8.5 Planning

Codes of conduct and licenses are a project's constitution, but how do contributors know what they should actually be doing on any given day? Whether we are working by ourselves or with a group of people, the best way to manage this is to use an **issue tracking system** to keep track of tasks we need to complete or problems we need to fix. **Issues** are sometimes called **tickets**, so issue tracking systems are sometimes called **ticketing systems**. They are also often called **bug trackers**, but they can be used to manage any kind of work, and are often a convenient way to manage discussions as well.

Like other **forges**, GitHub allows participants to create issues for a project, comment on existing issues, and search all available issues. Every issue can hold:

- A unique ID, such as `#123`, which is also part of its URL. This makes issues easy to find and refer to: GitHub automatically turns the expression `#123` in a **commit message** into a link to that issue.
- A one-line title to aid browsing and search.
- The issue's current status. In simple systems (like GitHub's) each issue is either open or closed, and by default, only open issues are displayed. Closed items are generally removed from default interfaces, so issues should only be closed when they no longer require any attention.
- The user ID of the issue's creator. Just as `#123` refers to a particular issue, `@name` is automatically translated into a link to that person. The IDs of people who have commented on it or modified it are embedded in the issue's history, which helps figure out who to talk to about what.
- The user ID of the person assigned to review the issue, if someone is assigned.
- A full description that may include screenshots, error messages, and anything else that can be put in a web page.
- Replies, counter-replies, and so on from people who are interested in the issue.

Broadly speaking, people create three kinds of issues:

1. **Bug reports** to describe problems they have encountered.
2. **Feature requests** describing what could be done next, such as "add this function to this package" or "add a menu to the website."
3. Questions about how to use the software, how parts of the project work, or its future directions. These can eventually turn into bug reports or feature requests, and can often be recycled as documentation.

Helping Users Find Information

Many projects encourage people to ask questions on a mailing list or in a chat channel. However, answers given there can be hard to find later, which leads to the same questions coming up over and over again. If people can be persuaded to ask questions by filing issues, and to respond to issues of this kind, then the project's old issues become a customized Stack Overflow[10] for the project. Some projects go so far as to create a page of links to old questions and answers that are particularly helpful.

8.6 Bug Reports

One steady supply of work in any active project is **bug reports**. Unsurprisingly, a well-written bug report is more likely to get a fast response that actually addresses the problem (Bettenburg et al. 2008). To write a good bug report:

1. Make sure the problem actually *is* a bug. It's always possible that we have called a function the wrong way or done an analysis using the wrong configuration file. If we take a minute to double-check, or ask someone else on our team to check our logic, we could well fix the problem ourselves.
2. Try to come up with a **reproducible example** or "reprex" that includes only the steps needed to make the problem happen, and that (if possible) uses simplified data rather than a complete dataset. Again, we can often solve the problem ourselves as we trim down the steps to create one.
3. Write a one-line title for the issue and a longer (but still brief) description that includes relevant details.
4. Attach any screenshots that show the problem, resulting errors, or (slimmed-down) input files needed to re-create it.
5. Describe the version of the software we were using, the operating system we were running on, which version of the programming language we ran it with, and anything else that might affect behavior. If the software in question uses a logging framework (Section 12.4), turn debugging output on and include it with the issue.

[10] `https://stackoverflow.com/`

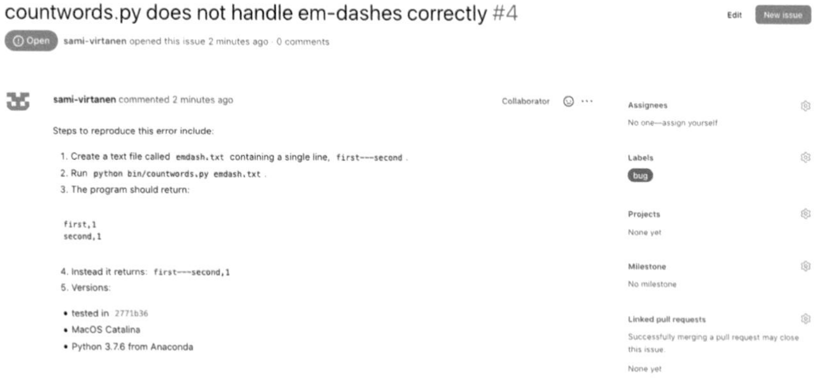

FIGURE 8.1: Example of a bug report.

6. Describe each problem separately so that each one can be tackled on its own. This parallels the rule about creating a branch in version control for each bug fix or feature discussed in Section 7.

An example of a well-written bug report with all of the components mentioned above is shown in Figure 8.1.

It takes time and energy to write a good error report. If the report is being filed by a member of the development team, the incentive to document errors well is that resolving the issue later is easier. You can encourage users from outside the project to write thorough error reports by including an issue template for your project. An issue template is a file included in your GitHub repository that proliferates each new issue with text that describes expectations for content that should be submitted. You can't force new issues to be as complete as you might like, but you can use an issue template to make it easier for contributors to remember and complete documentation about bug reports.

Sometimes the person creating the issue may not know or have the right answer for some of these things, and will be doing their best with limited information about the error. Responding with kindness and encouragement is important to maintain a healthy community, and should be enforced by the project's Code of Conduct (Section 8.3).

8.7 Labeling Issues

The bigger or older a project gets, the harder it is to find things—unless, that is, the project's members put in a bit of work to make things findable (Lin, Ali, and Wilson in press). Issue trackers let project members add **labels** to issues to make things easier to search and organize. Labels are also often called **tags**; whatever term is used, each one is just a descriptive word or two.

GitHub allows project owners to define any labels they want. A small project should always use some variation on these three:

- *Bug*: something should work but doesn't.
- *Enhancement*: something that someone wants added to the software.
- *Task*: something needs to be done, but won't show up in code (e.g., organizing the next team meeting).

Projects also often use:

- *Question*: where is something or how is something supposed to work? As noted above, issues with this label can often be recycled as documentation.
- *Discussion* or *Proposal*: something the team needs to make a decision about or a concrete proposal to resolve such a discussion. All issues can have discussion: this category is for issues that start that way. (Issues that are initially labeled *Question* are often relabeled *Discussion* or *Proposal* after some back and forth.)
- *Suitable for Newcomer* or *Beginner-Friendly*: to identify an easy starting point for someone who has just joined the project. If we help potential new contributors find places to start, they are more likely to do so (Steinmacher et al. 2014).

The labels listed above identify the kind of work an issue describes. A separate set of labels can be used to indicate the state of an issue:

- *Urgent*: work needs to be done right away. (This label is typically reserved for security fixes).
- *Current*: this issue is included in the current round of work.
- *Next*: this issue is (probably) going to be included in the next round.

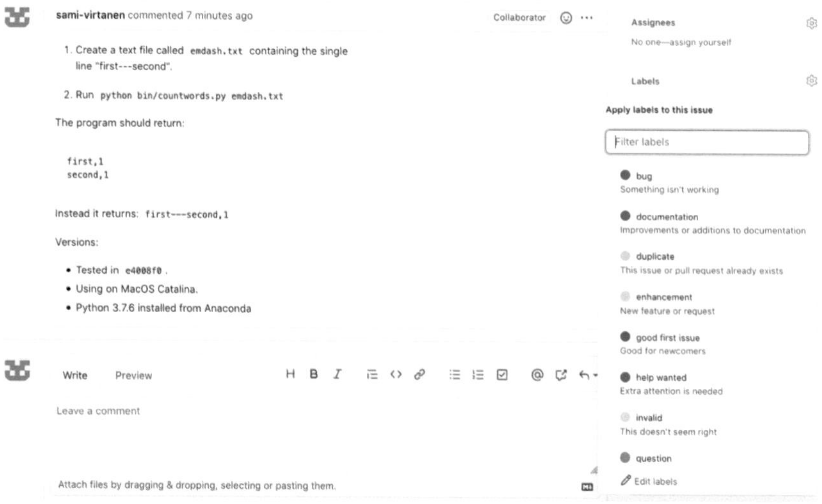

FIGURE 8.2: Labels for GitHub issues.

- *Eventually*: someone has looked at the issue and believes it needs to be tackled, but there's no immediate plan to do it.
- *Won't Fix*: someone has decided that the issue isn't going to be addressed, either because it's out of scope or because it's not actually a bug. Once an issue has been marked this way, it is usually then closed. When this happens, send the issue's creator a note explaining why the issue won't be addressed and encourage them to continue working with the project.
- *Duplicate*: this issue is a duplicate of one that's already in the system. Issues marked this way are usually also then closed; this is another opportunity to encourage people to stay involved.

Some of the labels that GitHub creates for repositories by default are shown in Figure 8.2. These labels can be modified or otherwise customized for each repository.

Some projects use labels corresponding to upcoming software releases, journal issues, or conferences instead of *Current*, *Next*, and *Eventually*. This approach works well in the short term, but becomes unwieldy as labels with names like `sprint-2020-08-01` and `spring-2020-08-16` pile up.

Instead, a project team will usually create a **milestone**, which is a set of issues and pull requests in a single project repository. GitHub milestones can have a due date and display aggregate progress toward completion, so the team can easily see when work is due and how much is left to be done. Teams can also create projects, which can include issues and pull requests from several repositories as well as notes and reminders for miscellaneous tasks.

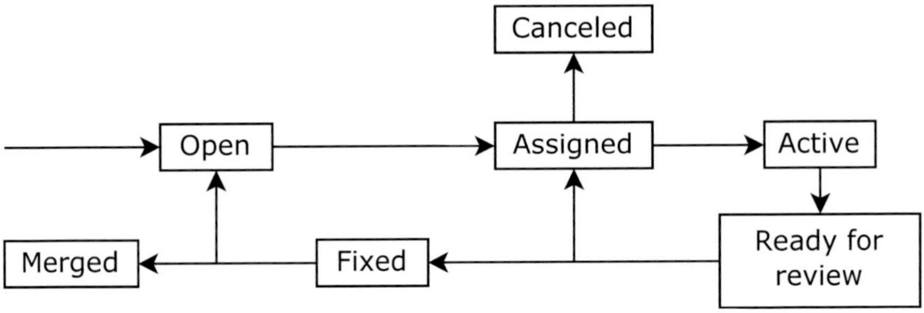

FIGURE 8.3: Example of issue lifecycle.

8.7.1 Standardizing workflows

Adding labels to issues also helps us standardize a workflow for the project. Conventions about who can do what to issues with various labels, and who can change those labels, let us define a workflow like the one shown in Figure 8.3.

- An *Open* issue becomes *Assigned* when someone is made responsible for it.
- They can then move the issue to *Canceled* if they think it was filed mistakenly or doesn't need to be fixed. (This is different from closing the issue after working on it.)
- An *Assigned* issue becomes *Active* when someone starts to work on it.
- When work is done it becomes *Ready for Review.*
- From there it is either *Assigned* again because it needs more work or moved to *Fixed*, which means the change is ready to be incorporated into the project.
- When the change is actually merged, the issue's state is changed to reflect that.

Small projects do not need this much formality, but when the team is distributed, contributors need to be able to find out what's going on without having to wait for someone to respond to email (or wondering who they *should* have emailed).

8.8 Prioritizing

Between bug reports, feature requests, and general cleanup, there is always more work to do than time to do it, so every project needs some way to figure out what to focus on. Labeling issues helps with **triage**, which is the process of deciding what is a priority and what isn't. This is never an easy job for software projects that need to balance fixing bugs with creating new features, and is even more challenging for research projects for which "done" is hard to define or whose team members are widely distributed or do not all work for the same institution.

Many commercial and open source teams have adopted **agile development** as a solution to these problems. Instead of carefully formulating long-term plans that could be derailed by changing circumstances, agile development uses a sequence of short development **sprints**, each typically one or two weeks long. Each sprint starts with a planning session lasting one or two hours in which the successes and failures of the previous sprint are reviewed and issues to be resolved in the current sprint are selected. If team members believe an issue is likely to take longer than a single sprint to complete, it should be broken into smaller pieces that *can* be finished so that the team can track progress more accurately. (Something large can be "90% done" for weeks; with smaller items, it's easier to see how much headway is being made.)

To decide which issues to work on in the next sprint, a team can construct an **impact/effort matrix** (Figure 8.4). Impact measures how important the issue is to reaching the team's goals, and is typically measured on a low–medium–high scale. (Some teams use ratings from 1 to 10, but this just leads to arguing over whether something is a 4 or a 5.) Effort measures how much work the issue requires. Since this can't always be estimated accurately, it's common to classify things as "an hour," "a day," or "multiple days." Again, anything that's likely to take longer than multiple days should be broken down so that planning and progress tracking can be more accurate.

The impact/effort matrix makes the priorities for the coming sprint clear: anything that is of high importance and requires little effort should be included, while things of low importance that require a lot of effort should not. The team must still make hard decisions, though:

- Should a single large high-priority item be done, or should several smaller low-priority items be tackled instead?
- What should be done about medium-priority items that keep being put off?

Each team has to answer these questions for each sprint, but that begs the question of exactly who has the final say in answering them. In a large project, a **product manager** decides how important items are, while a **project manager** is responsible for estimating effort and tracking progress. In a typical research software project the principal investigator either makes the decision or delegates that responsibility (and authority) to the lead developer.

Regardless of who is ultimately responsible, it is essential to include project participants in the planning and decision making. This may be as simple as

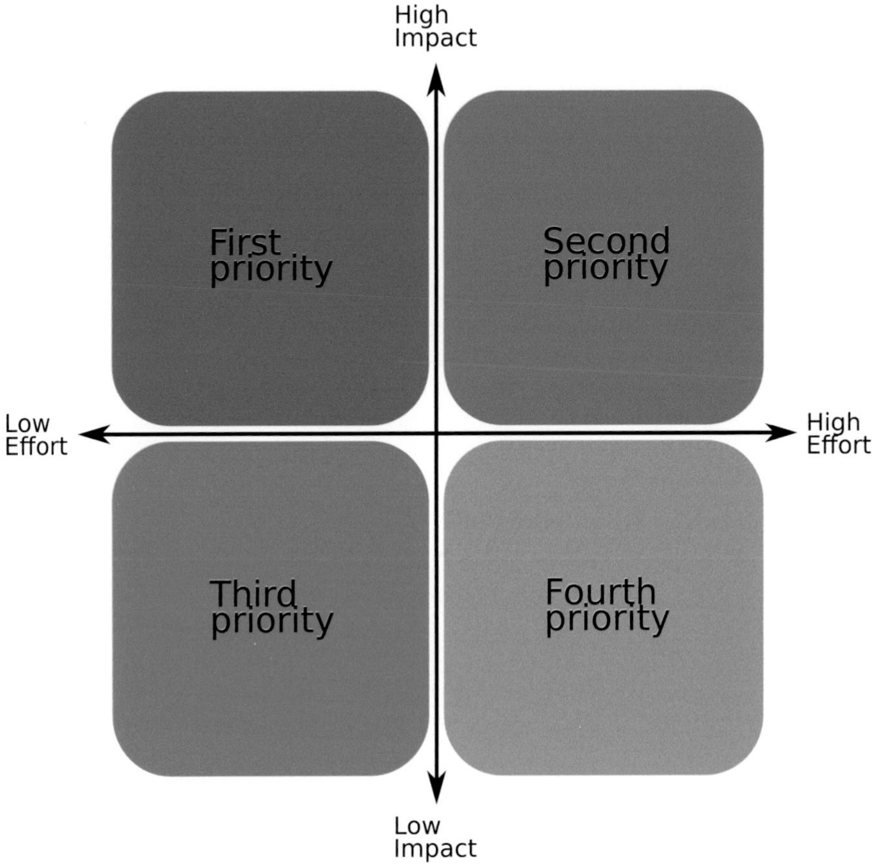

FIGURE 8.4: An impact/effort matrix.

having them add **up-votes** and **down-votes** to issues to indicate their opinions on importance, or as complex as asking them to propose a multi-sprint breakdown of a particularly complex feature. Doing this shows people that their contributions are valued, which in turn increases their commitment to doing the work. It also produces better plans, since everyone knows something that someone else doesn't.

8.9 Meetings

Pull requests and GitHub issues are good tools for asynchronous work, but team meetings are often a more efficient way to make decisions, and help build a sense of community. Knowing how to run a meeting well is as important as knowing how to use version control; the rules doing so are simple but rarely followed:

Decide if there actually needs to be a meeting. If the only purpose is to share information, have everyone send a brief email instead. Remember, people can read faster than they can speak: if someone has facts for the rest of the team to absorb, the most polite way to communicate them is to type them in.

Write an agenda. If nobody cares enough about the meeting to prepare a point-form list of what's to be discussed, the meeting itself probably doesn't need to happen. Note that "the agenda is all the open issues in our GitHub repo" doesn't count.

Include timings in the agenda. Timings help prevent early items stealing time from later ones. The first estimates with any new group are inevitably optimistic, so we should revise them upward for subsequent meetings. However, we shouldn't have a second or third meeting just because the first one ran over-time: instead, we should try to figure out *why* we're running over and fix the underlying problem.

Prioritize. Tackle issues that will have high impact but take little time first, and things that will take more time but have less impact later. That way, if the first things run over time, the meeting will still have accomplished something.

Make one person responsible for keeping things moving. One person should be made moderator and be responsible for keeping items to time, chiding people who are having side conversations or checking email, and asking people who are talking too much to get to the point. The moderator should *not* do all the talking: in fact, whoever is in charge will talk less in a well-run meeting than most other participants. This should be a rotating duty among members.

Require politeness. No one gets to be rude, no one gets to ramble, and if someone goes off topic, it's the moderator's job to say, "Let's discuss that elsewhere."

No interruptions. Participants should raise a finger, hand, put up a sticky note, or make another well understood gesture to indicate when they want to speak. The moderator should keep track of who wants to speak and give them time in turn.

No distractions. Side conversations make meetings less efficient because nobody can actually pay attention to two things at once. Similarly, if someone is checking their email or texting a friend during a meeting, it's a clear signal that they don't think the speaker or their work is important. This doesn't mean a complete ban on technology—people may need accessibility aids, or may be waiting for a call from a dependent—but by default, phones should be face down and laptops should be closed during in-person meetings.

Take minutes. Someone other than the moderator should take point-form notes about the most important information that was shared, and about every decision that was made or every task that was assigned to someone. This should be a rotating duty among members.

End early. If the meeting is scheduled for 10:00–11:00, aim to end at 10:50 to give people a break before whatever they're doing next.

As soon as the meeting is over, circulate the minutes by emailing them to everyone or adding a text file to the project's repository:

People who weren't at the meeting can follow what's going on. We all have to juggle tasks from several projects or courses, which means that sometimes we can't make it to meetings. Checking a written record is a more accurate and efficient way to catch up than asking a colleague, "So, what did I miss?"

Everyone can check what was actually said or promised. More than once, one of us has looked over the minutes of a meeting and thought, "Did I say that?" or, "I didn't promise to have it ready then!" Accidentally or not, people will often remember things differently; writing them down gives everyone a chance to correct mistakes, misinterpretations, or misrepresentations.

People can be held accountable at subsequent meetings. There's no point making lists of questions and action items if we don't follow up on them later. If we are using an issue-tracking system, we should create a ticket for each new question or task right after the meeting and update those that are being carried forward. This helps a lot when the time comes to draw up the agenda for the next meeting.

8.9.1 Air time

One of the problems in a synchronous meeting is the tendency of some people to speak far more than others. Other meeting members may be so accustomed to this that they don't speak up even when they have valuable points to make.

One way to combat this is to give everyone **three sticky notes** at the start of the meeting. Every time they speak, they have to give up one sticky note. When they're out of stickies, they aren't allowed to speak until everyone has used at least one, at which point everyone gets all of their sticky notes back. This ensures that nobody talks more than three times as often as the quietest person in the meeting, and completely changes group dynamics. People who have given up trying to be heard suddenly have space to contribute, and the overly frequent speakers realize how unfair they have been.

Another useful technique is called **interruption bingo**. Draw a grid and label the rows and columns with the participants' names. Each time one person interrupts another, add a tally mark to the appropriate cell; halfway through the meeting, take a moment to look at the results. In most cases it will be clear that one or two people are doing all of the interrupting. After that, saying, "All right, I'm adding another tally to the bingo card," is often enough to get them to throttle back.

8.9.2 Online meetings

Online meetings provide special challenges, both in the context of regulating how often individuals speak, as well as running meetings in general. Troy (2018) discusses why online meetings are often frustrating and unproductive and points out that in most online meetings, the first person to speak during a pause gets the floor. As a result, "If you have something you want to say, you have to stop listening to the person currently speaking and instead focus on when they're gonna pause or finish so you can leap into that nanosecond of silence and be the first to utter something. The format...encourages participants who want to contribute to say more and listen less."

The solution is to run a text chat beside the video conference where people can signal that they want to speak. The moderator can then select people from the waiting list. This practice can be reinforced by having everyone mute themselves, and only allowing the moderator to unmute people. Brookfield and Preskill (2016) has many other useful suggestions for managing meetings.

8.10 Making Decisions

The purpose of a well-run meeting is to make decisions, so sooner or later, the members of a project must decide who has a say in what. The first step is to acknowledge that every team has a power structure: the question is whether it's formal or informal—in other words, whether it's accountable or unaccountable (Freeman 1972). The latter can work for groups of up to half a dozen people in which everyone knows everyone else. Beyond that, groups need to spell out who has the authority to make which decisions and how to achieve consensus. In short, they need explicit **governance**.

Martha's Rules are a practical way to do this in groups of up to a few dozen members (Minahan 1986):

1. Before each meeting, anyone who wishes may sponsor a proposal. Proposals must be filed at least 24 hours before a meeting in order to be considered at that meeting, and must include:

- a one-line summary
- the full text of the proposal
- any required background information
- pros and cons
- possible alternatives

2. A quorum is established in a meeting if half or more of voting members are present.
3. Once a person has sponsored a proposal, they are responsible for it. The group may not discuss or vote on the issue unless the sponsor or their delegate is present. The sponsor is also responsible for presenting the item to the group.
4. After the sponsor presents the proposal, a **sense vote** is cast for the proposal prior to any discussion:

   - Who likes the proposal?
   - Who can live with the proposal?
   - Who is uncomfortable with the proposal?

5. If all of the group likes or can live with the proposal, it passes with no further discussion.
6. If most of the group is uncomfortable with the proposal, it is sent back to its sponsor for further work. (The sponsor may decide to drop it if it's clear that the majority isn't going to support it.)
7. If some members are uncomfortable with the proposal, a timer is set for a brief discussion moderated by the meeting moderator. After 10 minutes or when no one has anything further to add, the moderator calls for a straight yes-or-no vote on the question: "Should we implement this decision over the stated objections?" If a majority votes "yes" the proposal is implemented. Otherwise, it is returned to the sponsor for further work.

Every group that uses Martha's Rules must make two procedural decisions:

How are proposals put forward? In a software development project, the easiest way is to file an issue in the project's GitHub repository tagged *Proposal*, or to create a pull request containing a single file with the text of the proposal. Team members can then comment on the proposal, and the sponsor can revise it before bringing it to a vote.

Who gets to vote? The usual answer is "whoever is working on the project," but as it attracts more volunteer contributors, a more explicit rule is needed. One common method is for existing members to nominate new ones, who are then voted in (or not) using the process described above.

8.11 Make All This Obvious to Newcomers

Rules that people don't know about can't help them. Once your team agrees on a project structure, a workflow, how to get items on a meeting agenda, or how to make decisions, you should therefore take the time to document this for newcomers. This information may be included as sections in the existing `README` file or put into files of their own:

- `CONTRIBUTING` explains how to contribute, i.e., what naming conventions to use for functions, what tags to put on issues (Section 8.5), or how to install and configure the software needed to start work on the project. These instructions can also be included as a section in `README`; wherever they go, remember that the easier it is for people to get set up and contribute, the more likely they are to do so (Steinmacher et al. 2014).
- `GOVERNANCE` explains how the project is run (Section 8.10). It is still uncommon for this to be in a file of its own—it is more often included in `README` or `CONTRIBUTING`—but open communities have learned the hard way that *not* being explicit about who has a voice in decisions and how contributors can tell what decisions have been made causes trouble sooner or later.

Having these files helps new contributors orient themselves, and also signals that the project is well run.

8.12 Handling Conflict

You just missed an important deadline, and people are unhappy. The sick feeling in the pit of your stomach has turned to anger: you did *your* part, but Sylvie didn't finish her stuff until the very last minute, which meant that no one else had time to spot the two big mistakes she'd made. As for Cho, he didn't deliver at all—again. If something doesn't change, contributors are going to start looking for a new project.

Conflicts like this come up all the time. Broadly speaking, there are four ways we can deal with them:

1. Cross our fingers and hope that things will get better on their own, even though they didn't the last three times.

2. Do extra work to make up for others' shortcomings. This saves us the mental anguish of confronting others in the short run, but the time for that "extra" has to come from somewhere. Sooner or later, our personal lives or other parts of the project will suffer.
3. Lose our temper. People often wind up displacing anger into other parts of their life: they may yell at someone for taking an extra thirty seconds to make a change when what they really need to do is tell their boss that they won't work through another holiday weekend to make up for management's decision to short-staff the project.
4. Take constructive steps to fix the underlying problem.

Most of us find the first three options easiest, even though they don't actually fix the problem. The fourth option is harder because we don't like confrontation. If we manage it properly, though, it is a lot less bruising, which means that we don't have to be as afraid of initiating it. Also, if people believe that we will take steps when they bully, lie, procrastinate, or do a half-assed job, they will usually avoid making it necessary.

Make sure we are not guilty of the same sin. We won't get very far complaining about someone else interrupting in meetings if we do it just as frequently.

Check expectations. Are we sure the offender knows what standards they are supposed to be meeting? This is where things like job descriptions or up-front discussion of who's responsible for what come in handy.

Check the situation. Is someone dealing with an ailing parent or immigration woes? Have they been put to work on three other projects that we don't know about? Use open questions like, "Can you help me understand this?" when checking in. This gives them the freedom to explain something you may not have expected, and avoids the pressure of being asked directly about something they don't want to explain.

Document the offense. Write down what the offender has actually done and why it's not good enough. Doing this helps us clarify what we're upset about and is absolutely necessary if we have to escalate.

Check with other team members. Are we alone in feeling that the offender is letting the team down? If so, we aren't necessarily wrong, but it'll be a lot easier to fix things if we have the support of the rest of the team. Finding out who else on the team is unhappy can be the hardest part of the whole process, since we can't even ask the question without letting on that we are upset and word will almost certainly get back to whoever we are asking about, who might then accuse us of stirring up trouble.

Talk with the offender. This should be a team effort: put it on the agenda for a team meeting, present the complaint, and make sure that the offender understands it. Doing this once is often enough: if someone realizes that they're going to be called on their hitchhiking or bad manners, they will usually change their ways.

Escalate as soon as there's a second offense. People who don't have good intentions count on us giving them one last chance after another until the project is finished and they can go suck the life out of their next victim. *Don't fall into this trap.* If someone stole a laptop, we would report it right away. If someone steals time, we are being foolish if we give them a chance to do it again and again.

In academic research projects, "escalation" means "taking the issue to the project's principal investigator." Of course, the PI has probably had dozens of students complain to her over the years about teammates not doing their share, and it isn't uncommon to have both halves of a pair tell the supervisor that they're doing all the work. (This is yet another reason to use version control: it makes it easy to check who's actually written what.) In order to get her to take us seriously and help us fix our problem, we should send her an email signed by several people that describes the problem and the steps we have already taken to resolve it. Make sure the offender gets a copy as well, and ask the supervisor to arrange a meeting to resolve the issue.

Hitchhikers

Hitchhikers who show up but never actually do anything are particularly difficult to manage, in part because they are usually very good at appearing reasonable. They will nod as we present our case, then say, "Well, yes, but…" and list a bunch of minor exceptions or cases where others on the team have also fallen short of expectations. Having collaborator guidelines (Section 8.3) and tracking progress (Section 8.7.1) are essential for handling them. If we can't back up our complaint, our supervisor will likely be left with the impression that the whole team is dysfunctional.

What can we do if conflict becomes more personal and heated, especially if it relates to violations of our Code of Conduct? A few simple guidelines will go a long way:

1. Be short, simple, and firm.
2. Don't try to be funny. It almost always backfires, or will later be used against us.
3. Play for the audience. We probably won't change the person we are calling out, but we might change the minds or strengthen the resolve of people who are observing.

4. Pick our battles. We can't challenge everyone, every time, without exhausting ourselves and deafening our audience. An occasional sharp retort will be much more effective than constant criticism.
5. Don't shame or insult one group when trying to help another. For example, don't call someone stupid when what we really mean is that they're racist or homophobic.

Captain Awkward[11] has useful advice for discussions like these, and Charles' Rules of Argument[12] are very useful online.

Finally, it's important to recognize that good principles sometimes conflict. For example, consider this scenario:

> A manager consistently uses male pronouns to refer to software and people of unknown gender. When you tell them it makes you uncomfortable to treat maleness as the norm, they say that male is the default gender in their first language and you should be more considerate of people from other cultures.

On the one hand, we want to respect other people's cultures; on the other hand, we want to be inclusive of women. In this case, the manager's discomfort about changing pronouns matters less than the career harm caused by them being exclusionary, but many cases are not this clear cut.

8.13 Summary

This chapter was the hardest in this book to write, but is probably also the most important. A project can survive bad code or stumbles with Git, but not confusion and interpersonal conflict. Collaboration and management become easier with practice, and everything you learn from taking part in research software projects will help other things you do as well.

[11] `https://captainawkward.com/`

[12] `https://geekfeminism.wikia.com/wiki/Charles%27_Rules_of_Argument`

8.14 Exercises

8.14.1 Finding information

Take a look at the GitHub repository for this book[13]. Where is the information for licensing and contributing?

8.14.2 Add a Code of Conduct to your project

Add a `CONDUCT.md` file to your Zipf's Law project repository. Use the Contributor Covenant[14] Code of Conduct template and modify the places that need updating (e.g., who to contact). Be sure to edit the contact information in both before committing the files.

8.14.3 Add a license to your project

Add either an MIT or a GPL `LICENSE.md` file to your Zipf's Law project repository. Modify the contents to include your full name and year.

8.14.4 Adding contribution guidelines

1. Add a section to the `README.md` file in the Zipf's Law project to tell people where to find out more about contributing.
2. Add a `CONTRIBUTING.md` file in the Zipf's Law project to describe how other people can contribute to it.

Be sure to add it to the root directory of your Git repository, so that when someone opens a pull request or creates an issue on GitHub, they will be presented with a link to the CONTRIBUTING file (see the GitHub contributors guide[15] for details).

8.14.5 File an issue

Create a feature request issue in your Zipf's Law project repository to ask that unit tests be written for `countwords.py` (we will do this in Chapter 11).

---

[13] `https://github.com/merely-useful/py-rse/`

[14] `https://www.contributor-covenant.org`

[15] `https://docs.github.com/en/github/building-a-strong-community/setting-guidelines-for-repository-contributors`

8.14.6 Label issues

1. Create the labels *current* and *discussion* to help organize and prioritize your issues.
2. Delete at least one of the labels that GitHub automatically created for you.
3. Apply at least one label to each issue in your repository.

8.14.7 Balancing individual and team needs

A new member of your team has a medically diagnosed attention disorder. In order to help themselves focus, they need to talk to themselves while coding. Several other members of your team have come to you privately to say that they find this distracting. What steps would you take?

8.14.8 Crediting invisible contributions

Your team has a rule: if someone's name appears in the Git history for a project, they are listed as a co-author on papers for that project. A new member of your team has complained that this is unfair: people who haven't contributed for over two years are still being included as authors, while they aren't included because the many hours they have spent doing code reviews don't show up in the Git history. How would you address this issue?

8.14.9 Who are you?

Which (if any) of the following profiles describes you best? How would you help each of these people if they were on your team?

- *Anna* thinks she knows more about every subject than everyone else on the team put together. No matter what you say, she'll correct you; no matter what you know, she knows better. If you keep track in team meetings of how often people interrupt one another, her score is usually higher than everyone else's put together.

- *Bao* is a contrarian: no matter what anyone says, he'll take the opposite side. This is healthy in small doses, but when Bao does it, there's always another objection lurking behind the first half dozen.

- *Frank* believes that knowledge is power. He enjoys knowing things that other people don't—or to be more accurate, he enjoys it when people know he knows things they don't. Frank can actually make things work, but when asked how he did it, he'll grin and say, "Oh, I'm sure you can figure it out."

- *Hediyeh* is quiet. Very quiet. She never speaks up in meetings, even when she knows that what other people are saying is wrong. She might contribute to the mailing list, but she's very sensitive to criticism, and will always back down rather than defending her point of view.

- *Kenny* is a hitchhiker. He has discovered that most people would rather shoulder some extra work than snitch, and he takes advantage of it at every turn. The frustrating thing is that he's so damn *plausible* when someone finally does confront him. "There have been mistakes on all sides," he says, or, "Well, I think you're nit-picking."

- *Melissa* would easily have made the varsity procrastination team if she'd bothered to show up to tryouts. She means well—she really does feel bad about letting people down—but somehow her tasks are never finished until the last possible moment. Of course, that means that everyone who is depending on her can't do their work until *after* the last possible moment.

- *Petra*'s favorite phrase is "why don't we." Why don't we write a GUI to help people edit the program's configuration files? Hey, why don't we invent our own little language for designing GUIs?

- *Raj* is rude. "It's just the way I talk," he says. "If you can't hack it, maybe you should find another team." His favorite phrase is, "That's stupid," and he uses obscenity in every second sentence.

8.15 Key Points

- Welcome and nurture community members proactively.
- Create an explicit Code of Conduct for your project modeled on the Contributor Covenant[16].
- Include a license in your project so that it's clear who can do what with the material.
- Create **issues** for bugs, enhancement requests, and discussions.
- **Label issues** to identify their purpose.
- **Triage** issues regularly and group them into **milestones** to track progress.
- Include contribution guidelines in your project that specify its workflow and its expectations of participants.
- Make rules about **governance** explicit.
- Use common-sense rules to make project meetings fair and productive.
- Manage conflict between participants rather than hoping it will take care of itself.

[16] `https://www.contributor-covenant.org`

9

Automating Analyses with Make

> The three rules of the Librarians of Time and Space are: 1) Silence; 2) Books must be returned no later than the last date shown; and 3) Do not interfere with the nature of causality.
>
> — Terry Pratchett

It's easy to run one program to process a single data file, but what happens when our analysis depends on many files, or when we need to re-do the analysis every time new data arrives? What should we do if the analysis has several steps that we have to do in a particular order?

If we try to keep track of this ourselves, we will inevitably forget some crucial steps and it will be hard for other people to pick up our work. Instead, we should use a **build manager** to keep track of what depends on what and run our analysis programs automatically. These tools were invented to help programmers compile complex software, but can be used to automate any workflow.

Our Zipf's Law project currently includes these files:

```
zipf/
├── .gitignore
├── CONDUCT.md
├── CONTRIBUTING.md
├── LICENSE.md
├── README.md
├── bin
│   ├── book_summary.sh
│   ├── collate.py
│   ├── countwords.py
│   ├── plotcounts.py
│   ├── script_template.py
│   └── utilities.py
├── data
│   ├── README.md
│   ├── dracula.txt
```

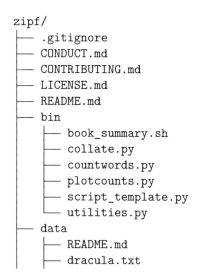

```
│   ├── frankenstein.txt
│   └── ...
└── results
    ├── dracula.csv
    ├── dracula.png
    └── ...
```

Now that the project's main building blocks are in place, we're ready to automate our analysis using a build manager. We will use a program called Make[1] to do this so that every time we add a new book to our data, we can create a new plot of the word count distribution with a single command. Make works as follows:

1. Every time the **operating system** creates, reads, or changes a file, it updates a **timestamp** on the file to show when the operation took place. Make can compare these timestamps to figure out whether files are newer or older than one another.
2. A user can describe which files depend on each other by writing **rules** in a **Makefile**. For example, one rule could say that `results/moby_dick.csv` depends on `data/moby_dick.txt`, while another could say that the plot `results/comparison.png` depends on all of the CSV files in the `results` directory.
3. Each rule also tells Make how to update an out-of-date file. For example, the rule for *Moby Dick* could tell Make to run `bin/countwords.py` if the result file is older than either the raw data file or the program.
4. When the user runs Make, the program checks all of the rules in the Makefile and runs the commands needed to update any that are out of date. If there are **transitive dependencies**—i.e., if A depends on B and B depends on C—then Make will trace them through and run all of the commands it needs to in the right order.

This chapter uses a version of Make called GNU Make[2]. It comes with MacOS and Linux; please see Section 1.3 for Windows installation instructions.

Keep Tracking with Version Control

We encourage you to use the Git workflow from Chapters 6 and 7 throughout the rest of this book, though we won't continue to remind you.

[1] `https://www.gnu.org/software/make/`
[2] `https://www.gnu.org/software/make/`

9.1 Updating a Single File

To start automating our analysis, let's create a file called `Makefile` in the root of our project and add the following:

```
# Regenerate results for "Moby Dick"
results/moby_dick.csv : data/moby_dick.txt
	python bin/countwords.py \
		data/moby_dick.txt > results/moby_dick.csv
```

As in the shell and many other programming languages, `#` indicates that the first line is a comment. The second and third lines form a **build rule**: the **target** of the rule is `results/moby_dick.csv`, its single **prerequisite** is the file `data/moby_dick.txt`, and the two are separated by a single colon `:`. There is no limit on the length of statement lines in Makefiles, but to aid readability we have used a backslash (\) character to split the lengthy third line in this example.

The target and prerequisite tell Make what depends on what. The line below them describes the **recipe** that will update the target if it is out of date. The recipe consists of one or more shell commands, each of which *must* be prefixed by a single tab character. Spaces cannot be used instead of tabs here, which can be confusing as they are interchangeable in most other programming languages. In the rule above, the recipe is "run `bin/countwords.py` on the raw data file and put the output in a CSV file in the `results` directory."

To test our rule, run this command in the shell:

```
$ make
```

Make automatically looks for a file called `Makefile`, follows the rules it contains, and prints the commands that were executed. In this case it displays:

```
python bin/countwords.py \
  data/moby_dick.txt > results/moby_dick.csv
```

Indentation Errors

If a `Makefile` indents a rule with spaces rather than tabs, Make produces an error message like this:

```
Makefile:3: *** missing separator.  Stop.
```

When Make follows the rules in our Makefile, one of three things will happen:

1. If `results/moby_dick.csv` doesn't exist, Make runs the recipe to create it.
2. If `data/moby_dick.txt` is newer than `results/moby_dick.csv`, Make runs the recipe to update the results.
3. If `results/moby_dick.csv` is newer than its prerequisite, Make does nothing.

In the first two cases, Make prints the commands it runs, along with anything those commands print to the screen via **standard output** or **standard error**. There is no screen output in this case, so we only see the command.

No matter what happened the first time we ran `make`, if we run it again right away it does nothing because our rule's target is now up to date. It tells us this by displaying the message:

```
make: `results/moby_dick.csv' is up to date.
```

We can check that it is telling the truth by listing the files with their timestamps, ordered by how recently they have been updated:

```
$ ls -l -t data/moby_dick.txt results/moby_dick.csv
```

```
-rw-r--r-- 1 amira staff  274967 Nov 29 12:58
  results/moby_dick.csv
-rw-r--r-- 1 amira staff 1253891 Nov 27 20:56
  data/moby_dick.txt
```

As a further test:

1. Delete `results/moby_dick.csv` and run `make` again. This is case #1, so Make runs the recipe.
2. Use `touch data/moby_dick.txt` to update the timestamp on the data file, then run `make`. This is case #2, so again, Make runs the recipe.

Managing Makefiles

We don't have to call our file `Makefile`: if we prefer something like `workflows.mk`, we can tell Make to read recipes from that file using `make -f workflows.mk`.

9.2 Managing Multiple Files

Our Makefile documents exactly how to reproduce one specific result. Let's add another rule to reproduce another result:

```
# Regenerate results for "Moby Dick"
results/moby_dick.csv : data/moby_dick.txt
	python bin/countwords.py \
		data/moby_dick.txt > results/moby_dick.csv

# Regenerate results for "Jane Eyre"
results/jane_eyre.csv : data/jane_eyre.txt
	python bin/countwords.py \
		data/jane_eyre.txt > results/jane_eyre.csv
```

When we run `make` it tells us:

```
make: `results/moby_dick.csv' is up to date.
```

By default, Make only attempts to update the first target it finds in the Makefile, which is called the **default target**. In this case, the first target is `results/moby_dick.csv`, which is already up to date. To update something else, we need to tell Make specifically what we want:

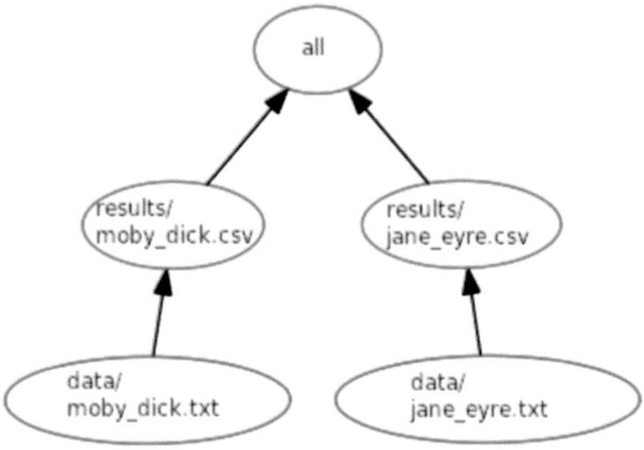

FIGURE 9.1: Dependency graph when making everything.

```
$ make results/jane_eyre.csv
```

```
python bin/countwords.py \
  data/jane_eyre.txt > results/jane_eyre.csv
```

If we have to run `make` once for each result, we're right back where we started. However, we can add a rule to our Makefile to update all of our results at once. We do this by creating a **phony target** that doesn't correspond to an actual file. Let's add this line to the top of our Makefile:

```
# Regenerate all results.
all : results/moby_dick.csv results/jane_eyre.csv
```

There is no file called `all`, and this rule doesn't have any recipes of its own, but when we run `make all`, Make finds everything that `all` depends on, then brings each of those prerequisites up to date (Figure 9.1).

The order in which rules appear in the Makefile does not necessarily determine the order in which recipes are run. Make is free to run commands in any order so long as nothing is updated before its prerequisites are up to date.

We can use phony targets to automate and document other steps in our workflow. For example, let's add another target to our Makefile to delete all of the result files we have generated so that we can start afresh. By convention this target is called `clean`, and we'll place it below the two existing targets:

```
# Remove all generated files.
clean :
	rm -f results/*.csv
```

The `-f` flag to `rm` means "force removal": if it is present, `rm` won't complain if the files we have told it to remove are already gone. If we now run:

```
$ make clean
```

Make will delete any results files we have. This is a lot safer than typing `rm -f results/*.csv` at the command line each time, because if we mistakenly put a space before the `*` we would delete all of the CSV files in the project's root directory.

Phony targets are very useful, but there is a catch. Try doing this:

```
$ mkdir clean
$ make clean
```

```
make: `clean' is up to date.
```

Since there is a directory called `clean`, Make thinks the target `clean` in the Makefile refers to this directory. Since the rule has no prerequisites, it can't be out of date, so no recipes are executed.

We can unconfuse Make by putting this line at the top of Makefile to explicitly state which targets are phony:

```
.PHONY : all clean
```

9.3 Updating Files When Programs Change

Our current Makefile says that each result file depends on the corresponding data file. That's not entirely true: each result also depends on the program used to generate it. If we change our program, we should regenerate our results. To get Make to do that, we can change our prerequisites to include the program:

```
# Regenerate results for "Moby Dick"
results/moby_dick.csv : data/moby_dick.txt bin/countwords.py
	python bin/countwords.py \
		data/moby_dick.txt > results/moby_dick.csv

# Regenerate results for "Jane Eyre"
results/jane_eyre.csv : data/jane_eyre.txt bin/countwords.py
	python bin/countwords.py \
		data/jane_eyre.txt > results/jane_eyre.csv
```

To run both of these rules, we can type `make all`. Alternatively, since `all` is the first target in our Makefile, Make will use it if we just type `make` on its own:

```
$ touch bin/countwords.py
$ make
```

```
python bin/countwords.py \
  data/moby_dick.txt > results/moby_dick.csv
python bin/countwords.py \
  data/jane_eyre.txt > results/jane_eyre.csv
```

The exercises will explore how we can write a rule to tell us whether our results will be different after a change to a program without actually updating them. Rules like this can help us test our programs: if we don't think an addition or modification ought to affect the results, but it would, we may have some debugging to do.

9.4 Reducing Repetition in a Makefile

Our Makefile now mentions `bin/countwords.py` four times. If we ever change the name of the program or move it to a different location, we will have to find and replace each of those occurrences. More importantly, this redundancy makes our Makefile harder to understand, just as scattering **magic numbers** through programs makes them harder to understand.

The solution is the same one we use in programs: define and use variables. Let's modify the results regeneration code by creating targets for the word-counting script and the command used to run it. The entire file should now read:

```
.PHONY : all clean

COUNT=bin/countwords.py
RUN_COUNT=python $(COUNT)

# Regenerate all results.
all : results/moby_dick.csv results/jane_eyre.csv

# Regenerate results for "Moby Dick"
results/moby_dick.csv : data/moby_dick.txt $(COUNT)
	$(RUN_COUNT) data/moby_dick.txt > results/moby_dick.csv

# Regenerate results for "Jane Eyre"
results/jane_eyre.csv : data/jane_eyre.txt $(COUNT)
	$(RUN_COUNT) data/jane_eyre.txt > results/jane_eyre.csv

# Remove all generated files.
clean :
	rm -f results/*.csv
```

Each definition takes the form `NAME=value`. Variables are written in upper case by convention so that they'll stand out from filenames (which are usually in lower case), but Make doesn't require this. What *is* required is using parentheses to refer to the variable, i.e., to use `$(NAME)` and not `$NAME`.

Why the Parentheses?

For historical reasons, Make interprets `$NAME` to be a variable called `N` followed by the three characters "AME." If no variable called `N` exists, `$NAME` becomes `AME`, which is almost certainly not what we want.

As in programs, variables don't just cut down on typing. They also tell readers that several things are always and exactly the same, which reduces **cognitive load**.

9.5 Automatic Variables

We could add a third rule to analyze a third novel and a fourth to analyze a fourth, but that won't scale to hundreds or thousands of novels. Instead, we can write a generic rule that does what we want for every one of our data files.

To do this, we need to understand Make's **automatic variables**. The first step is to use the very cryptic expression `$@` in the rule's recipe to mean "the target of the rule." It lets us turn this:

```
# Regenerate results for "Moby Dick"
results/moby_dick.csv : data/moby_dick.txt $(COUNT)
    $(RUN_COUNT) data/moby_dick.txt > results/moby_dick.csv
```

into this:

```
# Regenerate results for "Moby Dick"
results/moby_dick.csv : data/moby_dick.txt $(COUNT)
    $(RUN_COUNT) data/moby_dick.txt > $@
```

Make defines a value of `$@` separately for each rule, so it always refers to that rule's target. And yes, `$@` is an unfortunate name: something like `$TARGET` would have been easier to understand, but we're stuck with it now.

The next step is to replace the explicit list of prerequisites in the recipe with the automatic variable `$^`, which means "all the prerequisites in the rule":

```
# Regenerate results for "Moby Dick"
results/moby_dick.csv : data/moby_dick.txt $(COUNT)
    $(RUN_COUNT) $^ > $@
```

However, this doesn't work. The rule's prerequisites are the novel and the word-counting program. When Make expands the recipe, the resulting command tries to process the program `bin/countwords.py` as if it was a data file:

```
python bin/countwords.py data/moby_dick.txt bin/countwords.py >
  results/moby_dick.csv
```

Make solves this problem with another automatic variable `$<`, which means "only the first prerequisite." Using it lets us rewrite our rule as:

```
# Regenerate results for "Moby Dick"
results/moby_dick.csv : data/moby_dick.txt $(COUNT)
	$(RUN_COUNT) $< > $@
```

If you use this approach, the rule for *Jane Eyre* should be updated as well. The next section, however, includes instructions for generalizing rules.

9.6 Generic Rules

`$< > $@` is even harder to read than `$@` on its own, but we can now replace all the rules for generating results files with one **pattern rule** using the **wildcard** `%`, which matches zero or more characters in a filename. Whatever matches `%` in the target also matches in the prerequisites, so the rule:

```
results/%.csv : data/%.txt $(COUNT)
	$(RUN_COUNT) $< > $@
```

will handle *Jane Eyre*, *Moby Dick*, *The Time Machine*, and every other novel in the `data/` directory. `%` cannot be used in recipes, which is why `$<` and `$@` are needed.

With this rule in place, our entire Makefile is reduced to:

```
.PHONY: all clean

COUNT=bin/countwords.py
RUN_COUNT=python $(COUNT)

# Regenerate all results.
```

```
all : results/moby_dick.csv results/jane_eyre.csv \
  results/time_machine.csv

# Regenerate result for any book.
results/%.csv : data/%.txt $(COUNT)
	$(RUN_COUNT) $< > $@

# Remove all generated files.
clean :
	rm -f results/*.csv
```

We now have fewer lines of text, but we've also included a third book. To test our shortened Makefile, let's delete all of the results files:

```
$ make clean
```

```
rm -f results/*.csv
```

and then re-create them:

```
$ make  # Same as `make all` as "all" is the first target
```

```
python bin/countwords.py data/moby_dick.txt >
  results/moby_dick.csv
python bin/countwords.py data/jane_eyre.txt >
  results/jane_eyre.csv
python bin/countwords.py data/time_machine.txt >
  results/time_machine.csv
```

We can still rebuild individual files if we want, since Make will take the target filename we give on the command line and see if a pattern rule matches it:

```
$ touch data/jane_eyre.txt
$ make results/jane_eyre.csv
```

```
python bin/countwords.py data/jane_eyre.txt >
  results/jane_eyre.csv
```

9.7 Defining Sets of Files

Our analysis is still not fully automated: if we add another book to `data`, we have to remember to add its name to the `all` target in the Makefile as well. Once again we will fix this in steps.

To start, imagine that all the results files already exist and we just want to update them. We can define a variable called `RESULTS` to be a list of all the results files using the same wildcards we would use in the shell:

```
RESULTS=results/*.csv
```

We can then rewrite `all` to depend on that list:

```
# Regenerate all results.
all : $(RESULTS)
```

However, this only works if the results files already exist. If one doesn't, its name won't be included in `RESULTS` and Make won't realize that we want to generate it.

What we really want is to generate the list of results files based on the list of books in the `data/` directory. We can create that list using Make's `wildcard` function:

```
DATA=$(wildcard data/*.txt)
```

This calls the function `wildcard` with the argument `data/*.txt`. The result is a list of all the text files in the `data` directory, just as we would get with `data/*.txt` in the shell. The syntax is odd because functions were added to Make long after it was first written, but at least they have readable names.

To check that this line does the right thing, we can add another phony target called `settings` that uses the shell command `echo` to print the names and values of our variables:

```
.PHONY: all clean settings

# ...everything else...

# Show variables' values.
settings :
    echo COUNT: $(COUNT)
    echo DATA: $(DATA)
```

Let's run this:

```
$ make settings
```

```
echo COUNT: bin/countwords.py
COUNT: bin/countwords.py
echo DATA: data/dracula.txt data/frankenstein.txt
  data/jane_eyre.txt data/moby_dick.txt
  data/sense_and_sensibility.txt
  data/sherlock_holmes.txt data/time_machine.txt
DATA: data/dracula.txt data/frankenstein.txt
  data/jane_eyre.txt data/moby_dick.txt
  data/sense_and_sensibility.txt data/sherlock_holmes.txt
  data/time_machine.txt
```

The output appears twice because Make shows us the command it's going to run before running it. Putting `@` before the command in the recipe prevents this, which makes the output easier to read:

```
settings :
    @echo COUNT: $(COUNT)
    @echo DATA: $(DATA)
```

```
$ make settings
```

```
COUNT: bin/countwords.py
DATA: data/dracula.txt data/frankenstein.txt
  data/jane_eyre.txt data/moby_dick.txt
  data/sense_and_sensibility.txt data/sherlock_holmes.txt
  data/time_machine.txt
```

We now have the names of our input files. To create a list of corresponding output files, we use Make's `patsubst` function (short for **pat**tern **subst**itution):

```
RESULTS=$(patsubst data/%.txt,results/%.csv,$(DATA))
```

The first argument to `patsubst` is the pattern to look for, which in this case is a text file in the `data` directory. We use `%` to match the **stem** of the file's name, which is the part we want to keep.

The second argument is the replacement we want. As in a pattern rule, Make replaces `%` in this argument with whatever matched `%` in the pattern, which creates the name of the result file we want. Finally, the third argument is what to do the substitution in, which is our list of books' names.

Let's check the `RESULTS` variable by adding another command to the `settings` target:

```
settings :
	@echo COUNT: $(COUNT)
	@echo DATA: $(DATA)
	@echo RESULTS: $(RESULTS)
```

```
$ make settings
```

```
COUNT: bin/countwords.py
DATA: data/dracula.txt data/frankenstein.txt data/jane_eyre.txt
  data/moby_dick.txt data/sense_and_sensibility.txt
  data/sherlock_holmes.txt data/time_machine.txt
RESULTS: results/dracula.csv results/frankenstein.csv
  results/jane_eyre.csv results/moby_dick.csv
  results/sense_and_sensibility.csv
  results/sherlock_holmes.csv results/time_machine.csv
```

Excellent: `DATA` has the names of the files we want to process and `RESULTS` automatically has the names of the corresponding result files.

Why haven't we included `RUN_COUNT` when assessing our variables' values? This is another place we can streamline our script, by removing `RUN_COUNT` from the list of variables and changing our regeneration rule:

```
# Regenerate result for any book.
results/%.csv : data/%.txt $(COUNT)
    python $(COUNT) $< > $@
```

Since the phony target `all` depends on `$(RESULTS)` (i.e., all the files whose names appear in the variable `RESULTS`) we can regenerate all the results in one step:

```
$ make clean
```

```
rm -f results/*.csv
```

```
$ make  # Same as `make all` since "all" is the first target
```

```
python bin/countwords.py data/dracula.txt > results/dracula.csv
python bin/countwords.py data/frankenstein.txt >
  results/frankenstein.csv
python bin/countwords.py data/jane_eyre.txt >
  results/jane_eyre.csv
python bin/countwords.py data/moby_dick.txt >
  results/moby_dick.csv
python bin/countwords.py data/sense_and_sensibility.txt >
  results/sense_and_sensibility.csv
python bin/countwords.py data/sherlock_holmes.txt >
  results/sherlock_holmes.csv
python bin/countwords.py data/time_machine.txt >
  results/time_machine.csv
```

Our workflow is now just two steps: add a data file and run Make. This is a big improvement over running things manually, particularly as we start to add more steps like merging data files and generating plots.

9.8 Documenting a Makefile

Every well-behaved program should tell people how to use it (Taschuk and Wilson 2017). If we run `make --help`, we get a (very) long list of options that

Make understands, but nothing about our specific workflow. We could create another phony target called `help` that prints a list of available commands:

```
.PHONY: all clean help settings

# ...other definitions...

# Show help.
help :
	@echo "all : regenerate all results."
	@echo "results/*.csv : regenerate result for any book."
	@echo "clean : remove all generated files."
	@echo "settings : show variables' values."
	@echo "help : show this message."
```

but sooner or later we will add a target or rule and forget to update this list.

A better approach is to format some comments in a special way and then extract and display those comments when asked to. We'll use `##` (a double comment marker) to indicate the lines we want displayed and `grep` (Section 4.5) to pull these lines out of the file:

```
.PHONY: all clean help settings

COUNT=bin/countwords.py
DATA=$(wildcard data/*.txt)
RESULTS=$(patsubst data/%.txt,results/%.csv,$(DATA))

## all : regenerate all results.
all : $(RESULTS)

## results/%.csv : regenerate result for any book.
results/%.csv : data/%.txt $(COUNT)
	python $(COUNT) $< > $@

## clean : remove all generated files.
clean :
	rm -f results/*.csv

## settings : show variables' values.
settings :
	@echo COUNT: $(COUNT)
	@echo DATA: $(DATA)
```

```
	@echo RESULTS: $(RESULTS)

## help : show this message.
help :
	@grep '^##' ./Makefile
```

Let's test:

```
$ make help
```

```
## all : regenerate all results.
## results/%.csv : regenerate result for any book.
## clean : remove all generated files.
## settings : show variables' values.
## help : show this message.
```

The exercises will explore how to format this more readably.

9.9 Automating Entire Analyses

To finish our discussion of Make, let's automatically generate a collated list of word frequencies. The target is a file called `results/collated.csv` that depends on the results generated by `countwords.py`. To create it, we add or change these lines in our Makefile:

```
# ...phony targets and previous variable definitions...

COLLATE=bin/collate.py

## all : regenerate all results.
all : results/collated.csv

## results/collated.csv : collate all results.
results/collated.csv : $(RESULTS) $(COLLATE)
	mkdir -p results
	python $(COLLATE) $(RESULTS) > $@
```

```
# ...other rules...

## settings : show variables' values.
settings :
	@echo COUNT: $(COUNT)
	@echo DATA: $(DATA)
	@echo RESULTS: $(RESULTS)
	@echo COLLATE: $(COLLATE)

# ...help rule...
```

The first two lines tell Make about the collation program, while the change to `all` tells it what the final target of our pipeline is. Since this target depends on the results files for single novels, `make all` will regenerate all of those automatically.

The rule to regenerate `results/collated.csv` should look familiar by now: it tells Make that all of the individual results have to be up-to-date and that the final result should be regenerated if the program used to create it has changed. One difference between the recipe in this rule and the recipes we've seen before is that this recipe uses `$(RESULTS)` directly instead of an automatic variable. We have written the rule this way because there isn't an automatic variable that means "all but the last prerequisite," so there's no way to use automatic variables that wouldn't result in us trying to process our program.

Likewise, we can also add the `plotcounts.py` script to this workflow and update the `all` and `settings` rules accordingly. Note that there is no `>` needed before the `$@` because the default action of `plotcounts.py` is to write to a file rather than to standard output.

```
# ...phony targets and previous variable definitions...

PLOT=bin/plotcounts.py

## all : regenerate all results.
all : results/collated.png

## results/collated.png: plot the collated results.
results/collated.png : results/collated.csv
	python $(PLOT) $< --outfile $@

# ...other rules...
```

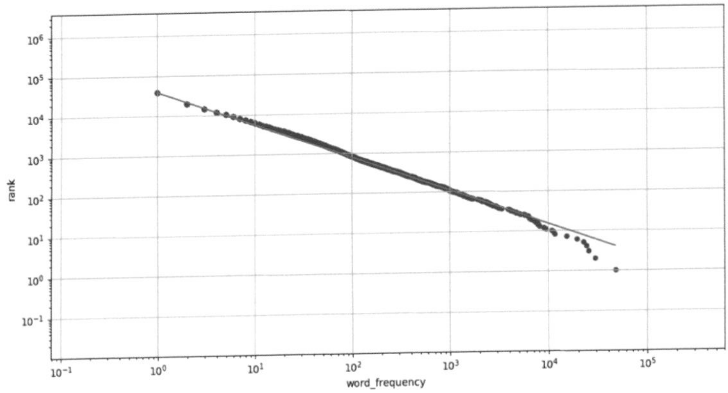

FIGURE 9.2: Word count distribution for all books combined.

```
## settings : show variables' values.
settings :
	@echo COUNT: $(COUNT)
	@echo DATA: $(DATA)
	@echo RESULTS: $(RESULTS)
	@echo COLLATE: $(COLLATE)
	@echo PLOT: $(PLOT)

# ...help...
```

Running `make all` should now generate the new `collated.png` plot (Figure 9.2):

```
$ make all
```

```
python bin/collate.py results/time_machine.csv
  results/moby_dick.csv results/jane_eyre.csv
  results/dracula.csv results/sense_and_sensibility.csv
  results/sherlock_holmes.csv results/frankenstein.csv >
  results/collated.csv
python bin/plotcounts.py results/collated.csv --outfile
  results/collated.png
alpha: 1.1712445413685917
```

Finally, we can update the `clean` target to only remove files created by the Makefile. It is a good habit to do this rather than using the asterisk wildcard to remove all files, since you might manually place files in the results directory and forget that these will be cleaned up when you run `make clean`.

```
# ...phony targets and previous variable definitions...

## clean : remove all generated files.
clean :
	rm $(RESULTS) results/collated.csv results/collated.png
```

9.10 Summary

Make's reliance on shell commands instead of direct calls to functions in Python sometimes makes it clumsy to use. However, that also makes it very flexible: a single Makefile can run shell commands and programs written in a variety of languages, which makes it a great way to assemble pipelines out of whatever is lying around.

Programmers have created many replacements for Make in the 45 years since it was first created—so many, in fact, that none have attracted enough users to displace it. If you would like to explore them, check out Snakemake[3] (for Python). If you want to go deeper, Smith (2011) describes the design and implementation of several build managers.

9.11 Exercises

Our `Makefile` currently reads as follows:

```
.PHONY: all clean help settings

COUNT=bin/countwords.py
COLLATE=bin/collate.py
```

[3] `https://snakemake.readthedocs.io/`

```
PLOT=bin/plotcounts.py
DATA=$(wildcard data/*.txt)
RESULTS=$(patsubst data/%.txt,results/%.csv,$(DATA))

## all : regenerate all results.
all : results/collated.png

## results/collated.png: plot the collated results.
results/collated.png : results/collated.csv
	python $(PLOT) $< --outfile $@

## results/collated.csv : collate all results.
results/collated.csv : $(RESULTS) $(COLLATE)
	@mkdir -p results
	python $(COLLATE) $(RESULTS) > $@

## results/%.csv : regenerate result for any book.
results/%.csv : data/%.txt $(COUNT)
	python $(COUNT) $< > $@

## clean : remove all generated files.
clean :
	rm $(RESULTS) results/collated.csv results/collated.png

## settings : show variables' values.
settings :
	@echo COUNT: $(COUNT)
	@echo DATA: $(DATA)
	@echo RESULTS: $(RESULTS)
	@echo COLLATE: $(COLLATE)
	@echo PLOT: $(PLOT)

## help : show this message.
help :
	@grep '^##' ./Makefile
```

A number of the exercises below ask you to make further edits to `Makefile`.

9.11.1 Report results that would change

How can you get `make` to show the commands it would run without actually running them? (Hint: look at the manual page.)

9.11.2 Useful options

1. What does Make's `-B` option do and when is it useful?
2. What about the `-C` option?
3. What about the `-f` option?

9.11.3 Make sure the output directory exists

One of our **build recipes** includes `mkdir -p`. What does this do and why is it useful?

9.11.4 Print the title and author

The build rule for regenerating the result for any book is currently:

```
## results/%.csv : regenerate result for any book.
results/%.csv : data/%.txt $(COUNT)
    python $(COUNT) $< > $@
```

Add an extra line to the recipe that uses the `book_summary.sh` script to print the title and author of the book to the screen. Use `@bash` so that the command itself isn't printed to the screen and don't forget to update the settings build rule to include the `book_summary.sh` script.

If you've successfully made those changes, you should get the following output for *Dracula*:

```
$ make -B results/dracula.csv
```

```
Title: Dracula
Author: Bram Stoker
python bin/countwords.py data/dracula.txt > results/dracula.csv
```

9.11.5 Create all results

The default target of our final `Makefile` re-creates `results/collated.csv`. Add a target to `Makefile` so that `make results` creates or updates any result files that are missing or out of date, but does *not* regenerate `results/collated.csv`.

9.11.6 The perils of shell wildcards

What is wrong with writing the rule for `results/collated.csv` like this:

```
results/collated.csv : results/*.csv
    python $(COLLATE) $^ > $@
```

(The fact that the result no longer depends on the program used to create it isn't the biggest problem.)

9.11.7 Making documentation more readable

We can format the documentation in our Makefile more readably using this command:

```
## help : show all commands.
help :
    @grep -h -E '^##' ${MAKEFILE_LIST} | sed -e 's/## //g' \
    | column -t -s ':'
```

Using `man` and online search, explain what every part of this recipe does.

9.11.8 Configuration

A next step in automating this analysis might include moving the definitions of the `COUNT`, `COLLATE`, and `PLOT` variables into a separate file called `config.mk`:

```
COUNT=bin/countwords.py
COLLATE=bin/collate.py
PLOT=bin/plotcounts.py
```

and using the `include` command to access those definitions in the existing `Makefile`:

```
.PHONY: results all clean help settings

include config.mk

# ... the rest of the Makefile ...
```

Under what circumstances would this strategy be useful?

9.12 Key Points

- Make[4] is a widely used build manager.
- A **build manager** re-runs commands to update files that are out of date.
- A **build rule** has **targets**, **prerequisites**, and a **recipe**.
- A target can be a file or a **phony target** that simply triggers an action.
- When a target is out of date with respect to its prerequisites, Make executes the recipe associated with its rule.
- Make executes as many rules as it needs to when updating files, but always respects prerequisite order.
- Make defines **automatic variables** such as `$@` (target), `$^` (all prerequisites), and `$<` (first prerequisite).
- **Pattern rules** can use `%` as a placeholder for parts of filenames.
- Makefiles can define variables using `NAME=value`.
- Make also has functions such as `$(wildcard...)` and `$(patsubst...)`.
- Use specially formatted comments to create self-documenting Makefiles.

[4] `https://www.gnu.org/software/make/`

10

Configuring Programs

> Always be wary of any helpful item that weighs less than its operating manual.
>
> — Terry Pratchett

In previous chapters we used command-line options to control our scripts and programs. If they are more complex, we may want to use up to four layers of configuration:

1. A system-wide configuration file for general settings.
2. A user-specific configuration file for personal preferences.
3. A job-specific file with settings for a particular run.
4. Command-line options to change things that commonly change.

This is sometimes called **overlay configuration** because each level overrides the ones above it: the user's configuration file overrides the system settings, the job configuration overrides the user's defaults, and the command-line options overrides that. This is more complex than most research software needs initially (Xu et al. 2015), but being able to read a complete set of options from a file is a big boost to reproducibility.

In this chapter, we'll explore a number approaches for configuring our Zipf's Law project, and ultimately decide to apply one of them. That project should now contain:

```
zipf/
├── .gitignore
├── CONDUCT.md
├── CONTRIBUTING.md
├── LICENSE.md
├── Makefile
├── README.md
├── bin
│   ├── book_summary.sh
│   ├── collate.py
```

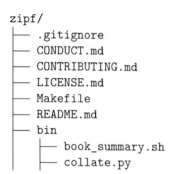

```
│   ├── countwords.py
│   ├── plotcounts.py
│   ├── script_template.py
│   └── utilities.py
├── data
│   ├── README.md
│   ├── dracula.txt
│   └── ...
└── results
    ├── collate.csv
    ├── collate.png
    ├── dracula.csv
    ├── dracula.png
    └── ...
```

Be Careful When Applying Settings outside Your Project

This chapter's examples modify files outside of the Zipf's Law project in order to illustrate some concepts. If you alter these files while following along, remember to change them back later.

10.1 Configuration File Formats

Programmers have invented far too many formats for configuration files; rather than creating one of your own, you should adopt some widely used approach. One is to write the configuration as a Python module and load it as if it was a library. This is clever, but means that tools in other languages can't process it.

A second option is Windows INI format[1], which is laid out like this:

```
[section1]
key1=value1
key2=value2

[section2]
```

[1]https://en.wikipedia.org/wiki/INI_file

```
key3=value3
key4=value4
```

INI files are simple to read and write, but the format is slowly falling out of use in favor of **YAML**. A simple YAML configuration file looks like this:

```yaml
# Standard settings for thesis.
logfile: "/tmp/log.txt"
quiet: false
overwrite: false
fonts:
- Verdana
- Serif
```

Here, the keys `logfile`, `quiet`, and `overwrite` have the values `/tmp/log.txt`, `false`, and `false` respectively, while the value associated with the key `fonts` is a list containing `Verdana` and `Serif`. For more discussion of YAML, see Appendix H.

10.2 Matplotlib Configuration

To see overlay configuration in action, let's consider a common task in data science: changing the size of the labels on a plot. The labels on our *Jane Eyre* word frequency plot are fine for viewing on screen (Figure 10.1), but they will need to be bigger if we want to include the figure in a slideshow or report.

We could use any of the overlay options described above to change the size of the labels:

- Edit the system-wide Matplotlib configuration file (which would affect everyone using this computer).
- Create a user-specific Matplotlib style sheet.
- Create a job-specific configuration file to set plotting options in `plotcounts.py`.
- Add some new command-line options to `plotcounts.py` .

Let's consider these options one by one.

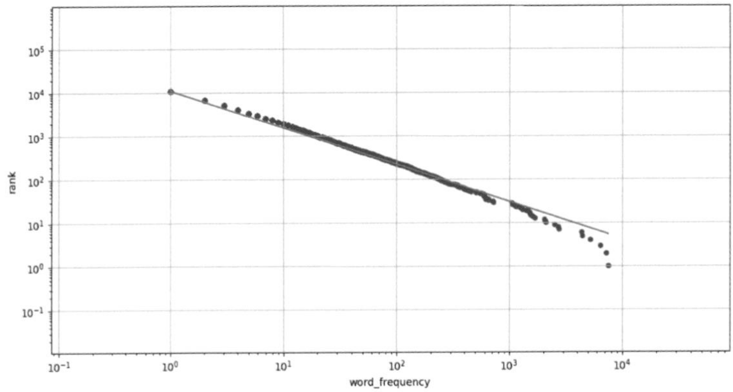

FIGURE 10.1: Word frequency distribution for Jane Eyre with default label sizes.

10.3 The Global Configuration File

Our first configuration possibility is to edit the system-wide Matplotlib run-time configuration file, which is called `matplotlibrc`. When we import Matplotlib, it uses this file to set the default characteristics of the plot. We can find it on our system by running this command:

```
import matplotlib as mpl
mpl.matplotlib_fname()
```

```
/Users/amira/anaconda3/lib/python3.7/site-packages/matplotlib/
mpl-data/matplotlibrc
```

In this case the file is located in the Python installation directory (`anaconda3`). All the different Python packages installed with Anaconda live in a `python3.7/site-packages` directory, including Matplotlib.

`matplotlibrc` lists all the default settings as comments. The default size of the X and Y axis labels is "medium," as is the size of the tick labels:

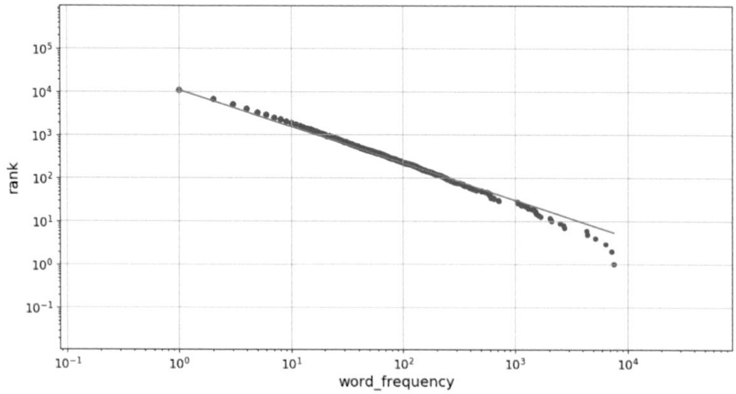

FIGURE 10.2: Word frequency distribution for Jane Eyre with larger labels.

```
#axes.labelsize     : medium  ## fontsize of the x and y labels
#xtick.labelsize    : medium  ## fontsize of the tick labels
#ytick.labelsize    : medium  ## fontsize of the tick labels
```

We can uncomment those lines and change the sizes to "large" and "extra large":

```
axes.labelsize      : x-large  ## fontsize of the x and y labels
xtick.labelsize     : large    ## fontsize of the tick labels
ytick.labelsize     : large    ## fontsize of the tick labels
```

and then re-generate the *Jane Eyre* plot with bigger labels (Figure 10.2):

```
$ python bin/plotcounts.py results/jane_eyre.csv --outfile
  results/jane_eyre.png
```

This does what we want, but is usually the wrong approach. Since the `matplotlibrc` file sets system-wide defaults, we will now have big labels by default for all plotting we do in the future, which we may not want. Secondly, we want to package our Zipf's Law code and make it available to other people (Chapter 14). That package won't include our `matplotlibrc` file, and we don't have access to the one on their computer, so this solution isn't as reproducible as others.

A global options file *is* useful, though. If we are using Matplotlib with **LaTeX** to generate reports and the latter is installed in an unusual place on our computing cluster, a one-line change in `matplotlibrc` can prevent a lot of failed jobs.

10.4 The User Configuration File

If we don't want to change the configuration for everyone, we can change it for just ourself. Matplotlib defines several carefully designed styles for plots:

```
import matplotlib.pyplot as plt
print(plt.style.available)
```

```
['seaborn-dark', 'seaborn-darkgrid', 'seaborn-ticks',
'fivethirtyeight', 'seaborn-whitegrid', 'classic',
'_classic_test', 'fast', 'seaborn-talk', 'seaborn-dark-palette',
'seaborn-bright', 'seaborn-pastel', 'grayscale',
'seaborn-notebook', 'ggplot', 'seaborn-colorblind',
'seaborn-muted', 'seaborn', 'Solarize_Light2', 'seaborn-paper',
'bmh', 'tableau-colorblind10', 'seaborn-white',
'dark_background', 'seaborn-poster', 'seaborn-deep']
```

In order to make the labels bigger in all of our Zipf's Law plots, we could create a custom Matplotlib style sheet. The convention is to store custom style sheets in a `stylelib` sub-directory in the Matplotlib configuration directory. That directory can be located by running the following command:

```
mpl.get_configdir()
```

```
/Users/amira/.matplotlib
```

Once we've created the new sub-directory:

```
$ mkdir /Users/amira/.matplotlib/stylelib
```

we can add a new file called `big-labels.mplstyle` that has the same YAML format as the `matplotlibrc` file:

```
axes.labelsize   : x-large  ## fontsize of the x and y labels
xtick.labelsize  : large    ## fontsize of the tick labels
ytick.labelsize  : large    ## fontsize of the tick labels
```

To use this new style, we would just need to add one line to `plotcounts.py`:

```
plt.style.use('big-labels')
```

Using a custom style sheet leaves the system-wide defaults unchanged, and it's a good way to achieve a consistent look across our personal data visualization projects. However, since each user has their own `stylelib` directory, it doesn't solve the problem of ensuring that other people can reproduce our plots.

10.5 Adding Command-Line Options

A third way to change the plot's properties is to add some new command-line arguments to `plotcounts.py`. The `choices` parameter of `add_argument` lets us tell `argparse` that the user is only allowed to specify a value from a predefined list:

```
mpl_sizes = ['xx-small', 'x-small', 'small', 'medium',
             'large', 'x-large', 'xx-large']
parser.add_argument('--labelsize', type=str, default='x-large',
                    choices=mpl_sizes,
                    help='fontsize of the x and y labels')
parser.add_argument('--xticksize', type=str, default='large',
                    choices=mpl_sizes,
                    help='fontsize of the x tick labels')
parser.add_argument('--yticksize', type=str, default='large',
                    choices=mpl_sizes,
                    help='fontsize of the y tick labels')
```

We can then add a few lines after the `ax` variable is defined in `plotcounts.py` to update the label sizes according to the user input:

```
ax.xaxis.label.set_fontsize(args.labelsize)
ax.yaxis.label.set_fontsize(args.labelsize)
ax.xaxis.set_tick_params(labelsize=args.xticksize)
ax.yaxis.set_tick_params(labelsize=args.yticksize)
```

Alternatively, we can change the default runtime configuration settings before the plot is created. These are stored in a variable called `matplotlib.rcParams`:

```
mpl.rcParams['axes.labelsize'] = args.labelsize
mpl.rcParams['xtick.labelsize'] = args.xticksize
mpl.rcParams['ytick.labelsize'] = args.yticksize
```

Adding extra command-line arguments is a good solution if we only want to change a small number of plot characteristics. It also makes our work more reproducible: if we use a Makefile to regenerate our plots (Chapter 9), the settings will all be saved in one place. However, `matplotlibrc` has hundreds of parameters we could change, so the number of new arguments can quickly get out of hand if we want to tweak other aspects of the plot.

10.6 A Job Control File

The final option for configuring our plots—the one we will actually adopt in this case—is to pass a YAML file full of Matplotlib parameters to `plotcounts.py`. First, we save the parameters we want to change in a file inside our project directory. We can call it anything, but `plotparams.yml` seems like it will be easy to remember. We'll store it in `bin` with the scripts that will use it:

```
# Plot characteristics
axes.labelsize   : x-large  ## fontsize of the x and y labels
xtick.labelsize  : large    ## fontsize of the tick labels
ytick.labelsize  : large    ## fontsize of the tick labels
```

Because this file is located in our project directory instead of the user-specific style sheet directory, we need to add one new option to `plotcounts.py` to load it:

```
parser.add_argument('--plotparams', type=str, default=None,
                    help='matplotlib parameters (YAML file)')
```

We can use Python's `yaml` library to read that file:

```
with open('plotparams.yml', 'r') as reader:
    plot_params = yaml.load(reader, Loader=yaml.BaseLoader)
print(plot_params)
```

```
{'axes.labelsize': 'x-large',
 'xtick.labelsize': 'large',
 'ytick.labelsize': 'large'}
```

and then loop over each item in `plot_params` to update Matplotlib's `mpl.rcParams`:

```
for (param, value) in param_dict.items():
    mpl.rcParams[param] = value
```

`plotcounts.py` now looks like this:

```
"""Plot word counts."""

import argparse

import yaml
import numpy as np
import pandas as pd
import matplotlib as mpl
from scipy.optimize import minimize_scalar


def nlog_likelihood(beta, counts):
    # ...as before...


def get_power_law_params(word_counts):
    # ...as before...
```

```
def set_plot_params(param_file):
    """Set the matplotlib parameters."""
    if param_file:
        with open(param_file, 'r') as reader:
            param_dict = yaml.load(reader,
                                   Loader=yaml.BaseLoader)
    else:
        param_dict = {}
    for param, value in param_dict.items():
        mpl.rcParams[param] = value


def plot_fit(curve_xmin, curve_xmax, max_rank, beta, ax):
    # ...as before...


def main(args):
    """Run the command line program."""
    set_plot_params(args.plotparams)
    df = pd.read_csv(args.infile, header=None,
                     names=('word', 'word_frequency'))
    df['rank'] = df['word_frequency'].rank(ascending=False,
                                           method='max')
    ax = df.plot.scatter(x='word_frequency',
                         y='rank', loglog=True,
                         figsize=[12, 6],
                         grid=True,
                         xlim=args.xlim)

    word_counts = df['word_frequency'].to_numpy()
    alpha = get_power_law_params(word_counts)
    print('alpha:', alpha)

    # Since the ranks are already sorted, we can take the last
    # one instead of computing which row has the highest rank
    max_rank = df['rank'].to_numpy()[-1]

    # Use the range of the data as the boundaries
    # when drawing the power law curve
    curve_xmin = df['word_frequency'].min()
    curve_xmax = df['word_frequency'].max()
```

```
    plot_fit(curve_xmin, curve_xmax, max_rank, alpha, ax)
    ax.figure.savefig(args.outfile)


if __name__ == '__main__':
    parser = argparse.ArgumentParser(description=__doc__)
    parser.add_argument('infile', type=argparse.FileType('r'),
                        nargs='?', default='-',
                        help='Word count csv file name')
    parser.add_argument('--outfile', type=str,
                        default='plotcounts.png',
                        help='Output image file name')
    parser.add_argument('--xlim', type=float, nargs=2,
                        metavar=('XMIN', 'XMAX'),
                        default=None, help='X-axis limits')
    parser.add_argument('--plotparams', type=str, default=None,
                        help='matplotlib parameters (YAML file)')
    args = parser.parse_args()
    main(args)
```

10.7 Summary

Programs are only useful if they can be controlled, and work is only reproducible if those controls are explicit and shareable. If the number of controls needed is small, adding command-line options to programs and setting those options in Makefiles is usually the best solution. As the number of options grows, so too does the value of putting options in files of their own. And if we are installing the software on large systems that are used by many people, such as a research cluster, system-wide configuration files let us hide the details from people who just want to get their science done.

More generally, the problem of configuring a program illustrates the difference between "works for me on my machine" and "works for everyone, everywhere." From reproducible workflows (Chapter 9) to logging (Section 12.4), this difference influences every aspect of a research software engineer's work. We don't always have to design for large-scale re-use, but knowing what it entails allows us to make a conscious, thoughtful choice.

10.8 Exercises

10.8.1 Building with plotting parameters

In the `Makefile` created in Chapter 9, the build rule involving `plotcounts.py` was defined as:

```
## results/collated.png: plot the collated results.
results/collated.png : results/collated.csv
	python $(PLOT) $< --outfile $@
```

Update that build rule to include the new `--plotparams` option. Make sure `plotparams.yml` is added as a second prerequisite in the updated build rule so that the appropriate commands will be re-run if the plotting parameters change.

Hint: We use the automatic variable `$<` to access the first prerequisite, but you'll need `$(word 2,$^)` to access the second. Read about automatic variables (Section 9.5) and functions for string substitution and analysis[2] to understand what that command is doing.

10.8.2 Using different plot styles

There are many pre-defined matplotlib styles (Section 10.4), as illustrated at the Python Graph Gallery[3].

1. Add a new option `--style` to `plotcounts.py` that allows the user to pick a style from the list of pre-defined matplotlib styles.

Hint: Use the `choices` parameter discussed in Section 10.5 to define the valid choices for the new `--style` option.

2. Re-generate the plot of the *Jane Eyre* word count distribution using a bunch of different styles to decide which you like best.
3. Matplotlib style sheets are designed to be composed together. (See the style sheets tutorial[4] for details.) Use the `nargs` parameter to

[2]`https://www.gnu.org/software/make/manual/html_node/Text-Functions.html#Text-Functions`

[3]`https://python-graph-gallery.com/199-matplotlib-style-sheets/`

[4]`https://matplotlib.org/tutorials/introductory/customizing.html`

allow the user to pass any number of styles when using the `--style` option.

10.8.3 Saving configurations

1. Add an option `--saveconfig filename` to `plotcounts.py` that writes all of its configuration to a file. Make sure this option saves *all* of the configuration, including any defaults that the user hasn't changed.
2. Add a new target `test-saveconfig` to the `Makefile` created in Chapter 9 to test that the new option is working.
3. How would this new `--saveconfig` option make your work more reproducible?

10.8.4 Using INI syntax

If we used Windows INI format[5] instead of YAML for our plot parameters configuration file (i.e., `plotparams.ini` instead of `plotparams.yml`) that file would read as follows:

```
[AXES]
axes.labelsize=x-large

[TICKS]
xtick.labelsize=large
ytick.labelsize=large
```

The `configparser`[6] library can be used to read and write INI files. Install that library by running `pip install configparser` at the command line.

Using `configparser`, rewrite the `set_plot_params` function in `plotcounts.py` to handle a configuration file in INI rather than YAML format.

1. Which file format do you find easier to work with?
2. What other factors should influence your choice of a configuration file syntax?

Note: the code modified in this exercise is not required for the rest of the book.

[5] `https://en.wikipedia.org/wiki/INI_file`
[6] `https://docs.python.org/3/library/configparser.html`

10.8.5 Configuration consistency

In order for a data processing pipeline to work correctly, some of the configuration parameters for Program A and Program B must be the same. However, the programs were written by different teams, and each has its own configuration file. What steps could you take to ensure the required consistency?

10.9 Key Points

- **Overlay configuration** specifies settings for a program in layers, each of which overrides previous layers.
- Use a system-wide configuration file for general settings.
- Use a user-specific configuration file for personal preferences.
- Use a job-specific configuration file with settings for a particular run.
- Use command-line options to change things that commonly change.
- Use **YAML** or some other standard syntax to write configuration files.
- Save configuration information to make your research **reproducible**.

11

Testing Software

> Opera happens because a large number of things amazingly fail to go wrong.
>
> — Terry Pratchett

We have written software to count and analyze the words in classic texts, but how can we be sure it's producing reliable results? The short is answer is that we can't—not completely—but we can test its behavior against our expectations to decide if we are sure enough. This chapter explores ways to do this, including assertions, unit tests, integration tests, and regression tests.

A Scientist's Nightmare

Why is testing research software important? A successful early career researcher in protein crystallography, Geoffrey Chang, had to retract five published papers—three from the journal *Science*—because his code had inadvertently flipped two columns of data (G. Miller 2006). More recently, a simple calculation mistake in a paper by Reinhart and Rogoff contributed to making the financial crash of 2008 even worse for millions of people (Borwein and Bailey 2013). Testing helps to catch errors like these.

Here's the current structure of our Zipf's Law project files:

```
zipf/
├── .gitignore
├── CONDUCT.md
├── CONTRIBUTING.md
├── LICENSE.md
├── Makefile
├── README.md
├── bin
│   ├── book_summary.sh
│   ├── collate.py
│   ├── countwords.py
```

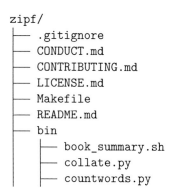

```
│   ├── plotcounts.py
│   ├── plotparams.yml
│   ├── script_template.py
│   └── utilities.py
├── data
│   ├── README.md
│   ├── dracula.txt
│   └── ...
└── results
    ├── collated.csv
    ├── collated.png
    ├── dracula.csv
    ├── dracula.png
    └── ...
```

11.1 Assertions

The first step in building confidence in our programs is to assume that mistakes will happen and guard against them. This is called **defensive programming**, and the most common way to do it is to add **assertions** to our code so that it checks itself as it runs. An assertion is a statement that something must be true at a certain point in a program. When Python sees an assertion, it checks the assertion's condition. If it's true, Python does nothing; if it's false, Python halts the program immediately and prints a user-defined error message. For example, this code halts as soon as the loop encounters an impossible word frequency:

```
frequencies = [13, 10, 2, -4, 5, 6, 25]
total = 0.0
for freq in frequencies[:5]:
    assert freq >= 0.0, 'Word frequencies must be non-negative'
    total += freq
print('total frequency of first 5 words:', total)
```

```
---------------------------------------------------------------------
AssertionError                    Traceback (most recent call last)
<ipython-input-19-33d87ea29ae4> in <module>()
      2 total = 0.0
      3 for freq in frequencies[:5]:
```

```
----> 4     assert freq >= 0.0, 'Word frequencies must be
            non-negative'
      5     total += freq
      6 print('total frequency of first 5 words:', total)

AssertionError: Word frequencies must be non-negative
```

Programs intended for widespread use are full of assertions: 10%–20% of the code they contain are there to check that the other 80%–90% are working correctly. Broadly speaking, assertions fall into three categories:

- A **precondition** is something that must be true at the start of a function in order for it to work correctly. For example, a function might check that the list it has been given has at least two elements and that all of its elements are integers.
- A **postcondition** is something that the function guarantees is true when it finishes. For example, a function could check that the value being returned is an integer that is greater than zero, but less than the length of the input list.
- An **invariant** is something that is true for every iteration in a loop. The invariant might be a property of the data (as in the example above), or it might be something like, "the value of `highest` is less than or equal to the current loop index."

The function `get_power_law_params` in our `plotcounts.py` script is a good example of the need for a precondition. Its docstring does not say that its `word_counts` parameter must be a list of numeric word counts; even if we add that, a user might easily pass in a list of the words themselves instead. Adding an assertion makes the requirement clearer, and also guarantees that the function will fail as soon as it is called rather than returning an error from `scipy.optimize.minimize_scalar` that would be more difficult to interpret/debug.

```
def get_power_law_params(word_counts):
    """
    Get the power law parameters.

    References
    ----------
    Moreno-Sanchez et al (2016) define alpha (Eq. 1),
      beta (Eq. 2) and the maximum likelihood estimation (mle)
```

```
    of beta (Eq. 6).

  Moreno-Sanchez I, Font-Clos F, Corral A (2016)
    Large-Scale Analysis of Zipf's Law in English Texts.
    PLoS ONE 11(1): e0147073.
    https://doi.org/10.1371/journal.pone.0147073
  """
  assert type(word_counts) == np.ndarray, \
      'Input must be a numerical (numpy) array of word counts'
  mle = minimize_scalar(nlog_likelihood,
                        bracket=(1 + 1e-10, 4),
                        args=word_counts,
                        method='brent')
  beta = mle.x
  alpha = 1 / (beta - 1)
  return alpha
```

Update `plotcounts.py` with the assertion described above. You'll see additional examples of assertions throughout this chapter.

11.2 Unit Testing

Catching errors is good, but preventing them is better, so responsible programmers test their code. As the name suggests, a **unit test** checks the correctness of a single unit of software. Exactly what constitutes a "unit" is subjective, but it typically means the behavior of a single function in one situation. In our Zipf's Law software, the `count_words` function in `countwords.py` is a good candidate for unit testing:

```
def count_words(reader):
    """Count the occurrence of each word in a string."""
    text = reader.read()
    chunks = text.split()
    npunc = [word.strip(string.punctuation) for word in chunks]
    word_list = [word.lower() for word in npunc if word]
    word_counts = Counter(word_list)
    return word_counts
```

A single unit test will typically have:

- a **fixture**, which is the thing being tested (e.g., an array of numbers);
- an **actual result**, which is what the code produces when given the fixture; and
- an **expected result** that the actual result is compared to.

The fixture is typically a subset or smaller version of the data the function will typically process. For instance, in order to write a unit test for the `count_words` function, we could use a piece of text small enough for us to count word frequencies by hand. Let's add the poem *Risk* by Anaïs Nin to our data:

```
$ cd ~/zipf
$ mkdir test_data
$ cat test_data/risk.txt
```

```
And then the day came,
when the risk
to remain tight
in a bud
was more painful
than the risk
it took
to blossom.
```

We can then count the words by hand to construct the expected result:

```
from collections import Counter


risk_poem_counts = {'the': 3, 'risk': 2, 'to': 2, 'and': 1,
  'then': 1, 'day': 1, 'came': 1, 'when': 1, 'remain': 1,
  'tight': 1, 'in': 1, 'a': 1, 'bud': 1, 'was': 1,
  'more': 1, 'painful': 1, 'than': 1, 'it': 1, 'took': 1,
  'blossom': 1}
expected_result = Counter(risk_poem_counts)
```

We then generate the actual result by calling `count_words`, and use an assertion to check if it is what we expected:

```
import sys
sys.path.append('/Users/amira/zipf/bin')
import countwords


with open('test_data/risk.txt', 'r') as reader:
    actual_result = countwords.count_words(reader)
assert actual_result == expected_result
```

There's no output, which means the assertion (and test) passed. (Remember, assertions only do something if the condition is false.)

Appending the Import Path

The last code chunk included `sys.path.append`, which allowed us to import scripts from elsewhere than our current working directory. This isn't a perfect solution, but works well enough for quick tests while developing code, especially since the tool we'll use for the rest of the chapter handles this issue for us. Import statements[1] are notoriously tricky, but we'll learn better methods for organizing our code for function imports in Chapter 14.

11.3 Testing Frameworks

Writing one unit test is easy enough, but we should check other cases as well. To manage them, we can use a **test framework** (also called a **test runner**). The most widely used test framework for Python is called `pytest`[2], which structures tests as follows:

1. Tests are put in files whose names begin with `test_`.
2. Each test is a function whose name also begins with `test_`.
3. These functions use `assert` to check results.

Following these rules, we can create a `test_zipfs.py` script in your `bin` directory that contains the test we just developed:

[1] https://chrisyeh96.github.io/2017/08/08/definitive-guide-python-imports.html
[2] https://pytest.org/

```
from collections import Counter

import countwords


def test_word_count():
    """Test the counting of words.

    The example poem is Risk, by Anaïs Nin.
    """
    risk_poem_counts = {'the': 3, 'risk': 2, 'to': 2, 'and': 1,
      'then': 1, 'day': 1, 'came': 1, 'when': 1, 'remain': 1,
      'tight': 1, 'in': 1, 'a': 1, 'bud': 1, 'was': 1,
      'more': 1, 'painful': 1, 'than': 1, 'it': 1, 'took': 1,
      'blossom': 1}
    expected_result = Counter(risk_poem_counts)
    with open('test_data/risk.txt', 'r') as reader:
        actual_result = countwords.count_words(reader)
    assert actual_result == expected_result
```

The `pytest` library comes with a command-line tool that is also called `pytest`. When we run it with no options, it searches for all files in or below the working directory whose names match the pattern `test_*.py`. It then runs the tests in these files and summarizes their results. (If we only want to run the tests in a particular file, we can use the command `pytest path/to/test_file.py`.)

```
$ pytest
```

```
==================== test session starts ======================
platform darwin -- Python 3.7.6, pytest-6.2.0, py-1.10.0,
pluggy-0.13.1
rootdir: /Users/amira
collected 1 item

bin/test_zipfs.py .                                      [100%]

===================== 1 passed in 0.02s =======================
```

To add more tests, we simply write more `test_` functions in `test_zipfs.py`. For instance, besides counting words, the other critical part of our code is the calculation of the α parameter. Earlier we defined a power law relating α to the word frequency f, the word rank r, and a constant of proportionality c (Section 7.3):

$$r = cf^{\frac{-1}{\alpha}}$$

We also noted that Zipf's Law holds exactly when α is equal to one. Setting α to one and re-arranging the power law gives us:

$$c = f/r$$

We can use this formula to generate synthetic word count data (i.e., our test fixture) with a constant of proportionality set to a hypothetical maximum word frequency of 600 (and thus r ranges from 1 to 600):

```
import numpy as np


max_freq = 600
counts = np.floor(max_freq / np.arange(1, max_freq + 1))
print(counts)
```

```
[600. 300. 200. 150. 120. 100.  85.  75.  66.  60.  54.  50.
  46.  42.  40.  37.  35.  33.  31.  30.  28.  27.  26.  25.
 ...
   1.   1.   1.   1.   1.   1.   1.   1.   1.   1.   1.   1.]
```

(We use `np.floor` to round down to the nearest whole number, because we can't have fractional word counts.) Passing this test fixture to `get_power_law_params` in `plotcounts.py` should give us a value of 1.0. To test this, we can add a second test to `test_zipfs.py`:

```
from collections import Counter

import numpy as np

import plotcounts
import countwords


def test_alpha():
    """Test the calculation of the alpha parameter.
```

```
    The test word counts satisfy the relationship,
      r = cf**(-1/alpha), where
      r is the rank,
      f the word count, and
      c is a constant of proportionality.

    To generate test word counts for an expected alpha of
      1.0, a maximum word frequency of 600 is used
      (i.e. c = 600 and r ranges from 1 to 600)
    """
    max_freq = 600
    counts = np.floor(max_freq / np.arange(1, max_freq + 1))
    actual_alpha = plotcounts.get_power_law_params(counts)
    expected_alpha = 1.0
    assert actual_alpha == expected_alpha


def test_word_count():
    #...as before...
```

Let's re-run both of our tests:

```
$ pytest
```

```
==================== test session starts ======================
platform darwin -- Python 3.7.6, pytest-6.2.1, py-1.10.0,
pluggy-0.13.1
rootdir: /Users/amira
collected 2 items

bin/test_zipfs.py F.                                    [100%]

========================= FAILURES ==========================
________________________ test_alpha _________________________

  def test_alpha():
      """Test the calculation of the alpha parameter.

      The test word counts satisfy the relationship,
        r = cf**(-1/alpha), where
        r is the rank,
```

```
        f the word count, and
        c is a constant of proportionality.

      To generate test word counts for an expected alpha of
        1.0, a maximum word frequency of 600 is used
        (i.e. c = 600 and r ranges from 1 to 600)
      """
      max_freq = 600
      counts = np.floor(max_freq / np.arange(1, max_freq + 1))
      actual_alpha = plotcounts.get_power_law_params(counts)
      expected_alpha = 1.0
>     assert actual_alpha == expected_alpha
E     assert 0.9951524579316625 == 1.0

bin/test_zipfs.py:26: AssertionError
================== short test summary info ===================
FAILED bin/test_zipfs.py::test_alpha - assert 0.99515246 == 1.0
================= 1 failed, 1 passed in 3.98s =================
```

The output tells us that one test failed but the other test passed. This is a very useful feature of test runners like `pytest`: they continue on and complete all the tests rather than stopping at the first assertion failure as a regular Python script would.

11.4 Testing Floating-Point Values

The output above shows that while `test_alpha` failed, the `actual_alpha` value of 0.9951524579316625 was very close to the expected value of 1.0. After a bit of thought, we decide that this isn't actually a failure: the value produced by `get_power_law_params` is an estimate, and being off by half a percent is good enough.

This example shows that testing scientific software almost always requires us to make the same kind of judgment calls that scientists have to make when doing any other sort of experimental work. If we are measuring the mass of a proton, we might expect ten decimal places of accuracy. If we are measuring the weight of a baby penguin, on the other hand, we'll probably be satisfied if we're within five grams. What matters most is that we are explicit about the bounds we used so that other people can tell what we actually did.

Degrees of Difficulty

There's an old joke that physicists worry about decimal places, astronomers worry about powers of ten, and economists are happy if they've got the sign right.

So how should we write tests when we don't know precisely what the right answer is? The best approach is to write tests that check if the actual value is within some **tolerance** of the expected value. The tolerance can be expressed as the **absolute error**, which is the absolute value of the difference between two, or the **relative error**, which the ratio of the absolute error to the value we're approximating (Goldberg 1991). For example, if we add 9+1 and get 11, the absolute error is 1 (i.e., $11 - 10$), and the relative error is 10%. If we add $99 + 1$ and get 101, on the other hand, the absolute error is still 1, but the relative error is only 1%.

For `test_alpha`, we might decide that an absolute error of 0.01 in the estimation of α is acceptable. If we are using `pytest`, we can check that values lie within this tolerance using `pytest.approx`:

```
from collections import Counter

import pytest
import numpy as np

import plotcounts
import countwords


def test_alpha():
    """Test the calculation of the alpha parameter.

    The test word counts satisfy the relationship,
      r = cf**(-1/alpha), where
      r is the rank,
      f the word count, and
      c is a constant of proportionality.

    To generate test word counts for an expected alpha of
      1.0, a maximum word frequency of 600 is used
      (i.e. c = 600 and r ranges from 1 to 600)
    """
    max_freq = 600
```

```
    counts = np.floor(max_freq / np.arange(1, max_freq + 1))
    actual_alpha = plotcounts.get_power_law_params(counts)
    expected_alpha = pytest.approx(1.0, abs=0.01)
    assert actual_alpha == expected_alpha


def test_word_count():
    #...as before...
```

When we re-run pytest, both tests now pass:

```
$ pytest
```

```
==================== test session starts ======================
platform darwin -- Python 3.7.6, pytest-6.2.0, py-1.10.0,
pluggy-0.13.1
rootdir: /Users/amira
collected 2 items

bin/test_zipfs.py ..                                     [100%]

===================== 2 passed in 0.69s =======================
```

Testing Visualizations

Testing visualizations is hard: any change to the dimension of the plot, however small, can change many pixels in a **raster image**, and cosmetic changes such as moving the legend up a couple of pixels will cause all of our tests to fail.

The simplest solution is therefore to test the data used to produce the image rather than the image itself. Unless we suspect that the plotting library contains bugs, the correct data should always produce the correct plot.

11.5 Integration Testing

Our Zipf's Law analysis has two steps: counting the words in a text and estimating the α parameter from the word count. Our unit tests give us some confidence that these components work in isolation, but do they work correctly together? Checking that is called **integration testing**.

Integration tests are structured the same way as unit tests: a fixture is used to produce an actual result that is compared against the expected result. However, creating the fixture and running the code can be considerably more complicated. For example, in the case of our Zipf's Law software, an appropriate integration test fixture might be a text file with a word frequency distribution that has a known α value. In order to create this text fixture, we need a way to generate random words.

Fortunately, a Python library called `RandomWordGenerator`[3] exists to do just that. We can install it using `pip`[4], the Python Package Installer:

```
$ pip install Random-Word-Generator
```

Borrowing from the word count distribution we created for `test_alpha`, we can use the following code to create a text file full of random words with a frequency distribution that corresponds to an α of approximately 1.0:

```
import numpy as np
from RandomWordGenerator import RandomWord


max_freq = 600
word_counts = np.floor(max_freq / np.arange(1, max_freq + 1))
rw = RandomWord()
random_words = rw.getList(num_of_words=max_freq)
writer = open('test_data/random_words.txt', 'w')
for index in range(max_freq):
    count = int(word_counts[index])
    word_sequence = f'{random_words[index]} ' * count
    writer.write(word_sequence + '\n')
writer.close()
```

[3] https://github.com/AbhishekSalian/Random-Word-Generator
[4] https://pypi.org/project/pip/

We can confirm this code worked by checking the resulting file:

```
tail -n 5 test_data/random_words.txt
```

```
lZnkzoBHRb
djiroplqrJ
HmAUGOncHg
DGLpfTIitu
KALSfPkrga
```

We can then add this integration test to `test_zipfs.py`:

```python
def test_integration():
    """Test the full word count to alpha parameter workflow."""
    with open('test_data/random_words.txt', 'r') as reader:
        word_counts_dict = countwords.count_words(reader)
    counts_array = np.array(list(word_counts_dict.values()))
    actual_alpha = plotcounts.get_power_law_params(counts_array)
    expected_alpha = pytest.approx(1.0, abs=0.01)
    assert actual_alpha == expected_alpha
```

Finally, we re-run `pytest` to check that the integration test passes:

```
$ pytest
```

```
===================== test session starts ======================
platform darwin -- Python 3.7.6, pytest-6.2.0, py-1.10.0,
pluggy-0.13.1
rootdir: /Users/amira
collected 3 items

bin/test_zipfs.py ...                                    [100%]

======================= 3 passed in 0.48s ======================
```

11.6 Regression Testing

So far we have tested two simplified texts: a short poem and a collection of random words with a known frequency distribution. The next step is to test with real data, i.e., an actual book. The problem is, we don't know the expected result: it's not practical to count the words in *Dracula* by hand, and even if we tried, the odds are good that we'd make a mistake.

For this kind of situation we can use **regression testing**. Rather than assuming that the test's author knows what the expected result should be, regression tests compare today's answer with a previous one. This doesn't guarantee that the answer is right—if the original answer is wrong, we could carry that mistake forward indefinitely—but it does draw attention to any changes (or "regressions").

In Section 7.4 we calculated an α of 1.0866646252515038 for *Dracula*. Let's use that value to add a regression test to `test_zipfs.py`:

```
def test_regression():
    """Regression test for Dracula."""
    with open('data/dracula.txt', 'r') as reader:
        word_counts_dict = countwords.count_words(reader)
    counts_array = np.array(list(word_counts_dict.values()))
    actual_alpha = plotcounts.get_power_law_params(counts_array)
    expected_alpha = pytest.approx(1.087, abs=0.001)
    assert actual_alpha == expected_alpha
```

```
$ pytest
```

```
==================== test session starts ====================
platform darwin -- Python 3.7.6, pytest-6.2.0, py-1.10.0,
pluggy-0.13.1
rootdir: /Users/amira
collected 4 items

bin/test_zipfs.py ....                                 [100%]

===================== 4 passed in 0.56s =====================
```

11.7 Test Coverage

How much of our code do the tests we have written check? More importantly, what parts of our code *aren't* being tested (yet)? To find out, we can use a tool to check their **code coverage**. Most Python programmers use the `coverage` library, which we can once again install using `pip`:

```
$ pip install coverage
```

Once we have it, we can use it to run `pytest` on our behalf:

```
$ coverage run -m pytest
```

```
==================== test session starts =====================
platform darwin -- Python 3.7.6, pytest-6.2.0, py-1.10.0,
pluggy-0.13.1
rootdir: /Users/amira
collected 4 items

bin/test_zipfs.py ....                                  [100%]

===================== 4 passed in 0.72s ======================
```

The `coverage` command doesn't display any information of its own, since mixing that in with our program's output would be confusing. Instead, it puts coverage data in a file called `.coverage` (with a leading `.`) in the current directory. To display that data, we run:

```
$ coverage report -m
```

```
bin/countwords.py      20      7     65%   25-26, 30-38
bin/plotcounts.py      58     37     36%   48-55, 75-77, 82-83,
                                           88-118, 122-140
bin/test_zipfs.py      31      0    100%
bin/utilities.py        8      5     38%   18-22
-------------------------------------------------
TOTAL                 117     49     58%
```

This summary shows us that some lines of `countwords.py` or `plotcounts.py` were not executed when we ran the tests: in fact, only 65% and 36% of the lines were run respectively. This makes sense, since much of the code in those scripts is devoted to handling command-line arguments or file I/O rather than the word counting and parameter estimation functionality that our unit, integration, and regression tests focus on.

To make sure that's the case, we can get a more complete report by running `coverage html` at the command line and opening `htmlcov/index.html`. Clicking on the name of our `countwords.py` script, for instance, produces the colorized line-by-line display shown in Figure 11.1.

This output confirms that all lines relating to word counting were tested, but not any of the lines related to argument handling or I/O.

Commit our Coverage?

At this point, you're probably wondering if you should use version control to track the files reporting your code's coverage. While it won't necessarily harm your code, the reports will become inaccurate unless you continue updating your coverage reports as your code changes. Therefore, we recommend adding `.coverage` and `htmlcov/` to your `.gitignore` file[5]. In the next section, we'll explore an approach that can help you automate tasks like assessing coverage.

Is this good enough? The answer depends on what the software is being used for and by whom. If it is for a safety-critical application such as a medical device, we should aim for 100% code coverage, i.e., every single line in the application should be tested. In fact, we should probably go further and aim for 100% **path coverage** to ensure that every possible path through the code has been checked. Similarly, if the software has become popular and is being used by thousands of researchers all over the world, we should probably check that it's not going to embarrass us.

But most of us don't write software that people's lives depend on, or that is in a "top 100" list, so requiring 100% code coverage is like asking for ten decimal places of accuracy when checking the voltage of a household electrical outlet. We always need to balance the effort required to create tests against the likelihood that those tests will uncover useful information. We also have to accept that no amount of testing can prove a piece of software is completely correct. A function with only two numeric arguments has 2^{128} possible inputs. Even if we could write the tests, how could we be sure we were checking the result of each one correctly?

[5] `https://github.com/github/gitignore`

Coverage for **bin/countwords.py** : 65%

20 statements | 13 run | 7 missing | 0 excluded

```
"""
Count the occurrences of all words in a text
and write them to a CSV-file.
"""

import argparse
import string
from collections import Counter

import utilities as util


def count_words(reader):
    """Count the occurrence of each word in a string."""
    text = reader.read()
    chunks = text.split()
    stripped = [word.strip(string.punctuation) for word in chunks]
    word_list = [word.lower() for word in stripped if word]
    word_counts = Counter(word_list)
    return word_counts


def main(args):
    """Run the command line program."""
    word_counts = count_words(args.infile)
    util.collection_to_csv(word_counts, num=args.num)


if __name__ == '__main__':
    parser = argparse.ArgumentParser(description=__doc__)
    parser.add_argument('infile', type=argparse.FileType('r'),
                        nargs='?', default='-',
                        help='Input file name')
    parser.add_argument('-n', '--num',
                        type=int, default=None,
                        help='Output only n most frequent words')
    args = parser.parse_args()
    main(args)
```

FIGURE 11.1: Example of Python code coverage report.

Luckily, we can usually put test cases into groups. For example, when testing a function that summarizes a table full of data, it's probably enough to check that it handles tables with:

- no rows
- only one row
- many identical rows
- rows having keys that are supposed to be unique, but aren't
- rows that contain nothing but missing values

Some projects develop **checklists** like this one to remind programmers what they ought to test. These checklists can be a bit daunting for newcomers, but they are a great way to pass on hard-earned experience.

11.8 Continuous Integration

Now that we have a set of tests, we could run `pytest` every now and again to check our code. This is probably sufficient for short-lived projects, but if several people are involved, or if we are making changes over weeks or months, we might forget to run the tests or it might be difficult to identify which change is responsible for a test failure.

The solution is **continuous integration** (CI), which runs tests automatically whenever a change is made. CI tells developers immediately if changes have caused problems, which makes them much easier to fix. CI can also be set up to run tests with several different configurations of the software or on several different operating systems, so that a programmer using Windows can be warned that a change breaks things for Mac users and vice versa.

One popular CI tool is Travis CI[6], which integrates well with GitHub[7]. If Travis CI has been set up, then every time a change is committed to a GitHub repository, Travis CI creates a fresh environment, makes a fresh clone of the repository (Section 7.8), and runs whatever commands the project's managers have set up.

Before setting up our account with Travis CI, however, we need to prepare our repository to be recognized by the tool. We'll first add a file called `.travis.yml` to our repository, which includes instructions for Travis CI. (The leading `.` in the name hides the file from casual listings on Mac or Linux, but not on Windows.) This file must be in the root directory of the repository,

[6] `https://travis-ci.com/`
[7] `https://github.com`

and is written in **YAML** (Section 10.1 and Appendix H). For our project, we add the following lines:

```
language: python

python:
- "3.6"

install:
- pip install -r requirements.txt

script:
- pytest
```

Python 3.6?

An early draft of the book used Python 3.6, and we later updated to 3.7. Over the life of a real project, software versions change, and it's not always feasible or possible to re-do all parts of the project. In this case, our first build in Travis CI used Python 3.6, so that's what we've shown here, and later checked that updating to 3.7 didn't break anything. We can't expect our software will be perfect the first time around, but we can do our best to document what's changed and confirm our results are still accurate.

The `language` key tells Travis CI which programming language to use, so that it knows which of its standard **virtual machines** to use as a starting point for the project. The `python` key specifies the version or versions of Python to use, while the `install` key indicates the name of the file (`requirements.txt`) listing the libraries that need to be installed. The `script` key lists the commands to run—in this case, `pytest`—which in turn executes our project scripts (i.e., `test_zipfs.py`, `plotcounts.py`, `countwords.py` and `utilities.py`). These scripts import a number of packages that don't come with the Python Standard Library[8], so these are what comprise `requirements.txt`:

```
numpy
pandas
matplotlib
scipy
pytest
pyyaml
```

[8] `https://docs.python.org/3/library/`

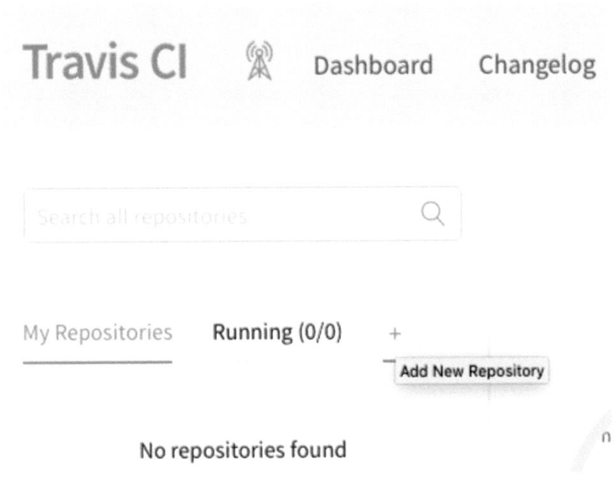

FIGURE 11.2: Click to add a new GitHub repository to Travis CI.

Be sure to save both `requirements.txt` and `.travis.yml` to your project's root directory, commit to your repository, and push to GitHub, or Travis CI will not be able to recognize your project.

We're now ready to set up CI for our project. The basic steps are as follows:

1. Create an account on Travis CI[9] (if we don't already have one).
2. Link our Travis CI account to our GitHub account (if we haven't done so already).
3. Tell Travis CI to watch the repository that contains our project.

Creating an account with an online service is probably a familiar process, but linking our Travis CI account to our GitHub account may be something new. We only have to do this once to allow Travis CI to access all our GitHub repositories, but we should always be careful when giving sites access to other sites, and only trust well-established and widely used services.

We can tell Travis CI which repository we want it to watch by clicking the "+" next to the "My Repositories" link on the left-hand side of the Travis CI homepage (Figure 11.2).

To add the GitHub repository we have been using throughout this book, find it in the repository list (Figure 11.3). If the repository doesn't show up, re-synchronize the list using the green "Sync account" button on the left sidebar. If it still doesn't appear, the repository may belong to someone else or be private, or your required files may be incorrect. Click the "Trigger a build" button to initiate your first test using Travis CI.

---
[9] `https://travis-ci.com/`

FIGURE 11.3: Find Zipf's Law repository and trigger a build.

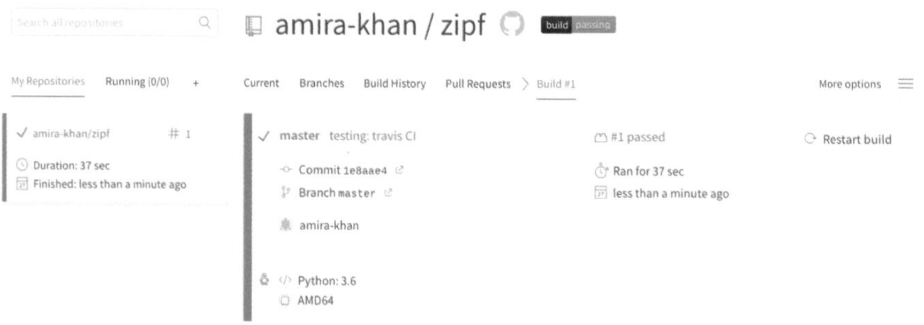

FIGURE 11.4: Travis build overview (build succeeded).

Once your repository has been activated, Travis CI follows the instructions in `.travis.yml` and reports whether the build passed (shown in green) or produced warnings or errors (shown in red). To create this report, Travis CI has done the following:

1. Created a new Linux virtual machine.
2. Installed the desired version of Python.
3. Run the commands below the `script` key.
4. Reported the results at `https://travis-ci.com/USER/REPO`, where `USER/REPO` identifies the repository for a given user.

Our tests pass and the build completes successfully. We can view additional details about the test by clicking on the repository name (Figure 11.4).

This example shows one of the other benefits of CI: it forces us to be explicit about what we are doing and how we do it, just as writing a Makefile forces us to be explicit about exactly how we produce results (Zampetti et al. 2020).

11.9 When to Write Tests

We have now met the three major types of test: unit, integration, and regression. At what point in the code development process should we write these? The answer depends on who you ask.

Many programmers are passionate advocates of a practice called **test-driven development** (TDD). Rather than writing code and then writing tests, they write the tests first and then write just enough code to make those tests pass. Once the code is working, they clean it up (Appendix F.4) and then move on to the next task. TDD's advocates claim that this leads to better code because:

1. Writing tests clarifies what the code is actually supposed to do.
2. It eliminates **confirmation bias**. If someone has just written a function, they are predisposed to want it to be right, so they will bias their tests towards proving that it is correct instead of trying to uncover errors.
3. Writing tests first ensures that they actually get written.

These arguments are plausible. However, studies such as Fucci et al. (2016) and Fucci et al. (2017) don't support them: in practice, writing tests first or last doesn't appear to affect productivity. What *does* have an impact is working in small, interleaved increments, i.e., writing just a few lines of code and testing it before moving on rather than writing several pages of code and then spending hours on testing.

So how do most data scientists figure out if their software is doing the right thing? The answer is spot checks: each time they produce an intermediate or final result, they scan a table, create a chart, or inspect some summary statistics to see if everything looks OK. Their heuristics are usually easy to state, like "there shouldn't be NAs at this point" or "the age range should be reasonable," but applying those heuristics to a particular analysis always depends on their evolving insight into the data in question.

By analogy with test-driven development, we could call this process "checking-driven development." Each time we add a step to our pipeline and look at its output, we can also add a check of some kind to the pipeline to ensure that what we are checking for remains true as the pipeline evolves or is run on other data. Doing this helps reusability—it's amazing how often a one-off analysis winds up being used many times—but the real goal is comprehensibility. If someone can get our code and data, then runs the code on the data, and gets the same result that we did, then our computation is reproducible, but that doesn't mean they can understand it. Comments help (either in the code or as blocks of prose in a **computational notebook**), but they won't check that assumptions and invariants hold. And unlike comments, runnable assertions can't fall out of step with what the code is actually doing.

11.10 Summary

Testing data analysis pipelines is often harder than testing mainstream software applications, since data analysts often don't know what the right answer is (Braiek and Khomh 2018). (If we did, we would have submitted our report and moved on to the next problem already.) The key distinction is the difference between **validation**, which asks whether the specification is correct, and **verification**, which asks whether we have met that specification. The difference between them is the difference between building the right thing and building something right; the practices introduced in this chapter will help with both.

11.11 Exercises

11.11.1 Explaining assertions

Given a list of numbers, the function `total` returns the total:

```
total([1, 2, 3, 4])
```

```
10
```

The function only works on numbers:

```
total(['a', 'b', 'c'])
```

```
ValueError: invalid literal for int() with base 10: 'a'
```

Explain in words what the assertions in this function check, and for each one, give an example of input that will make that assertion fail.

```
def total(values):
    assert len(values) > 0
    for element in values:
        assert int(element)
    values = [int(element) for element in values]
    total = sum(values)
    assert total > 0
    return total
```

11.11.2 Rectangle normalization

A rectangle can be described using a tuple of four cartesian coordinates `(x0, y0, x1, y1)`, where `(x0, y0)` represents the lower left corner and `(x1, y1)` the upper right. In order to do some calculations, suppose we need to be able to normalize rectangles so that the lower left corner is at the origin (i.e., `(x0, y0) = (0, 0)`) and the longest side is 1.0 units long. This function does that:

```
def normalize_rectangle(rect):
    """Normalizes a rectangle so that it is at the origin
    and 1.0 units long on its longest axis.  Input should be
    (x0, y0, x1, y1), where (x0, y0) and (x1, y1) define the
    lower left and upper right corners of the rectangle."""

    # insert preconditions
    x0, y0, x1, y1 = rect
    # insert preconditions

    dx = x1 - x0
    dy = y1 - y0
    if dx > dy:
        scaled = float(dx) / dy
        upper_x, upper_y = 1.0, scaled
    else:
        scaled = float(dx) / dy
        upper_x, upper_y = scaled, 1.0

    # insert postconditions here

    return (0, 0, upper_x, upper_y)
```

In order to answer the following questions, cut and paste the `normalize_rectangle` function into a new file called `geometry.py` (outside of your `zipf` project) and save that file in a new directory called `exercises`.

1. To ensure that the inputs to `normalize_rectangle` are valid, add **preconditions** to check that:

(a) `rect` contains 4 coordinates,
(b) the width of the rectangle is a positive, non-zero value (i.e., `x0 < x1`), and
(c) the height of the rectangle is a positive, non-zero value (i.e., `y0 < y1`).

2. If the normalization calculation has worked correctly, the new `x1` coordinate will lie between 0 and 1 (i.e., `0 < upper_x <= 1.0`). Add a **postcondition** to check that this is true. Do the same for the new `y1` coordinate, `upper_y`.

Running `normalize_rectangle` for a short, wide rectangle should pass your new preconditions and postconditions:

```
import geometry
```

```
geometry.normalize_rectangle([2, 5, 3, 10])
```

```
(0, 0, 0.2, 1.0)
```

but will fail for a tall, skinny rectangle:

```
geometry.normalize_rectangle([20, 15, 30, 20])
```

```
AssertionError                    Traceback (most recent call last)
<ipython-input-3-f4e8cdf7f69d> in <module>
----> 1 geometry.normalize_rectangle([20, 15, 30, 20])

~/Desktop/exercises/geometry.py in normalize_rectangle(rect)
     19
```

```
     20     assert 0 < upper_x <= 1.0, \
     21         'Calculated upper X coordinate invalid'
---> 22     assert 0 < upper_y <= 1.0, \
     23         'Calculated upper Y coordinate invalid'
     24
     25     return (0, 0, upper_x, upper_y)

AssertionError: Calculated upper Y coordinate invalid
```

3. Find and correct the source of the error in `normalize_rectangle`. Once fixed, you should be able to successfully run `geometry.normalize_rectangle([20, 15, 30, 20])`.
4. Write a unit test for tall, skinny rectangles and save it in a new file called `test_geometry.py`. Run `pytest` to make sure the test passes.
5. Add a couple more unit tests to `test_geometry.py`. Explain the rationale behind each test.

11.11.3 Testing with randomness

Programs that rely on random numbers are impossible to test because there's (deliberately) no way to predict their output. Luckily, computer programs don't actually use random numbers: they use a **pseudo-random number generator** (PRNG) that produces values in a repeatable but unpredictable way. Given the same initial **seed**, a PRNG will always produce the same sequence of values. How can we use this fact when testing programs that rely on pseudo-random numbers?

11.11.4 Testing with relative error

If E is the expected result of a function and A is the actual value it produces, the **relative error** is `abs((A-E)/E)`. This means that if we expect the results of tests to be 2, 1, and 0, and we actually get 2.1, 1.1, and 0.1 the relative errors are 5%, 10%, and infinity. Why does this seem counter-intuitive, and what might be a better way to measure error in this case?

11.12 Key Points

- Test software to convince people (including yourself) that software is correct enough and to make tolerances on "enough" explicit.
- Add **assertions** to code so that it checks itself as it runs.
- Write **unit tests** to check individual pieces of code.
- Write **integration tests** to check that those pieces work together correctly.
- Write **regression tests** to check if things that used to work no longer do.
- A **test framework** finds and runs tests written in a prescribed fashion and reports their results.
- Test **coverage** is the fraction of lines of code that are executed by a set of tests.
- **Continuous integration** re-builds and/or re-tests software every time something changes.

12

Handling Errors

> "When Mister Safety Catch Is Not On, Mister Crossbow Is Not Your Friend."
>
> — Terry Pratchett

We are imperfect people living in an imperfect world. People will misunderstand how to use our programs, and even if we test thoroughly as described in the previous chapter, those programs might still contain bugs. We should therefore plan from the start to detect and handle errors.

Something that goes wrong while a program is running is sometimes referred to as an **exception** from normal behavior. Generally speaking, we distinguish between two types of errors/exceptions. **Internal errors**, are mistakes in the program itself, such as calling a function with `None` instead of a list. **External errors** are usually caused by interactions between the program and the outside world: a user may mis-type a filename, the network might be down, and so on.

When an internal error occurs, the only thing we can do in most cases is report it and halt the program. If a function has been passed `None` instead of a valid list, for example, the odds are good that one of our data structures is corrupted. We can try to guess what the problem is and take corrective action, but our guess will often be wrong and our attempt to correct the problem might actually make things worse. When an external error occurs on the other hand, we don't always want the program to stop. If a user mis-types her password, handling the error by prompting her to try again would be friendlier than halting and requiring her to restart the program.

This chapter looks at how we can raise, catch and handle errors. We consider how to write useful error messages, and how to make our programs log those messages along with other useful information as they are running, so that it's easier to figure out what happened when something goes wrong.

The Zipf's Law project should now include:

```
zipf/
├── .gitignore
```

```
├── .travis.yml
├── CONDUCT.md
├── CONTRIBUTING.md
├── LICENSE.md
├── Makefile
├── README.md
├── requirements.txt
├── bin
│   ├── book_summary.sh
│   ├── collate.py
│   ├── countwords.py
│   ├── plotcounts.py
│   ├── plotparams.yml
│   ├── script_template.py
│   ├── test_zipfs.py
│   └── utilities.py
├── data
│   ├── README.md
│   ├── dracula.txt
│   └── ...
├── results
│   ├── dracula.csv
│   ├── dracula.png
│   └── ...
└── test_data
    ├── random_words.txt
    └── risk.txt
```

12.1 Exceptions

Most modern programming languages use exceptions for error handling. As the name suggests, an exception is a way to represent an exceptional or unusual occurrence that doesn't fit neatly into the program's expected operation. The code below uses exceptions to report attempts to divide by zero:

```
for denom in [-5, 0, 5]:
    try:
        result = 1/denom
        print(f'1/{denom} == {result}')
```

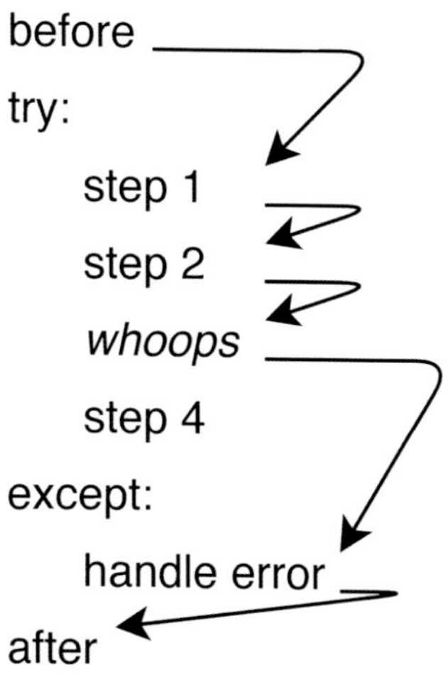

FIGURE 12.1: Exception control flow.

```
    except:
        print(f'Cannot divide by {denom}')
```

```
1/-5 == -0.2
Cannot divide by 0
1/5 == 0.2
```

`try/except` looks like `if/else` and works in a similar fashion. If nothing unexpected happens inside the `try` block, the `except` block isn't run (Figure 12.1). If something goes wrong inside the `try`, on the other hand, the program jumps immediately to the `except`. This is why the `print` statement inside the `try` doesn't run when `denom` is 0: as soon as Python tries to calculate `1/denom`, it skips directly to the code under `except`.

We often want to know exactly what went wrong, so Python and other languages store information about the error in an object (which is also called an exception). We can **catch** an exception and inspect it as follows:

```
for denom in [-5, 0, 5]:
    try:
        result = 1/denom
        print(f'1/{denom} == {result}')
```

```
    except Exception as error:
        print(f'{denom} has no reciprocal: {error}')
```

```
1/-5 == -0.2
0 has no reciprocal: division by zero
1/5 == 0.2
```

We can use any variable name we like instead of `error`; Python will assign the exception object to that variable so that we can do things with it in the `except` block.

Python also allows us to specify what kind of exception we want to catch. For example, we can write code to handle out-of-range indexing and division by zero separately:

```
numbers = [-5, 0, 5]
for i in [0, 1, 2, 3]:
    try:
        denom = numbers[i]
        result = 1/denom
        print(f'1/{denom} == {result}')
    except IndexError as error:
        print(f'index {i} out of range')
    except ZeroDivisionError as error:
        print(f'{denom} has no reciprocal: {error}')
```

```
1/-5 == -0.2
0 has no reciprocal: division by zero
1/5 == 0.2
index 3 out of range
```

Exceptions are organized in a hierarchy: for example, `FloatingPointError`, `OverflowError`, and `ZeroDivisionError` are all special cases of `ArithmeticError`, so an `except` that catches the latter will catch all three of the former, but an `except` that catches an `OverflowError` *won't* catch a `ZeroDivisionError`. The Python documentation describes all of the built-in exception types[1]; in practice, the ones that people handle most often are:

[1] `https://docs.python.org/3/library/exceptions.html#exception-hierarchy`

- `ArithmeticError`: something has gone wrong in a calculation.
- `IndexError` and `KeyError`: something has gone wrong indexing a list or lookup something up in a dictionary.
- `OSError`: thrown when a file is not found, the program doesn't have permission to read it, and so on.

So where do exceptions come from? The answer is that programmers can **raise** them explicitly:

```
for number in [1, 0, -1]:
    try:
        if number < 0:
            raise ValueError(f'no negatives: {number}')
        print(number)
    except ValueError as error:
        print(f'exception: {error}')
```

```
1
0
exception: no negatives: -1
```

We can define our own exception types, and many libraries do, but the built-in types are enough to cover common cases.

One final note is that exceptions don't have to be handled where they are raised. In fact, their greatest strength is that they allow long-range error handling. If an exception occurs inside a function and there is no `except` for it there, Python checks to see if whoever called the function is willing to handle the error. It keeps working its way up through the **call stack** until it finds a matching `except`. If there isn't one, Python takes care of the exception itself. The example below relies on this: the second call to `sum_reciprocals` tries to divide by zero, but the exception is caught in the calling code rather than in the function.

```
def sum_reciprocals(values):
    result = 0
    for v in values:
        result += 1/v
    return result

numbers = [-1, 0, 1]
try:
```

```
    one_over = sum_reciprocals(numbers)
except ArithmeticError as error:
    print(f'Error trying to sum reciprocals: {error}')
```

```
Error trying to sum reciprocals: division by zero
```

This behavior is designed to support a pattern called "throw low, catch high": write most of your code without exception handlers, since there's nothing useful you can do in the middle of a small utility function, but put a few handlers in the uppermost functions of your program to catch and report all errors.

We can now go ahead and add error handling to our Zipf's Law code. Some is already built in: for example, if we try to read a file that does not exist, the `open` function throws a `FileNotFoundError`:

```
$ python bin/collate.py results/none.csv results/dracula.csv
```

```
Traceback (most recent call last):
  File "bin/collate.py", line 35, in <module>
    main(args)
  File "bin/collate.py", line 23, in main
    with open(fname, 'r') as reader:
FileNotFoundError: [Errno 2] No such file or directory:
'results/none.csv'
```

But what happens if we try to read a file that exists, but was not created by `countwords.py`?

```
$ python bin/collate.py Makefile
```

```
Traceback (most recent call last):
  File "bin/collate.py", line 35, in <module>
    main(args)
  File "bin/collate.py", line 24, in main
    update_counts(reader, word_counts)
  File "bin/collate.py", line 15, in update_counts
    for word, count in csv.reader(reader):
ValueError: not enough values to unpack (expected 2, got 1)
```

This error is hard to understand, even if we are familiar with the code's internals. Our program should therefore check that the input files are CSV files, and if not, raise an error with a useful explanation of what went wrong. We could achieve this by wrapping the call to `open` in a `try/except` clause:

```
for fname in args.infiles:
    try:
        with open(fname, 'r') as reader:
            update_counts(reader, word_counts)
    except ValueError as e:
        print(f'{fname} is not a CSV file.')
        print(f'ValueError: {e}')
```

```
$ python bin/collate.py Makefile
```

```
Makefile is not a CSV file.
ValueError: not enough values to unpack (expected 2, got 1)
```

This is definitely more informative than before. However, *all* `ValueErrors` that are raised when trying to open a file will result in this error message, including those raised when we actually do use a CSV file as input. A more precise approach in this case would be to throw an exception only if some other kind of file is specified as an input:

```
for fname in args.infiles:
    if fname[-4:] != '.csv':
        raise OSError(f'{fname} is not a CSV file.')
    with open(fname, 'r') as reader:
        update_counts(reader, word_counts)
```

```
$ python bin/collate.py Makefile
```

```
Traceback (most recent call last):
  File "bin/collate.py", line 37, in <module>
    main(args)
  File "bin/collate.py", line 24, in main
    raise OSError(f'{fname} is not a CSV file.')
OSError: Makefile is not a CSV file.
```

FIGURE 12.2: An unhelpful error message.

This approach is still not perfect: we are checking that the file's suffix is `.csv` instead of checking the content of the file and confirming that it is what we require. What we *should* do is check that there are two columns separated by a comma, that the first column contains strings, and that the second is numerical.

Kinds of Errors

The "`if` then `raise`" approach is sometimes referred to as "look before you leap," while the `try/except` approach obeys the old adage that "it's easier to ask for forgiveness than permission." The first approach is more precise, but has the shortcoming that programmers can't anticipate everything that can go wrong when running a program, so there should always be an `except` somewhere to deal with unexpected cases.

The one rule we should *always* follow is to check for errors as early as possible so that we don't waste the user's time. Few things are as frustrating as being told at the end of an hour-long calculation that the program doesn't have permission to write to an output directory. It's a little extra work to check things like this up front, but the larger your program or the longer it runs, the more useful those checks will be.

12.2 Writing Useful Error Messages

The error message shown in Figure 12.2 is not helpful. Having `collate.py` print the message below would be equally unfriendly:

```
OSError: Something went wrong, try again.
```

This message doesn't provide any information on what went wrong, so it is difficult to know what to change for next time. A slightly better message would be:

```
OSError: Unsupported file type.
```

This tells us the problem is with the type of file we're trying to process, but it still doesn't tell us what file types are supported, which means we have to rely on guesswork or read the source code. Telling the user "*filename* is not a CSV file" (as we did in the previous section) makes it clear that the program only works with CSV files, but since we don't actually check the content of the file, this message could confuse someone who has comma-separated values saved in a `.txt` file. An even better message would therefore be:

```
OSError: File must end in .csv
```

This message tells us exactly what the criteria are to avoid the error.

Error messages are often the first thing people read about a piece of software, so they should therefore be the most carefully written documentation for that software. A web search for "writing good error messages" turns up hundreds of hits, but recommendations are often more like gripes than guidelines and are usually not backed up by evidence. What research there is gives us the following rules (Becker et al. 2016):

1. Tell the user what they did, not what the program did. Putting it another way, the message shouldn't state the effect of the error, it should state the cause.
2. Be spatially correct, i.e., point at the actual location of the error. Few things are as frustrating as being pointed at line 28 when the problem is really on line 35.
3. Be as specific as possible without being or seeming wrong from a user's point of view. For example, "file not found" is very different from "don't have permissions to open file" or "file is empty."

4. Write for your audience's level of understanding. For example, error messages should never use programming terms more advanced than those you would use to describe the code to the user.
5. Do not blame the user, and do not use words like fatal, illegal, etc. The former can be frustrating—in many cases, "user error" actually isn't—and the latter can make people worry that the program has damaged their data, their computer, or their reputation.
6. Do not try to make the computer sound like a human being. In particular, avoid humor: very few jokes are funny on the dozenth re-telling, and most users are going to see error messages at least that often.
7. Use a consistent vocabulary. This rule can be hard to enforce when error messages are written by several different people, but putting them all in one module makes review easier.

That last suggestion deserves a little elaboration. Most people write error messages directly in their code:

```
if fname[-4:] != '.csv':
    raise OSError(f'{fname}: File must end in .csv')
```

A better approach is to put all the error messages in a dictionary:

```
ERRORS = {
    'not_csv_suffix' : '{fname}: File must end in .csv',
    'config_corrupted' : '{config_name} corrupted',
    # ...more error messages...
    }
```

and then only use messages from that dictionary:

```
if fname[-4:] != '.csv':
    raise OSError(ERRORS['not_csv_suffix'].format(fname=fname))
```

Doing this makes it much easier to ensure that messages are consistent. It also makes it much easier to give messages in the user's preferred language:

```
ERRORS = {
  'en' : {
    'not_csv_suffix' : '{fname}: File must end in .csv',
    'config_corrupted' : '{config_name} corrupted',
    # ...more error messages in English...
  },
  'fr' : {
    'not_csv_suffix' : '{fname}: Doit se terminer par .csv',
    'config_corrupted' : f'{config_name} corrompu',
    # ...more error messages in French...
  }
  # ...other languages...
}
```

The error report is then looked up and formatted as:

```
ERRORS[user_language]['not_csv_suffix'].format(fname=fname)
```

where `user_language` is a two-letter code for the user's preferred language.

12.3 Testing Error Handling

An alarm isn't much use if it doesn't go off when it's supposed to. Equally, if a function doesn't raise an exception when it should then errors can easily slip past us. If we want to check that a function called `func` raises an `ExpectedError` exception, we can use the following template for a unit test:

```
#...set up fixture...
try:
    actual = func(fixture)
    assert False, 'Expected function to raise exception'
except ExpectedError as error:
    pass
except Exception as error:
    assert False, 'Function raised the wrong exception'
```

This template has three cases:

1. If the call to `func` returns a value without throwing an exception then something has gone wrong, so we `assert False` (which always fails).
2. If `func` raises the error it's supposed to, then we go into the first `except` branch *without* triggering the `assert` immediately below the function call. The code in this `except` branch could check that the exception contains the right error message, but in this case it does nothing (which in Python is written `pass`).
3. Finally, if the function raises the wrong kind of exception we also `assert False`. Checking this case might seem overly cautious, but if the function raises the wrong kind of exception, users could easily fail to catch it.

This pattern is so common that `pytest` provides support for it. Instead of the eight lines in our original example, we can write:

```
import pytest


#...set up fixture...
with pytest.raises(ExpectedError):
    actual = func(fixture)
```

The argument to `pytest.raises` is the type of exception we expect; the call to the function then goes in the body of the `with` statement. We will explore `pytest.raises` further in the exercises.

12.4 Reporting Errors

Programs should report things that go wrong; they should also sometimes report things that go right so that people can monitor their progress. Adding `print` statements is a common approach, but removing them or commenting them out when the code goes into production is tedious and error-prone.

A better approach is to use a **logging framework**, such as Python's `logging` library. This lets us leave debugging statements in our code and turn them on or off at will. It also lets us send output to any of several destinations, which is helpful when our data analysis pipeline has several stages and we are trying to figure out which one contains a bug.

To understand how logging frameworks work, suppose we want to turn `print` statements in our `collate.py` program on or off without editing the program's source code. We would probably wind up with code like this:

```
if LOG_LEVEL >= 0:
    print('Processing files...')
for fname in args.infiles:
    if LOG_LEVEL >= 1:
        print(f'Reading in {fname}...')
    if fname[-4:] != '.csv':
        msg = ERRORS['not_csv_suffix'].format(fname=fname)
        raise OSError(msg)
    with open(fname, 'r') as reader:
        if LOG_LEVEL >= 1:
            print(f'Computing word counts...')
        update_counts(reader, word_counts)
```

`LOG_LEVEL` acts as a threshold: any debugging output at a lower level than its value isn't printed. As a result, the first log message will always be printed, but the other two only in case the user has requested more details by setting `LOG_LEVEL` higher than zero.

A logging framework combines the `if` and `print` statements in a single function call and defines standard names for the **logging levels**. In order of increasing severity, the usual levels are:

- `DEBUG`: very detailed information used for localizing errors.
- `INFO`: confirmation that things are working as expected.
- `WARNING`: something unexpected happened, but the program will keep going.
- `ERROR`: something has gone badly wrong, but the program hasn't hurt anything.
- `CRITICAL`: potential loss of data, security breach, etc.

Each of these has a corresponding function: we can use `logging.debug`, `logging.info`, etc. to write messages at these levels. By default, only `WARNING` and above are displayed; messages appear on **standard error** so that the flow of data in pipes isn't affected. The logging framework also displays the source of the message, which is called `root` by default. Thus, if we run the small program shown below, only the warning message appears:

```
import logging

logging.warning('This is a warning.')
logging.info('This is just for information.')
```

```
WARNING:root:This is a warning.
```

Rewriting the `collate.py` example above using `logging` yields code that is less cluttered:

```
import logging

logging.info('Processing files...')
for fname in args.infiles:
    logging.debug(f'Reading in {fname}...')
    if fname[-4:] != '.csv':
        msg = ERRORS['not_csv_suffix'].format(fname=fname)
        raise OSError(msg)
    with open(fname, 'r') as reader:
        logging.debug('Computing word counts...')
        update_counts(reader, word_counts)
```

We can also configure logging to send messages to a file instead of standard error using `logging.basicConfig`. (This has to be done before we make any logging calls—it's not retroactive.) We can also use that function to set the logging level: everything at or above the specified level is displayed.

```
import logging

logging.basicConfig(level=logging.DEBUG, filename='logging.log')

logging.debug('This is for debugging.')
logging.info('This is just for information.')
logging.warning('This is a warning.')
logging.error('Something went wrong.')
logging.critical('Something went seriously wrong.')
```

```
DEBUG:root:This is for debugging.
INFO:root:This is just for information.
WARNING:root:This is a warning.
ERROR:root:Something went wrong.
CRITICAL:root:Something went seriously wrong.
```

By default, `basicConfig` re-opens the file we specify in **append mode**; we can use `filemode='w'` to overwrite the existing log data. Overwriting is useful during debugging, but we should think twice before doing it in production, since the information we throw away often turns out to be exactly what we need to find a bug.

Many programs allow users to specify logging levels and log filenames as command-line parameters. At its simplest, this is a single flag `-v` or `--verbose` that changes the logging level from `WARNING` (the default) to `DEBUG` (the noisiest level). There may also be a corresponding flag `-q` or `--quiet` that changes the level to `ERROR`, and a flag `-l` or `--logfile` that specifies the name of a log file. To log messages to a file while also printing them, we can tell `logging` to use two handlers simultaneously:

```python
import logging


logging.basicConfig(
    level=logging.DEBUG,
    handlers=[
        logging.FileHandler("logging.log"),
        logging.StreamHandler()])

logging.debug('This is for debugging.')
```

The string `'This is for debugging'` is both printed to standard error and appended to `logging.log`.

Libraries like `logging` can send messages to many destinations; in production, we might send them to a centralized logging server that collates logs from many different systems. We might also use **rotating files** so that the system always has messages from the last few hours but doesn't fill up the disk. We don't need any of these when we start, but the data engineers and system administrators who eventually have to install and maintain your programs will be very grateful if we use `logging` instead of `print` statements, because it allows them to set things up the way they want with very little work.

Logging Configuration

Chapter 10 explained why and how to save the configuration that produced a particular result. We clearly also want this information in the log, so we have three options:

1. Write the configuration values into the log one at a time.
2. Save the configuration as a single record in the log (e.g., as a single entry containing **JSON**).
3. Write the configuration to a separate file and save the filename in the log.

Option 1 usually means writing a lot of extra code to reassemble the configuration. Option 2 also often requires us to write extra code (since we need to be able to save and restore configurations as JSON as well as in whatever format we normally use), so on balance we recommend option 3.

12.5 Summary

Most programmers spend as much time debugging as they do writing new code, but most courses and textbooks only show working code, and never discuss how to prevent, diagnose, report, and handle errors. Raising our own exceptions instead of using the system's, writing useful error messages, and logging problems systematically can save us and our users a lot of needless work.

12.6 Exercises

This chapter suggested several edits to `collate.py`. Suppose our script now reads as follows:

```
"""
Combine multiple word count CSV-files
```

```
into a single cumulative count.
"""

import csv
import argparse
from collections import Counter
import logging

import utilities as util


ERRORS = {
    'not_csv_suffix' : '{fname}: File must end in .csv',
    }


def update_counts(reader, word_counts):
    """Update word counts with data from another reader/file."""
    for word, count in csv.reader(reader):
        word_counts[word] += int(count)


def main(args):
    """Run the command line program."""
    word_counts = Counter()
    logging.info('Processing files...')
    for fname in args.infiles:
        logging.debug(f'Reading in {fname}...')
        if fname[-4:] != '.csv':
            msg = ERRORS['not_csv_suffix'].format(fname=fname)
            raise OSError(msg)
        with open(fname, 'r') as reader:
            logging.debug('Computing word counts...')
            update_counts(reader, word_counts)
    util.collection_to_csv(word_counts, num=args.num)


if __name__ == '__main__':
    parser = argparse.ArgumentParser(description=__doc__)
    parser.add_argument('infiles', type=str, nargs='*',
                        help='Input file names')
    parser.add_argument('-n', '--num',
                        type=int, default=None,
                        help='Output n most frequent words')
```

```
    args = parser.parse_args()
    main(args)
```

The following exercises will ask you to make further edits to `collate.py`.

12.6.1 Set the logging level

Define a new command-line flag for `collate.py` called `--verbose` (or `-v`) that changes the logging level from `WARNING` (the default) to `DEBUG` (the noisiest level).

Hint: the following command changes the logging level to `DEBUG`:

```
logging.basicConfig(level=logging.DEBUG)
```

Once finished, running `collate.py` with and without the `-v` flag should produce the following output:

```
$ python bin/collate.py results/dracula.csv
  results/moby_dick.csv -n 5
```

```
the,22559
and,12306
of,10446
to,9192
a,7629
```

```
$ python bin/collate.py results/dracula.csv
  results/moby_dick.csv -n 5 -v
```

```
INFO:root:Processing files...
DEBUG:root:Reading in results/dracula.csv...
DEBUG:root:Computing word counts...
DEBUG:root:Reading in results/moby_dick.csv...
DEBUG:root:Computing word counts...
the,22559
and,12306
of,10446
to,9192
a,7629
```

12.6.2 Send the logging output to file

In Exercise 12.6.1, logging information is printed to the screen when the verbose flag is activated. This is problematic if we want to re-direct the output from `collate.py` to a CSV file, because the logging information will appear in the CSV file as well as the words and their counts.

1. Edit `collate.py` so that the logging information is sent to a log file called `collate.log` instead. (HINT: `logging.basicConfig` has an argument called `filename`.)
2. Create a new command-line option `-l` or `--logfile` so that the user can specify a different name for the log file if they don't like the default name of `collate.log`.

12.6.3 Handling exceptions

1. Modify the script `collate.py` so that it catches any exceptions that are raised when it tries to open files and records them in the log file. When you are finished, the program should collate all the files it can, rather than halting as soon as it encounters a problem.
2. Modify your first solution to handle nonexistent files and permission problems separately.

12.6.4 Testing error handling

In our suggested solution to the previous exercise, we modified `collate.py` to handle different types of errors associated with reading input files. If the `main` function in `collate.py` now reads:

```
def main(args):
    """Run the command line program."""
    log_lev = logging.DEBUG if args.verbose else logging.WARNING
    logging.basicConfig(level=log_lev, filename=args.logfile)
    word_counts = Counter()
    logging.info('Processing files...')
    for fname in args.infiles:
        try:
            logging.debug(f'Reading in {fname}...')
            if fname[-4:] != '.csv':
                msg = ERRORS['not_csv_suffix'].format(
                    fname=fname)
```

```
                raise OSError(msg)
            with open(fname, 'r') as reader:
                logging.debug('Computing word counts...')
                update_counts(reader, word_counts)
        except FileNotFoundError:
            msg = f'{fname} not processed: File does not exist'
            logging.warning(msg)
        except PermissionError:
            msg = f'{fname} not processed: No read permission'
            logging.warning(msg)
        except Exception as error:
            msg = f'{fname} not processed: {error}'
            logging.warning(msg)
    util.collection_to_csv(word_counts, num=args.num)
```

1. It is difficult to write a simple unit test for the lines of code dedicated to reading input files, because `main` is a long function that requires command-line arguments as input. Edit `collate.py` so that the six lines of code responsible for processing an input file appear in their own function that reads as follows (i.e., once you are done, `main` should call `process_file` in place of the existing code):

```
def process_file(fname, word_counts):
    """Read file and update word counts"""
    logging.debug(f'Reading in {fname}...')
    if fname[-4:] != '.csv':
        msg = ERRORS['not_csv_suffix'].format(
            fname=fname)
        raise OSError(msg)
    with open(fname, 'r') as reader:
        logging.debug('Computing word counts...')
        update_counts(reader, word_counts)
```

2. Add a unit test to `test_zipfs.py` that uses `pytest.raises` to check that the new `collate.process_file` function raises an `OSError` if the input file does not end in `.csv`. Run `pytest` to check that the new test passes.
3. Add a unit test to `test_zipfs.py` that uses `pytest.raises` to check that the new `collate.process_file` function raises a `FileNotFoundError` if the input file does not exist. Run `pytest` to check that the new test passes.

4. Use the `coverage` library (Section 11.7) to check that the relevant commands in `process_file` (specifically `raise OSError` and `open(fname, 'r')`) were indeed tested.

12.6.5 Error catalogs

In Section 12.2 we started to define an error catalog called `ERRORS`.

1. Read Appendix F.1 and explain why we have used capital letters for the name of the catalog.
2. Python has three ways to format strings: the `%` operator, the `str.format` method, and f-strings (where the "f" stands for "format"). Look up the documentation for each and explain why we have to use `str.format` rather than f-strings for formatting error messages in our catalog/lookup table.
3. There's a good chance we will eventually want to use the error messages we've defined in other scripts besides `collate.py`. To avoid duplication, move `ERRORS` to the `utilities` module that was first created in Section 5.8.

12.6.6 Tracebacks

Run the following code:

```
try:
    1/0
except Exception as e:
    help(e.__traceback__)
```

1. What kind of object is `e.__traceback__`?
2. What useful information can you get from it?

12.7 Key Points

- Signal errors by **raising exceptions**.
- Use `try/except` blocks to **catch** and handle exceptions.
- Python organizes its standard exceptions in a hierarchy so that programs can catch and handle them selectively.

- "Throw low, catch high," i.e., raise exceptions immediately but handle them at a higher level.
- Write error messages that help users figure out what to do to fix the problem.
- Store error messages in a lookup table to ensure consistency.
- Use a **logging framework** instead of `print` statements to report program activity.
- Separate logging messages into `DEBUG`, `INFO`, `WARNING`, `ERROR`, and `CRITICAL` levels.
- Use `logging.basicConfig` to define basic logging parameters.

13

Tracking Provenance

> The most important problem is that we are trying to understand the fundamental workings of the universe via a language devised for telling one another when the best fruit is.
>
> — Terry Pratchett

We have now developed, automated, and tested a workflow for plotting the word count distribution for classic novels. In the normal course of events, we would include the outputs from that workflow (e.g., our figures and α values) in a research paper or a report for a consulting client.

But modern publishing involves much more than producing a PDF and making it available on a preprint server such as arXiv[1] or bioRxiv[2]. It also entails providing the data underpinning the report and the code used to do the analysis:

> An article about computational science in a scientific publication is *not* the scholarship itself, it is merely *advertising* of the scholarship. The actual scholarship is the complete software development environment and the complete set of instructions which generated the figures.
>
> — Jonathan Buckheit and David Donoho, paraphrasing Jon Claerbout, in Buckheit and Donoho (1995)

While some reports, datasets, software packages, and analysis scripts can't be published without violating personal or commercial confidentiality, every researcher's default should be to make all these as widely available as possible. Publishing it under an open license (Section 8.4) is the first step; the sections below describe what else we can do to capture the **provenance** of our data analysis.

Our Zipf's Law project files are structured as they were at the end of the previous chapter:

[1] `https://arxiv.org/`
[2] `https://www.biorxiv.org/`

```
zipf/
├── .gitignore
├── .travis.yml
├── CONDUCT.md
├── CONTRIBUTING.md
├── LICENSE.md
├── Makefile
├── README.md
├── requirements.txt
├── bin
│   ├── book_summary.sh
│   ├── collate.py
│   ├── countwords.py
│   ├── plotcounts.py
│   ├── plotparams.yml
│   ├── script_template.py
│   ├── test_zipfs.py
│   └── utilities.py
├── data
│   ├── README.md
│   ├── dracula.txt
│   └── ...
├── results
│   ├── dracula.csv
│   ├── dracula.png
│   └── ...
└── test_data
    ├── random_words.txt
    └── risk.txt
```

Identifying Reports and Authors

Before publishing anything, we need to understand how authors and their works are identified. A **Digital Object Identifier** (DOI) is a unique identifier for a particular version of a particular digital artifact such as a report, a dataset, or a piece of software. DOIs are written as `doi:prefix/suffix`, but are often also represented as URLs like `http://dx.doi.org/prefix/suffix`. In order to be allowed to issue a DOI, an academic journal, data archive, or other organization must guarantee a certain level of security, longevity and access.

An **ORCID** is an Open Researcher and Contributor ID. Anyone can get an ORCID for free, and should include it in publications because people's names and affiliations change over time.

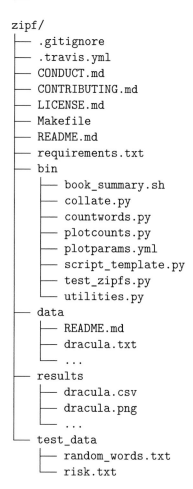

13.1 Data Provenance

The first step in documenting the data associated with a report is to determine what (if anything) needs to be published. If the report involved the analysis of a publicly available dataset that is maintained and documented by a third party, it's not necessary to publish a duplicate of the dataset: the report simply needs to document where to access the data and what version was analyzed. This is the case for our Zipf's Law analysis, since the texts we analyze are available at Project Gutenberg[3].

It's not strictly necessary to publish intermediate data produced during the analysis of a publicly available dataset either (e.g., the CSV files produced by `countwords.py`), so long as readers have access to the original data and the code/software used to process it. However, making intermediate data available can save people time and effort, particularly if it takes a lot of computing power to reproduce it (or if installing the required software is complicated). For example, NASA has published the Goddard Institute for Space Studies Surface Temperature Analysis[4], which estimates the global average surface temperature based on thousands of land and ocean weather observations, because a simple metric of global warming is expensive to produce and is useful in many research contexts.

If a report involves a new dataset, such as observations collected during a field experiment, then that data needs to be published in its raw (unprocessed) form. The publication of a dataset, whether raw or intermediate, should follow the FAIR Principles.

13.1.1 The FAIR Principles

The FAIR Principles[5] describe what research data should look like. They are still aspirational for most researchers, but tell us what to aim for (Goodman et al. 2014; Michener 2015; Hart et al. 2016; Brock 2019; Tierney and Ram 2020). The most immediately important elements of the FAIR Principles are outlined below.

13.1.1.1 Data should be *findable*

The first step in using or re-using data is to find it. We can tell we've done this if:

[3] `https://www.gutenberg.org/`
[4] `https://data.giss.nasa.gov/gistemp/`
[5] `https://www.go-fair.org/fair-principles/`

1. (Meta)data is assigned a globally unique and persistent identifier (i.e., a **DOI**).
2. Data is described with rich metadata.
3. Metadata clearly and explicitly includes the identifier of the data it describes.
4. (Meta)data is registered or indexed in a searchable resource, such as the data sharing platforms described in Section 13.1.2.

13.1.1.2 Data should be *accessible*

People can't use data if they don't have access to it. In practice, this rule means the data should be openly accessible (the preferred solution) or that authenticating in order to view or download it should be free. We can tell we've done this if:

1. (Meta)data is retrievable by its identifier using a standard communications protocol like HTTP.
2. Metadata is accessible even when the data is no longer available.

13.1.1.3 Data should be *interoperable*

Data usually needs to be integrated with other data, which means that tools need to be able to process it. We can tell we've done this if:

1. (Meta)data uses a formal, accessible, shared, and broadly applicable language for knowledge representation.
2. (Meta)data uses vocabularies that follow FAIR principles.
3. (Meta)data includes qualified references to other (meta)data.

13.1.1.4 Data should be *reusable*

This is the ultimate purpose of the FAIR Principles and much other work. We can tell we've done this if:

1. (Meta)data is described with accurate and relevant attributes.
2. (Meta)data is released with a clear and accessible data usage license.
3. (Meta)data has detailed **provenance**.
4. (Meta)data meets domain-relevant community standards.

13.1.2 Where to archive data

Small datasets (i.e., anything under 500 MB) can be stored in version control. If the data is being used in several projects, it may make sense to create one repository to hold only the data; these are sometimes referred to as **data packages**, and they are often accompanied by small scripts to clean up and query the data.

For medium-sized datasets (between 500 MB and 5 GB), it's better to put the data on platforms like the Open Science Framework[6], Dryad[7], Zenodo[8], and Figshare[9], which will give the dataset a DOI. Big datasets (i.e., anything more than 5 GB) may not be ours in the first place, and probably need the attention of a professional archivist.

Data Journals

While archiving data at a site like Dryad or Figshare (following the FAIR Principles) is usually the end of the data publishing process, there is the option of publishing a journal paper to describe the dataset in detail. Some research disciplines have journals devoted to describing particular types of data (e.g., the Geoscience Data Journal[10]) and there are also generic data journals (e.g., Scientific Data[11]).

13.2 Code Provenance

Our Zipf's Law analysis represents a typical data science project in that we've written some code that leverages other pre-existing software packages in order to produce the key results of a report. To make a computational workflow like this open, transparent, and reproducible we must archive three key items:

1. A copy of any **analysis scripts or notebooks** used to produce the key results presented in the report.
2. A detailed description of the **software environment** in which those analysis scripts or notebooks ran.
3. A description of the **data processing steps** taken in producing each key result, i.e., a step-by-step account of which scripts were executed in what order for each key result.

[6] `https://osf.io/`
[7] `https://datadryad.org/`
[8] `https://zenodo.org/`
[9] `https://figshare.com/`
[10] `https://rmets.onlinelibrary.wiley.com/journal/20496060`
[11] `https://www.nature.com/sdata/`

Unfortunately, librarians, publishers, and regulatory bodies are still trying to determine the best way to document and archive material like this, so a widely accepted set of FAIR Principles for research software is still under development (Lamprecht et al. 2020). In the meantime, the best advice we can give is presented below. It involves adding information about the software environment and data processing steps to a GitHub repository that contains the analysis scripts/notebooks, before creating a new release of that repository and archiving it (with a DOI) with Zenodo[12].

13.2.1 Software environment

In order to document the software packages used in our analysis, we should archive a list of the names and version numbers of each software package. We can get version information for the Python packages we are using by running:

```
$ pip freeze
```

```
alabaster==0.7.12
anaconda-client==1.7.2
anaconda-navigator==1.9.12
anaconda-project==0.8.3
appnope==0.1.0
appscript==1.0.1
asn1crypto==1.0.1
...
```

Other command-line tools will often have an option like `--version` or `--status` to access the version information.

Archiving a list of package names and version numbers would mean that our software environment is technically reproducible, but it would be left up to the reader of the report to figure out how to get all those packages installed and working together. This might be fine for a small number of packages with very few dependencies, but in more complex cases we probably want to make life easier for the reader (and for our future selves looking to re-run the analysis). One way to make things easier is to export a description of a complete conda environment (Section 14.2; Appendix I.2), which can be saved as YAML using:

[12] https://zenodo.org/

```
$ conda env export > environment.yml
$ cat environment.yml
```

```
name: base
channels:
  - conda-forge
  - defaults
dependencies:
  - _ipyw_jlab_nb_ext_conf=0.1.0=py37_0
  - alabaster=0.7.12=py37_0
  - anaconda=2019.10=py37_0
  - anaconda-client=1.7.2=py37_0
  - anaconda-navigator=1.9.12=py37_0
  - anaconda-project=0.8.3=py_0
  - appnope=0.1.0=py37_0
  - appscript=1.1.0=py37h1de35cc_0
  - asn1crypto=1.0.1=py37_0
...
```

That software environment can be re-created on another computer with one line of code:

```
$ conda env create -f environment.yml
```

Since this `environment.yml` file is an important part of reproducing the analysis, remember to add it to your GitHub repository.

Container Images

More complex tools like Docker[13] can install our entire environment (down to the precise version of the operating system) on a different computer (Nüst et al. 2020). However, their complexity can be daunting, and there is a lot of debate about how well (or whether) they actually make research more reproducible in practice.

13.2.2 Data processing steps

The second item that needs to be added to our GitHub repository is a description of the data processing steps involved in each key result. Assuming the

[13] `https://en.wikipedia.org/wiki/Docker_(software)`

author list on our report is Amira Khan and Sami Virtanen (Section 0.2), we could add a new Markdown file called `KhanVirtanen2020.md` to the repository to describe the steps:

```
The code in this repository was used in generating the results
for the following paper:

Khan A & Virtanen S, 2020. Zipf's Law in classic english texts.
*Journal of Important Research*, 27, 134-139.

The code was executed in the software environment described by
`environment.yml`. It can be installed using
[conda](https://docs.conda.io/en/latest/):
$ conda env create -f environment.yml

Figure 1 in the paper was created by running the following at
the command line:
$ make all
```

We should also add this information as an appendix to the report itself.

13.2.3 Analysis scripts

Later in this book we will package and release our Zipf's Law code so that it can be downloaded and installed by the wider research community, just like any other Python package (Chapter 14). Doing this is especially helpful if other people might be interested in using and/or extending it, but often the scripts and notebooks we write to produce a particular figure or table are too case-specific to be of broad interest. To fully capture the provenance of the results presented in a report, these analysis scripts and/or notebooks (along with the details of the associated software environment and data processing steps) can simply be archived with a repository like Figshare[14] or Zenodo[15], which specialize in storing the long tail of research projects (i.e., supplementary figures, data, and code). Uploading a zip file of analysis scripts to the repository is a valid option, but more recently the process has been streamlined via direct integration between GitHub and Zenodo. As described in this tutorial[16], the process involves creating a new release of our repository in GitHub that Zenodo copies and then issues a DOI for (Figure 13.1).

[14]https://figshare.com/
[15]https://zenodo.org/
[16]https://guides.github.com/activities/citable-code/

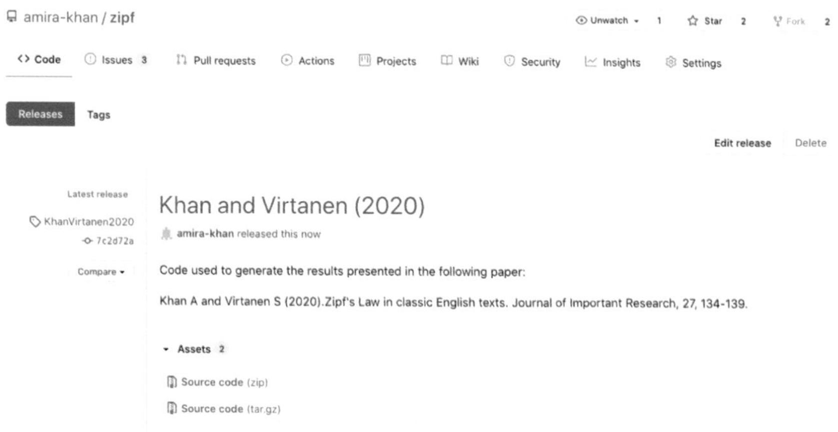

FIGURE 13.1: A new code release on GitHub.

13.2.4 Reproducibility versus inspectability

In most cases, documenting our software environment, analysis scripts, and data processing steps will ensure that our computational analysis is reproducible/repeatable at the time our report is published. But what about five or ten years later? As we have discussed, data analysis workflows usually depend on a hierarchy of packages. Our Zipf's Law analysis depends on a collection of Python libraries, which in turn depend on the Python language itself. Some workflows also depend critically on a particular operating system. Over time some of these dependencies will inevitably be updated or no longer supported, meaning our workflow will be documented but not reproducible.

Fortunately, most readers are not looking to exactly re-run a decade-old analysis: they just want to be able to figure out what was run and what the important decisions were, which is sometimes referred to as **inspectability** (Gil et al. 2016; T. Brown 2017). While exact repeatability has a short shelf-life, inspectability is the enduring legacy of a well-documented computational analysis.

Your Future Self Will Thank You

Data and code provenance is often promoted for the good of people trying to reproduce your work, who were not part of creating the work in the first place. Prioritizing their needs can be difficult: how can we justify spending time for other people when our current projects need work for the good of the people working on them right now?

Instead of thinking about people who are unknown and unrelated, we can think about newcomers to our team and the time we will save ourselves in onboarding them. We can also think about the time we will save ourselves when we come back to this project five months or five years from now. Documentation that serves these two groups well will almost certainly serve the needs of strangers as well.

13.3 Summary

The Internet started a revolution in scientific publishing that shows no sign of ending. Where an inter-library loan once took weeks to arrive and data had to be transcribed from published papers (if it could be found at all), we can now download one another's work in minutes: *if* we can find it and make sense of it. Organizations like Our Research[17] are building tools to help with both; by using DOIs and ORCIDs, publishing on preprint servers, following the FAIR Principles, and documenting our workflow, we help ensure that everyone can pursue their ideas as we did.

13.4 Exercises

13.4.1 ORCID

If you don't already have an **ORCID**, go to the website and register now. If you do have an ORCID, log in and make sure that your details and publication record are up-to-date.

[17] `http://ourresearch.org/`

13.4.2 A FAIR test

An online questionnaire[18] for measuring the extent to which datasets are FAIR has been created by the Australian Research Data Commons. Fill in the questionnaire for a dataset you have published or that you use often.

13.4.3 Evaluate a project's data provenance

This exercise is modified from Wickes and Stein (2016) and explores the dataset from Meili (2016). Go to the dataset's page `http://doi.org/10.3886/E17507V2` and download the files. You will need to make an ICPSER account and agree to their data agreement before you can download.

Review the dataset's main page to get a sense of the study, then review the spreadsheet file and the coded response file.

1. Who are the participants of this study?
2. What types of data were collected and used for analysis?
3. Can you find information on the demographics of the interviewees?
4. This dataset is clearly in support of an article. What information can you find about it, and can you find a link to it?

13.4.4 Evaluate a project's code provenance

The GitHub repository `borstlab/reversephi_paper`[19] provides the code and data for the paper Leonhardt et al. (2017). Browse the repository and answer the following questions:

1. Where is the software environment described? What files would you need to re-create the software environment?
2. Where are the data processing steps described? How could you re-create the results included in the manuscript?
3. How are the scripts and data archived? That is, where can you download the version of the code and data as it was when the manuscript was published?

To get a feel for the different approaches to code provenance, repeat steps 1-3 with the following:

[18] `https://ardc.edu.au/resources/working-with-data/fair-data/fair-self-assessment-tool`

[19] `https://github.com/borstlab/reversephi_paper/`

- The figshare page[20] that accompanies the paper Irving, Wijffels, and Church (2019).
- The GitHub repo `blab/h3n2-reassortment`[21] that accompanies the paper Potter et al. (2019).

13.4.5 Making permanent links

The link to the UK Home Office's accessibility guideline posters[22] might change in future. Use the Wayback Machine[23] to find a link that is more likely to be usable in the long run.

13.4.6 Create an archive of your Zipf's analysis

A slightly less permanent alternative to having a DOI for your analysis code is to provide a link to a GitHub release. Follow the instructions on GitHub[24] to create a release for the current state of your `zipf/` project.

Once you've created the release, read about how to link to it[25]. What is the URL that allows direct download of the zip archive of your release?

What about getting a DOI?

Creating a GitHub release is also a necessary step to get a DOI through the Zenodo/GitHub integration (Section 13.2.3). We are stopping short of getting the DOI here, since nobody reading this book needs to formally cite or archive the example Zipf's Law software we've been developing. Also, if every reader of the book generated a DOI, we'd have many DOIs pointing to the same code!

[20] `https://doi.org/10.6084/m9.figshare.7575830`
[21] `https://github.com/blab/h3n2-reassortment`
[22] `https://ukhomeoffice.github.io/accessibility-posters/posters/accessibility-posters.pdf`
[23] `https://web.archive.org/`
[24] `https://docs.github.com/en/github/administering-a-repository/managing-releases-in-a-repository`
[25] `https://docs.github.com/en/github/administering-a-repository/linking-to-releases`

13.4.7 Publishing your code

Think about a project that you're currently working on. How would you go about publishing the code associated with that project (i.e., the software description, analysis scripts, and data processing steps)?

13.5 Key Points

- Publish data and code as well as papers.
- Use **DOIs** to identify reports, datasets, and software releases.
- Use an **ORCID** to identify yourself as an author of a report, dataset, or software release.
- Data should be FAIR[26]: findable, accessible, interoperable, and reusable.
- Put small datasets in version control repositories; store large ones on data sharing sites.
- Describe your software environment, analysis scripts, and data processing steps in **reproducible** ways.
- Make your analyses **inspectable** as well as reproducible.

[26] `https://www.go-fair.org/fair-principles/`

14

Creating Packages with Python

> Another response of the wizards, when faced with a new and unique situation, was to look through their libraries to see if it had ever happened before. This was...a good survival trait. It meant that in times of danger you spent the day sitting very quietly in a building with very thick walls.
>
> — Terry Pratchett

The more software we write, the more we come to think of programming languages as a way to build and combine libraries. Every widely used language now has an online repository from which people can download and install those libraries. This lesson shows you how to use Python's tools to create and share libraries of your own.

We will continue with our Zipf's Law project, which should include the following files:

```
zipf/
├── .gitignore
├── .travis.yml
├── CONDUCT.md
├── CONTRIBUTING.md
├── KhanVirtanen2020.md
├── LICENSE.md
├── Makefile
├── README.md
├── environment.yml
├── requirements.txt
├── bin
│   ├── book_summary.sh
│   ├── collate.py
│   ├── countwords.py
│   ├── plotcounts.py
│   ├── plotparams.yml
│   ├── script_template.py
│   ├── test_zipfs.py
```

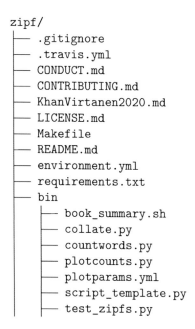

```
│   └── utilities.py
├── data
│   ├── README.md
│   ├── dracula.txt
│   └── ...
├── results
│   ├── dracula.csv
│   ├── dracula.png
│   └── ...
└── test_data
    ├── random_words.txt
    └── risk.txt
```

14.1 Creating a Python Package

A package consists of one or more Python source files in a specific directory structure combined with installation instructions for the computer. Python packages can come from various sources: some are distributed with Python itself as part of the language's standard library[1], but anyone can create one, and there are thousands that can be downloaded and installed from online repositories.

Terminology

People sometimes refer to packages as modules. Strictly speaking, a module is a single source file, while a package is a directory structure that contains one or more modules.

A generic package folder hierarchy looks like this:

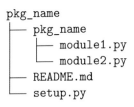

```
pkg_name
├── pkg_name
│   ├── module1.py
│   └── module2.py
├── README.md
└── setup.py
```

[1] https://docs.python.org/3/library/

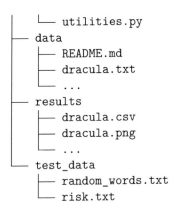

The top-level directory is named after the package. It contains a directory that is also named after the package, and that contains the package's source files. It is initially a little confusing to have two directories with the same name, but most Python projects follow this convention because it makes it easier to set up the project for installation.

> **`__init__.py`**
>
> Python packages often contain a file with a special name: `__init__.py` (two underscores before and after `init`). Just as importing a module file executes the code in the module, importing a package executes the code in `__init__.py`. Packages *had* to have this file before Python 3.3, even if it was empty, but since Python 3.3 it is only needed if we want to run some code as the package is being imported.

If we want to make our Zipf's Law software available as a Python package, we need to follow the generic folder hierarchy. A quick search of the Python Package Index (PyPI)[2] reveals that the package name `zipf` is already taken, so we will need to use something different. Let's use `pyzipf` and update our directory names accordingly:

```
$ mv ~/zipf ~/pyzipf
$ cd ~/pyzipf
$ mv bin pyzipf
```

> **Updating GitHub's Repository Name**
>
> We won't do it in this case (because it would break links/references from earlier in the book), but now that we've decided to name our package `pyzipf`, we would normally update the name of our GitHub repository to match. After changing the name at the GitHub website, we would need to update our `git remote` so that our local repository could still be synchronized with GitHub:
>
> ```
> $ git remote set-url origin
>   https://github.com/amira-khan/pyzipf.git
> ```

[2]https://pypi.org/

Python has several ways to build an installable package. We will show how to use `setuptools`[3], which is the lowest common denominator and will allow everyone, regardless of what Python distribution they have, to use our package. To use `setuptools`, we must create a file called `setup.py` in the directory *above* the root directory of the package. (This is why we require the two-level directory structure described earlier.) `setup.py` must have exactly that name, and must contain lines like these:

```
from setuptools import setup


setup(
    name='pyzipf',
    version='0.1.0',
    author='Amira Khan',
    packages=['pyzipf'])
```

The `name` and `author` parameters are self-explanatory. Most software projects use **semantic versioning** for software releases. A version number consists of three integers X.Y.Z, where X is the major version, Y is the minor version, and Z is the **patch** version. Major version zero (0.Y.Z) is for initial development, so we have started with 0.1.0. The first stable public release would be version 1.0.0, and in general, the version number is incremented as follows:

- Increment `major` every time there's an incompatible externally visible change.
- Increment `minor` when adding new functionality in a backwards-compatible manner (i.e., without breaking any existing code).
- Increment `patch` for backwards-compatible bug fixes that don't add any new features.

Finally, we specify the name of the directory containing the code to be packaged with the `packages` parameter. This is straightforward in our case because we only have a single package directory. For more complex projects, the `find_packages`[4] function from `setuptools` can automatically find all packages by recursively searching the current directory.

[3] https://setuptools.readthedocs.io/

[4] https://setuptools.readthedocs.io/en/latest/setuptools.html#using-find-packages

14.2 Virtual Environments

We can add additional information to our package later, but this is enough to be able to build it for testing purposes. Before we do that, though, we should create a **virtual environment** to test how our package installs without breaking anything in our main Python installation. We exported details of our environment in Chapter 13 as a way to document the software we're using; in this section, we'll use environments to make the software we're creating more robust.

A virtual environment is a layer on top of an existing Python installation. Whenever Python needs to find a package, it looks in the virtual environment before checking the main Python installation. This gives us a place to install packages that only some projects need without affecting other projects.

Virtual environments also help with package development:

- We want to be able to easily test install and uninstall our package, without affecting the entire Python environment.
- We want to answer problems people have with our package with something more helpful than "I don't know, it works for me." By installing and running our package in a completely empty environment, we can ensure that we're not accidentally relying on other packages being installed.

We can manage virtual environments using `conda`[5] (Appendix I). To create a new virtual environment called `pyzipf` we run `conda create`, specifying the environment's name with the `-n` or `--name` flag and including `pip` and our current version of Python in the new environment:

```
$ conda create -n pyzipf pip python=3.7.6
```

```
Collecting package metadata (current_repodata.json): done
Solving environment: done

## Package Plan ##

  environment location: /Users/amira/anaconda3/envs/pyzipf

  added / updated specs:
```

[5]`https://conda.io/`

```
    - pip
    - python=3.7.6


The following packages will be downloaded:
...list of packages...

The following NEW packages will be INSTALLED:
...list of packages...

Proceed ([y]/n)? y


...

Preparing transaction: done
Verifying transaction: done
Executing transaction: done
#
# To activate this environment, use
#
#     $ conda activate pyzipf
#
# To deactivate an active environment, use
#
#     $ conda deactivate
```

`conda` creates the directory `/Users/amira/anaconda3/envs/pyzipf`, which contains the subdirectories needed for a minimal Python installation, such as `bin` and `lib`. It also creates `/Users/amira/anaconda3/envs/pyzipf/bin/python`, which checks for packages in these directories before checking the main installation.

conda Variations

As with many of the other tools we've explored in this book, the behavior of some `conda` commands differ depending on the operating system. There are multiple ways to accomplish some of the tasks we present in this chapter. The options we present here represent the approaches most likely to work across multiple platforms.

Additionally, the path for Anaconda differs among operating systems. Our examples show the default path for Anaconda installed via the Unix shell on MacOS (`/Users/amira/anaconda3`), but for the MacOS graphical installer it is `/Users/amira/opt/anaconda3`, for Linux it is `/home/amira/anaconda3`, and on Windows it is `C:\Users\amira\Anaconda3`. During the installation process, users can also choose a custom location if they like (Section 1.3).

We can switch to the `pyzipf` environment by running:

```
$ conda activate pyzipf
```

Once we have done this, the `python` command runs the interpreter in `pyzipf/bin`:

```
(pyzipf)$ which python
```

```
/Users/amira/anaconda3/envs/pyzipf/bin/python
```

Notice that every shell command displays `(pyzipf)` when that virtual environment is active. Between Git branches and virtual environments, it can be very easy to lose track of what exactly we are working on and with. Prompts like this can make it a little less confusing; using virtual environment names that match the names of your projects (and branches, if you're testing different environments on different branches) quickly becomes essential.

We can now install packages safely. Everything we install will go into the `pyzipf` virtual environment without affecting the underlying Python installation. When we are done, we can switch back to the default environment using `conda deactivate`:

```
(pyzipf)$ conda deactivate
```

```
$ which python
```

```
/usr/bin/python
```

14.3 Installing a Development Package

Let's install our package in this virtual environment. First we re-activate it:

```
$ conda activate pyzipf
```

Next, we go into the upper `pyzipf` directory that contains our `setup.py` file and install our package:

```
(pyzipf)$ cd ~/pyzipf
(pyzipf)$ pip install -e .
```

```
Obtaining file:///Users/amira/pyzipf
Installing collected packages: pyzipf
  Running setup.py develop for pyzipf
Successfully installed pyzipf
```

The `-e` option indicates that we want to install the package in "editable" mode, which means that any changes we make in the package code are directly available to use without having to reinstall the package; the `.` means "install from the current directory."

If we look in the location containing package installations (e.g., `/Users/amira/anaconda3/envs/pyzipf/lib/python3.7/site-packages/`), we can see the `pyzipf` package beside all the other locally installed packages. If we try to use the package at this stage, though, Python will complain that some of the packages it depends on, such as `pandas`, are not installed.

We could install these manually, but it is more reliable to automate this process by listing everything that our package depends on using the `install_requires` parameter in `setup.py`:

```
from setuptools import setup


setup(
    name='pyzipf',
    version='0.1',
    author='Amira Khan',
    packages=['pyzipf'],
    install_requires=[
        'matplotlib',
        'pandas',
        'scipy',
        'pyyaml',
        'pytest'])
```

We don't have to list `numpy` explicitly because it will be installed as a dependency for `pandas` and `scipy`.

Versioning Dependencies

It is good practice to specify the versions of our dependencies and even better to specify version ranges. For example, if we have only tested our package on pandas version 1.1.2, we could put `pandas==1.1.2` or `pandas>=1.1.2` instead of just `pandas` in the list argument passed to the `install_requires` parameter.

Next, we can install our package using the modified `setup.py` file:

```
(pyzipf)$ cd ~/pyzipf
(pyzipf)$ pip install -e .
```

```
Obtaining file:///Users/amira/pyzipf
Collecting matplotlib
  Downloading matplotlib-3.3.3-cp37-cp37m-macosx_10_9_x86_64.whl
     |                    | 8.5 MB 3.1 MB/s
Collecting cycler>=0.10
  Using cached cycler-0.10.0-py2.py3-none-any.whl
```

```
Collecting kiwisolver>=1.0.1
  Downloading kiwisolver-1.3.1-cp37-cp37m-macosx_10_9_x86_64.whl
     |                    | 61 kB 2.0 MB/s
Collecting numpy>=1.15
  Downloading numpy-1.19.4-cp37-cp37m-macosx_10_9_x86_64.whl
     |                    | 15.3 MB 8.9 MB/s
Collecting pillow>=6.2.0
  Downloading Pillow-8.0.1-cp37-cp37m-macosx_10_10_x86_64.whl
     |                    | 2.2 MB 6.3 MB/s
Collecting pyparsing!=2.0.4,!=2.1.2,!=2.1.6,>=2.0.3
  Using cached pyparsing-2.4.7-py2.py3-none-any.whl
Collecting python-dateutil>=2.1
  Using cached python_dateutil-2.8.1-py2.py3-none-any.whl
Collecting six
  Using cached six-1.15.0-py2.py3-none-any.whl
Collecting pandas
  Downloading pandas-1.1.5-cp37-cp37m-macosx_10_9_x86_64.whl
     |                    | 10.0 MB 1.4 MB/s
Collecting pytz>=2017.2
  Using cached pytz-2020.4-py2.py3-none-any.whl
Collecting pytest
  Using cached pytest-6.2.1-py3-none-any.whl
Collecting attrs>=19.2.0
  Using cached attrs-20.3.0-py2.py3-none-any.whl
Collecting importlib-metadata>=0.12
  Downloading importlib_metadata-3.3.0-py3-none-any.whl
Collecting pluggy<1.0.0a1,>=0.12
  Using cached pluggy-0.13.1-py2.py3-none-any.whl
Collecting py>=1.8.2
  Using cached py-1.10.0-py2.py3-none-any.whl
Collecting typing-extensions>=3.6.4
  Downloading typing_extensions-3.7.4.3-py3-none-any.whl
Collecting zipp>=0.5
  Downloading zipp-3.4.0-py3-none-any.whl
Collecting iniconfig
  Using cached iniconfig-1.1.1-py2.py3-none-any.whl
Collecting packaging
  Using cached packaging-20.8-py2.py3-none-any.whl
Collecting pyyaml
  Using cached PyYAML-5.3.1.tar.gz
Collecting scipy
  Downloading scipy-1.5.4-cp37-cp37m-macosx_10_9_x86_64.whl
     |                    | 28.7 MB 10.7 MB/s
Collecting toml
  Using cached toml-0.10.2-py2.py3-none-any.whl
```

```
Building wheels for collected packages: pyyaml
  Building wheel for pyyaml (setup.py) ... done
  Created wheel for pyyaml:
    filename=PyYAML-5.3.1-cp37-cp37m-macosx_10_9_x86_64.whl
    size=44626
    sha256=5a59ccf08237931e7946ec6b526922e4
           f0c8ee903d43671f50289431d8ee689d
  Stored in directory: /Users/amira/Library/Caches/pip/wheels/
    5e/03/1e/e1e954795d6f35dfc7b637fe2277bff021303bd9570ecea653
Successfully built pyyaml
Installing collected packages: zipp, typing-extensions, six,
pyparsing, importlib-metadata, toml, pytz, python-dateutil, py,
pluggy, pillow, packaging, numpy, kiwisolver, iniconfig, cycler,
attrs, scipy, pyyaml, pytest, pandas, matplotlib, pyzipf
  Attempting uninstall: pyzipf
    Found existing installation: pyzipf 0.1.0
    Uninstalling pyzipf-0.1.0:
      Successfully uninstalled pyzipf-0.1.0
  Running setup.py develop for pyzipf
Successfully installed attrs-20.3.0 cycler-0.10.0
importlib-metadata-3.3.0 iniconfig-1.1.1 kiwisolver-1.3.1
matplotlib-3.3.3 numpy-1.19.4 packaging-20.8 pandas-1.1.5
pillow-8.0.1 pluggy-0.13.1 py-1.10.0 pyparsing-2.4.7
pytest-6.2.1 python-dateutil-2.8.1 pytz-2020.4 pyyaml-5.3.1
pyzipf scipy-1.5.4 six-1.15.0 toml-0.10.2
typing-extensions-3.7.4.3 zipp-3.4.0
```

(The precise output of this command will change depending on which versions of our dependencies get installed.)

We can now import our package in a script or a Jupyter notebook[6] just as we would any other package. For example, to use the function in `utilities`, we would write:

```
from pyzipf import utilities as util


util.collection_to_csv(...)
```

To allow our functions to continue accessing `utilities.py`, we need to change that line in both `countwords.py` and `collate.py`.

However, the useful command-line scripts that we used to count and plot

[6] `https://jupyter.org/`

word counts are no longer accessible directly from the Unix shell. Fortunately there is an alternative to changing the function import as described above. The `setuptools` package allows us to install programs along with the package. These programs are placed beside those of other packages. We tell `setuptools` to do this by defining **entry points** in `setup.py`:

```
from setuptools import setup


setup(
    name='pyzipf',
    version='0.1',
    author='Amira Khan',
    packages=['pyzipf'],
    install_requires=[
        'matplotlib',
        'pandas',
        'scipy',
        'pyyaml',
        'pytest'],
    entry_points={
        'console_scripts': [
            'countwords = pyzipf.countwords:main',
            'collate = pyzipf.collate:main',
            'plotcounts = pyzipf.plotcounts:main']})
```

The right side of the `=` operator is the location of a function, written as `package.module:function`; the left side is the name we want to use to call this function from the command line. In this case we want to call each module's `main` function; right now, it requires an input argument `args` containing the command-line arguments given by the user (Section 5.2). For example, the relevant section of our `countwords.py` program is:

```
def main(args):
    """Run the command line program."""
    word_counts = count_words(args.infile)
    util.collection_to_csv(word_counts, num=args.num)


if __name__ == '__main__':
    parser = argparse.ArgumentParser(description=__doc__)
    parser.add_argument('infile', type=argparse.FileType('r'),
```

```
                        nargs='?', default='-',
                        help='Input file name')
    parser.add_argument('-n', '--num',
                        type=int, default=None,
                        help='Output n most frequent words')
    args = parser.parse_args()
    main(args)
```

We can't pass any arguments to main when we define entry points in our setup.py file, so we need to change our script slightly:

```
def parse_command_line():
    """Parse the command line for input arguments."""
    parser = argparse.ArgumentParser(description=__doc__)
    parser.add_argument('infile', type=argparse.FileType('r'),
                        nargs='?', default='-',
                        help='Input file name')
    parser.add_argument('-n', '--num',
                        type=int, default=None,
                        help='Output n most frequent words')
    args = parser.parse_args()
    return args


def main():
    """Run the command line program."""
    args = parse_command_line()
    word_counts = count_words(args.infile)
    util.collection_to_csv(word_counts, num=args.num)


if __name__ == '__main__':
    main()
```

The new parse_command_line function handles the command-line arguments, so that main() no longer requires any input arguments.

Once we have made the corresponding change in collate.py and plotcounts.py, we can re-install our package:

```
(pyzipf)$ pip install -e .
```

```
Defaulting to user installation because normal site-packages is
  not writeable
Obtaining file:///Users/amira/pyzipf
Requirement already satisfied: matplotlib in
  /usr/lib/python3.7/site-packages (from pyzipf==0.1) (3.2.1)
Requirement already satisfied: pandas in
  /Users/amira/.local/lib/python3.7/site-packages
  (from pyzipf==0.1) (1.0.3)
Requirement already satisfied: scipy in
  /usr/lib/python3.7/site-packages (from pyzipf==0.1) (1.4.1)
Requirement already satisfied: pyyaml in
  /usr/lib/python3.7/site-packages (from pyzipf==0.1) (5.3.1)
Requirement already satisfied: cycler>=0.10 in
  /usr/lib/python3.7/site-packages
  (from matplotlib->pyzipf==0.1) (0.10.0)
Requirement already satisfied: kiwisolver>=1.0.1 in
  /usr/lib/python3.7/site-packages
  (from matplotlib->pyzipf==0.1) (1.1.0)
Requirement already satisfied: numpy>=1.11 in
  /usr/lib/python3.7/site-packages
  (from matplotlib->pyzipf==0.1) (1.18.2)
Requirement already satisfied: pyparsing!=2.0.4,!=2.1.2,
!=2.1.6,>=2.0.1 in
  /usr/lib/python3.7/site-packages
  (from matplotlib->pyzipf==0.1) (2.4.6)
Requirement already satisfied: python-dateutil>=2.1 in
  /usr/lib/python3.7/site-packages
  (from matplotlib->pyzipf==0.1) (2.8.1)
Requirement already satisfied: pytz>=2017.2 in
  /usr/lib/python3.7/site-packages
  (from pandas->pyzipf==0.1) (2019.3)
Requirement already satisfied: six in
  /usr/lib/python3.7/site-packages
  (from cycler>=0.10->matplotlib->pyzipf==0.1) (1.14.0)
Requirement already satisfied: setuptools in
  /usr/lib/python3.7/site-packages
  (from kiwisolver>=1.0.1->matplotlib->pyzipf==0.1) (46.1.3)
Installing collected packages: pyzipf
  Running setup.py develop for pyzipf
Successfully installed pyzipf
```

The output looks slightly different than the first run because `pip` could re-use some packages saved locally by the previous install rather than re-fetching them from online repositories. (If we hadn't used the `-e` option to make the package immediately editable, we would have to uninstall it before reinstalling it during development.)

We can now use our commands directly from the Unix shell without writing the full path to the file and without prefixing it with `python`.

```
(pyzipf)$ countwords data/dracula.txt -n 5
```

```
the,8036
and,5896
i,4712
to,4540
of,3738
```

Tracking `pyzipf.egg-info`?

Using `setuptools` automatically creates a new folder in your project directory named `pyzipf.egg-info`. This folder is another example[7] of information generated by a script that is also included in the repository, so it should be included in the `.gitignore` file to avoid tracking with Git.

14.4 What Installation Does

Now that we have created and installed a Python package, let's explore what actually happens during installation. The short version is that the contents of the package are copied into a directory that Python will search when it imports things. In theory we can "install" packages by manually copying source code into the right places, but it's much more efficient and safer to use a tool specifically made for this purpose, such as `conda` or `pip`.

Most of the time, these tools copy packages into the Python installation's `site-packages` directory, but this is not the only place Python searches. Just as the `PATH` environment in the shell contains a list of directories that the shell searches for programs it can execute (Section 4.6), the Python variable

[7] https://github.com/github/gitignore

`sys.path` contains a list of the directories it searches (Section 11.2). We can look at this list inside the interpreter:

```
import sys
sys.path
```

```
['',
'/Users/amira/anaconda3/envs/pyzipf/lib/python37.zip',
'/Users/amira/anaconda3/envs/pyzipf/lib/python3.7',
'/Users/amira/anaconda3/envs/pyzipf/lib/python3.7/lib-dynload',
'/Users/amira/.local/lib/python3.7/site-packages',
'/Users/amira/anaconda3/envs/pyzipf/lib/python3.7/
site-packages',
'/Users/amira/pyzipf']
```

The empty string at the start of the list means "the current directory." The rest are system paths for our Python installation, and will vary from computer to computer.

14.5 Distributing Packages

Look but Don't Execute

In this section we upload the `pyzipf` package to TestPyPI and PyPI:

`https://test.pypi.org/project/pyzipf`

`https://pypi.org/project/pyzipf/`

You won't be able to execute the `twine upload` commands below exactly as shown (because Amira has already uploaded the `pyzipf` package), but the general sequence of commands in this section is an excellent resource to refer to when you are uploading your own packages. If you want to try uploading your own `pyzipf` package via `twine`, you could edit the project name to include your name (e.g., `pyzipf-yourname`) and use the TestPyPI repository for the upload.

An installable package is most useful if we distribute it so that anyone who wants it can run `pip install pyzipf` and get it. To make this possible, we need to use `setuptools` to create a **source distribution** (known as an `sdist` in Python packaging jargon):

```
(pyzipf)$ python setup.py sdist
```

```
running sdist
running egg_info
creating pyzipf.egg-info
writing pyzipf.egg-info/PKG-INFO
writing dependency_links to pyzipf.egg-info/dependency_links.txt
writing entry points to pyzipf.egg-info/entry_points.txt
writing requirements to pyzipf.egg-info/requires.txt
writing top-level names to pyzipf.egg-info/top_level.txt
writing manifest file 'pyzipf.egg-info/SOURCES.txt'
package init file 'pyzipf/__init__.py' not found
(or not a regular file)
reading manifest file 'pyzipf.egg-info/SOURCES.txt'
writing manifest file 'pyzipf.egg-info/SOURCES.txt'
running check
warning: check: missing required meta-data: url

warning: check: missing meta-data: if 'author' supplied,
                'author_email' must be supplied too

creating pyzipf-0.1
creating pyzipf-0.1/pyzipf
creating pyzipf-0.1/pyzipf.egg-info
copying files to pyzipf-0.1...
copying README.md -> pyzipf-0.1
copying setup.py -> pyzipf-0.1
copying pyzipf/collate.py ->
  pyzipf-0.1/pyzipf
copying pyzipf/countwords.py ->
  pyzipf-0.1/pyzipf
copying pyzipf/plotcounts.py ->
  pyzipf-0.1/pyzipf
copying pyzipf/script_template.py ->
  pyzipf-0.1/pyzipf
copying pyzipf/test_zipfs.py ->
  pyzipf-0.1/pyzipf
copying pyzipf/utilities.py ->
  pyzipf-0.1/pyzipf
copying pyzipf.egg-info/PKG-INFO ->
```

```
  pyzipf-0.1/pyzipf.egg-info
copying pyzipf.egg-info/SOURCES.txt ->
  pyzipf-0.1/pyzipf.egg-info
copying pyzipf.egg-info/dependency_links.txt ->
  pyzipf-0.1/pyzipf.egg-info
copying pyzipf.egg-info/entry_points.txt ->
  pyzipf-0.1/pyzipf.egg-info
copying pyzipf.egg-info/requires.txt ->
  pyzipf-0.1/pyzipf.egg-info
copying pyzipf.egg-info/top_level.txt ->
  pyzipf-0.1/pyzipf.egg-info
Writing pyzipf-0.1/setup.cfg
creating dist
Creating tar archive
removing 'pyzipf-0.1' (and everything under it)
```

This creates a file named `pyzipf-0.1.tar.gz`, located in a new directory in our project, `dist/` (another directory to add to `.gitignore`[8]). These distribution files can now be distributed via PyPI[9], the standard repository for Python packages. Before doing that, though, we can put `pyzipf` on TestPyPI[10], which lets us test the distribution of our package without having things appear in the main PyPI repository. We must have an account, but they are free to create.

The preferred tool for uploading packages to PyPI is called twine[11], which we can install with:

```
(pyzipf)$ pip install twine
```

Following the Python Packaging User Guide[12], we upload our distribution from the `dist/` folder using the `--repository` option to specify the TestPyPI repository:

```
$ twine upload --repository testpypi dist/*
```

```
Uploading distributions to https://test.pypi.org/legacy/
```

[8] https://github.com/github/gitignore
[9] https://pypi.org/
[10] https://test.pypi.org
[11] https://twine.readthedocs.io/en/latest/
[12] https://packaging.python.org/guides/using-testpypi/

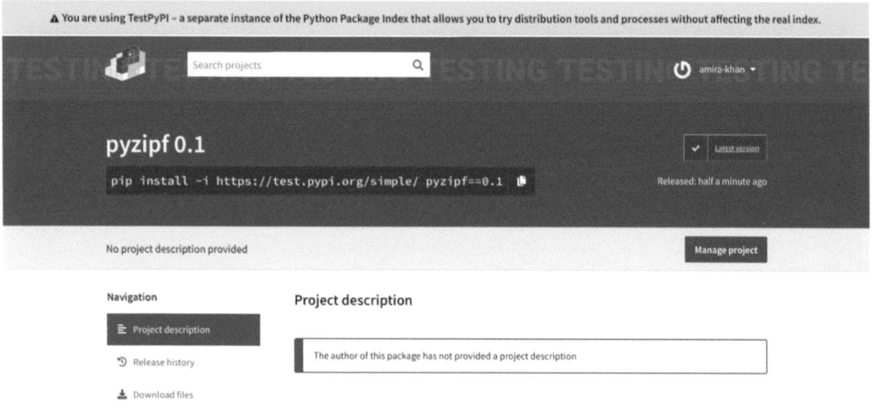

FIGURE 14.1: Our new project on TestPyPI.

```
Enter your username: amira-khan
Enter your password: *********
```

```
Uploading pyzipf-0.1.0.tar.gz
100%|          | 5.59k/5.59k [00:01<00:00, 3.27kB/s]

View at:
https://test.pypi.org/project/pyzipf/0.1/
```

and view the results at the new test project webpage (Figure 14.1). In the exercises, we will explore additional metadata that can be added to `setup.py` so that it appears on the project webpage.

We can test that everything works as expected by creating a virtual environment and installing our package from TestPyPI (the `--extra-index-url` reference to PyPI below accounts for the fact that not all of our package dependencies are available on TestPyPI):

```
(pyzipf)$ conda create -n pyzipf-test pip python=3.7.6
(pyzipf)$ conda activate pyzipf-test
(pyzipf-test)$ pip install --index-url
  https://test.pypi.org/simple
  --extra-index-url https://pypi.org/simple pyzipf
```

```
Looking in indexes: https://test.pypi.org/simple,
                    https://pypi.org/simple
Collecting pyzipf
  Downloading pyzipf-0.1.tar.gz (5.5 kB)
Collecting matplotlib
 Using cached matplotlib-3.3.3-cp37-cp37m-macosx_10_9_x86_64.whl
...collecting other packages...
Building wheels for collected packages: pyzipf
  Building wheel for pyzipf (setup.py) ... done
  Created wheel for pyzipf:
    filename=pyzipf-0.1-py3-none-any.whl
    size=6836
    sha256=62a23715379b71ad5a6b124444fab194
           596d094c7df293c4019d33bdd648aff1
  Stored in directory: /Users/amira/Library/Caches/pip/wheels/
   c6/d6/08/f16cf80ec82a9c70ab8a5d9c8acc7ab35c9a01009539aeb2be
Successfully built pyzipf
Installing collected packages: zipp, typing-extensions, six,
pyparsing, importlib-metadata, toml, pytz, python-dateutil, py,
pluggy, pillow, packaging, numpy, kiwisolver, iniconfig, cycler,
attrs, scipy, pyyaml, pytest, pandas, matplotlib, pyzipf
Successfully installed attrs-20.3.0 cycler-0.10.0
importlib-metadata-3.3.0 iniconfig-1.1.1 kiwisolver-1.3.1
matplotlib-3.3.3 numpy-1.19.4 packaging-20.8 pandas-1.1.5
pillow-8.0.1 pluggy-0.13.1 py-1.10.0 pyparsing-2.4.7
pytest-6.2.1 python-dateutil-2.8.1 pytz-2020.4 pyyaml-5.3.1
pyzipf-0.1 scipy-1.5.4 six-1.15.0 toml-0.10.2
typing-extensions-3.7.4.3 zipp-3.4.0
```

Once again, `pip` takes advantage of the fact that some packages already existing on our system (e.g., they are cached from our previous installs) and doesn't download them again. Once we are happy with our package at TestPyPI, we can go through the same process to put it on the main PyPI[13] repository.

[13] `https://pypi.org/`

Python Wheels

When we installed our package from TestPyPI, the output said that it collected our source distribution and then used it to build a **wheel** for `pyzipf`. This build takes time (especially for large, complex packages), so it can be a good idea for package authors to create and upload wheel files (`.whl`) to PyPI along with the source distribution. `pip` will use the appropriate wheel file if it's available at PyPI instead of building it from the source distribution, which makes the installation process faster and more efficient. Check out the Real Python guide to wheels[14] for details.

`conda` Installation Packages

Given the widespread use of `conda`[15] for package management, it can be a good idea to post a `conda` installation package to Anaconda Cloud[16]. The `conda` documentation has instructions[17] for quickly building a `conda` package for a Python module that is already available on PyPI. See Appendix I for more information about `conda` and Anaconda Cloud.

14.6 Documenting Packages

Now that our package has been distributed, we need to think about whether we have provided sufficient documentation. Docstrings (Section 5.3) and READMEs are sufficient to describe most simple packages, but as our code base grows larger, we will want to complement these manually written sections with automatically generated content, references between functions, and search functionality. For most large Python packages, such documentation is generated using a **documentation generator** called Sphinx[18], which is often used in combination with a free online hosting service called Read the Docs[19]. In this section we will update our README file with some basic package-level documentation, before using Sphinx and Read the Docs to host that information online along with more detailed function-level documentation. For further

[14] `https://realpython.com/python-wheels/`
[15] `https://conda.io/`
[16] `https://anaconda.org/`
[17] `https://docs.conda.io/projects/conda-build/en/latest/user-guide/tutorials/build-pkgs-skeleton.html`
[18] `https://www.sphinx-doc.org/en/master/`
[19] `https://docs.readthedocs.io/en/latest/`

advice on writing documentation for larger and more complex packages, see Appendix G.

14.6.1 Including package-level documentation in the README

When a user first encounters a package, they usually want to know what the package is meant to do, instructions on how to install it, and examples of how to use it. We can include these elements in the `README.md` file we started in Chapter 7. At the moment it reads as follows:

```
$ cat README.md
```

```
# Zipf's Law

These Zipf's Law scripts tally the occurrences of words in text
files and plot each word's rank versus its frequency.

...
```

This file is currently written in Markdown[20], but Sphinx uses a format called **reStructuredText** (reST), so we will switch to that. Like Markdown, reST is a plain-text markup format that can be rendered into HTML or PDF documents with complex indices and cross-links. GitHub recognizes files ending in `.rst` as reST files and displays them nicely, so our first task is to rename our existing file:

```
$ git mv README.md README.rst
```

We then make a few edits to the file formatting: titles are underlined and overlined, section headings are underlined, and code blocks are set off with two colons (`::`) and indented. We can also add some context about why to use the package, as well as updated information about package installation:

```
The ``pyzipf`` package tallies the occurrences of words in text
files and plots each word's rank versus its frequency together
with a line for the theoretical distribution for Zipf's Law.
```

[20]https://en.wikipedia.org/wiki/Markdown

```
Motivation
----------

Zipf's Law is often stated as an observational pattern in the
relationship between the frequency and rank of words in a text:

`"…the most frequent word will occur approximately twice as
often as the second most frequent word,
three times as often as the third most
frequent word, etc."`
- `wikipedia <https://en.wikipedia.org/wiki/Zipf%27s_law>`_

Many books are available to download in plain text format
from sites such as
`Project Gutenberg <https://www.gutenberg.org/>`_,
so we created this package to qualitatively explore how well
different books align with the word frequencies predicted by
Zipf's Law.

Installation
------------

``pip install pyzipf``

Usage
-----

After installing this package, the following three commands will
be available from the command line

- ``countwords`` for counting the occurrences of words in a text
- ``collate`` for collating multiple word count files together
- ``plotcounts`` for visualizing the word counts

A typical usage scenario would include running the following
from your terminal::

    countwords dracula.txt > dracula.csv
    countwords moby_dick.txt > moby_dick.csv
    collate dracula.csv moby_dick.csv > collated.csv
    plotcounts collated.csv --outfile zipf-drac-moby.jpg

Additional information on each function
can be found in their docstrings and appending the ``-h`` flag,
```

```
e.g., ``countwords -h``.

Contributing
------------

Interested in contributing?
Check out the CONTRIBUTING.md
file for guidelines on how to contribute.
Please note that this project is released with a
Contributor Code of Conduct (CONDUCT.md).
By contributing to this project,
you agree to abide by its terms.
Both of these files can be found in our
`GitHub repository. <https://github.com/amira-khan/zipf>`_
```

14.6.2 Creating a web page for documentation

Now that we've added package-level documentation to our README file, we need to think about function-level documentation. We want to provide users with a list of all the functions available in our package along with a short description of what they do and how to use them. We could achieve this by manually cutting and pasting function names and docstrings from our Python modules (i.e., `countwords.py`, `plotcounts.py`, etc.), but that would be a time-consuming process prone to errors as more functions are added over time. Instead, we can use a **documentation generator** called Sphinx[21] that is capable of scanning Python code for function names and docstrings and can export that information to HTML format for hosting on the web.

To start, let's install Sphinx and create a `docs/` directory at the top of our repository:

```
$ pip install sphinx
$ mkdir docs
$ cd docs
```

We can then run Sphinx's `quickstart` tool to create a minimal set of documentation that includes the package-level information in the `README.rst` file we just created and the function-level information in the docstrings we've written along the way. It asks us to specify the project's name, the name of the

[21] `https://www.sphinx-doc.org/en/master/`

project's author, and a release; we can use the default settings for everything else.

```
$ sphinx-quickstart
```

```
Welcome to the Sphinx 3.1.1 quickstart utility.

Please enter values for the following settings (just press Enter
to accept a default value, if one is given in brackets).

Selected root path: .

You have two options for placing the build directory for Sphinx
output. Either, you use a directory "_build" within the root
path, or you separate "source" and "build" directories within
the root path.
```

```
> Separate source and build directories (y/n) [n]: n
```

```
The project name will occur in several places in the built
documentation.
```

```
> Project name: pyzipf
> Author name(s): Amira Khan
> Project release []: 0.1
```

```
If the documents are to be written in a language other than
English, you can select a language here by its language code.
Sphinx will then translate text that it generates into that
language.

For a list of supported codes, see
https://www.sphinx-doc.org/en/master/usage/configuration.html
```

```
> Project language [en]:
```

```
Creating file /Users/amira/pyzipf/docs/conf.py.
Creating file /Users/amira/pyzipf/docs/index.rst.
Creating file /Users/amira/pyzipf/docs/Makefile.
Creating file /Users/amira/pyzipf/docs/make.bat.

Finished: An initial directory structure has been created.

You should now populate your master file
/Users/amira/pyzipf/docs/index.rst and create other documentation
source files. Use the Makefile to build the docs, like so:
   make builder
where "builder" is one of the supported builders, e.g. HTML,
LaTeX or linkcheck.
```

`quickstart` creates a file called `conf.py` in the `docs` directory that configures Sphinx. We will make two changes to that file so that another tool called `autodoc` can find our modules (and their docstrings). The first change relates to the "path setup" section near the head of the file:

```
# -- Path setup -----------------------------------------------

# If extensions (or modules to document with autodoc) are in
# another directory, add these directories to sys.path here. If
# the directory is relative to the documentation root, use
# os.path.abspath to make it absolute, like shown here.
```

Relative to the `docs/` directory, our modules (i.e., `countwords.py`, `utilities.py`, etc.) are located in the `../pyzipf` directory. We therefore need to uncomment the relevant lines of the path setup section in `conf.py` to tell Sphinx where those modules are:

```
import os
import sys
sys.path.insert(0, os.path.abspath('../pyzipf'))
```

We will also change the "general configuration" section to add `autodoc` to the list of Sphinx extensions we want:

```
extensions = ['sphinx.ext.autodoc']
```

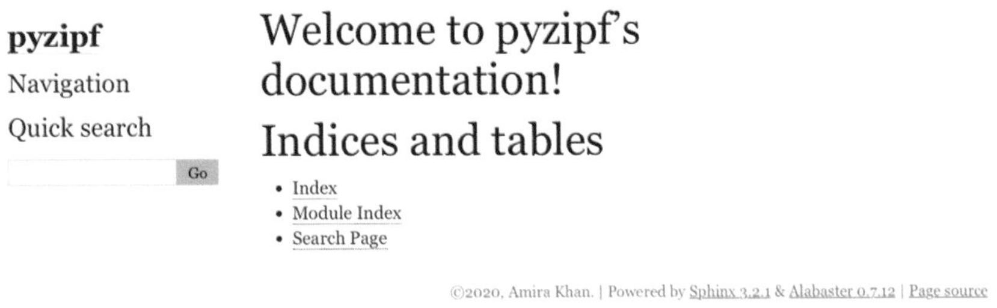

FIGURE 14.2: The default website landing page.

With those edits complete, we can now generate a Sphinx `autodoc` script that generates information about each of our modules and puts it in corresponding `.rst` files in the `docs/source` directory:

```
sphinx-apidoc -o source/ ../pyzipf
```

```
Creating file source/collate.rst.
Creating file source/countwords.rst.
Creating file source/plotcounts.rst.
Creating file source/test_zipfs.rst.
Creating file source/utilities.rst.
Creating file source/modules.rst.
```

At this point, we are ready to generate our webpage. The `docs` sub-directory contains a Makefile that was generated by `sphinx-quickstart`. If we run `make html` and open `docs/_build/index.html` in a web browser, we'll have a landing page with minimal documentation (Figure 14.2). If we click on the `Module Index` link we can access the documentation for the individual modules (Figures 14.3 and 14.4).

The landing page for the website is the perfect place for the content of our README file, so we can add the line `.. include:: ../README.rst` to the `docs/index.rst` file to insert it:

```
Welcome to pyzipf's documentation!
==================================

.. include:: ../README.rst
```

FIGURE 14.3: The module index.

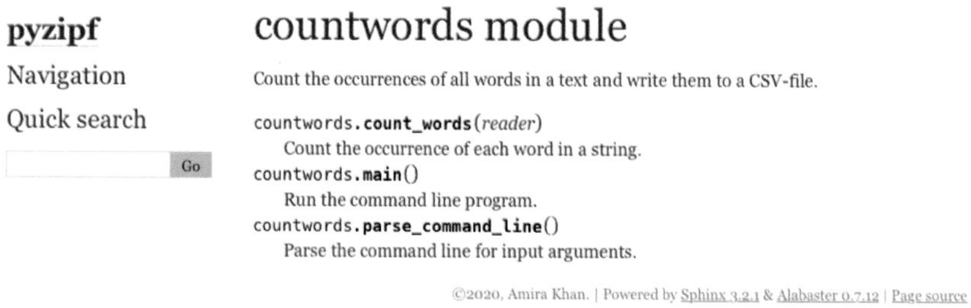

FIGURE 14.4: The countwords documentation.

```
.. toctree::
   :maxdepth: 2
   :caption: Contents:

Indices and tables
==================

* :ref:`genindex`
* :ref:`modindex`
* :ref:`search`
```

pyzipf
Navigation
Quick search
Go
Welcome to pyzipf's documentation!
Zipf's Law
The pyzipf package tallies the occurrences of words in text files and plots each word's rank versus its frequency together with a line for the theoretical distribution for Zipf's Law.
Motivation
Zipf's Law is often stated as an observational pattern seen in the relationship between the frequency and rank of words in a text:
"...the most frequent word will occur approximately twice as often as the second most frequent word, three times as often as the third most frequent word, etc." — wikipedia
Many books are available to download in plain text format from sites such as Project Gutenberg, so we created this package to qualitatively explore how well different books align with the word frequencies predicted by Zipf's Law.

FIGURE 14.5: The new landing page showing the contents of README.rst.

If we re-run `make html`, we now get an updated set of web pages that re-uses our README as the introduction to the documentation (Figure 14.5).

Before going on, note that Sphinx is not included in the installation requirements in `requirements.txt` (Section 11.8). Sphinx isn't needed to run, develop, or even test our package, but it is needed for building the documentation. To note this requirement, but without requiring everyone installing the package to install Sphinx, let's create a `requirements_docs.txt` file that contains this line (where the version number is found by running `pip freeze`):

```
Sphinx>=1.7.4
```

Anyone wanting to build the documentation (including us, on another computer) now only needs run `pip install -r requirements_docs.txt`

14.6.3 Hosting documentation online

Now that we have generated our package documentation, we need to host it online. A common option for Python projects is Read the Docs[22], which is a community-supported site that hosts software documentation free of charge.

Just as continuous integration systems automatically re-test things (Section 11.8), Read the Docs integrates with GitHub so that documentation is automatically re-built every time updates are pushed to the project's GitHub

[22]`https://docs.readthedocs.io/en/latest/`

repository. If we register for Read the Docs with our GitHub account, we can log in at the Read the Docs website and import a project from our GitHub repository. Read the Docs will then build the documentation (using `make html` as we did earlier) and host the resulting files.

For this to work, all of the source files generated by Sphinx need to be checked into your GitHub repository: in our case, this means `docs/source/*.rst`, `docs/Makefile`, `docs/conf.py`, and `docs/index.rst`. We also need to create and save a Read the Docs configuration file[23] in the root directory of our `pyzipf` package:

```
$ cd ~/pyzipf
$ cat .readthedocs.yml
```

```
# .readthedocs.yml
# Read the Docs configuration file
# See https://docs.readthedocs.io/en/stable/config-file/v2.html
# for details

# Required
version: 2

# Build documentation in the docs/ directory with Sphinx
sphinx:
  configuration: docs/conf.py

# Optionally set the version of Python and requirements required
# to build your docs
python:
  version: 3.7
  install:
    - requirements: requirements.txt
```

The configuration file uses the now-familiar **YAML** format (Section 10.1 and Appendix H) to specify the location of the Sphinx configuration script (`docs/conf.py`) and the dependencies for our package (`requirements.txt`).

Amira has gone through this process and the documentation is now available at:

`https://pyzipf.readthedocs.io/`

[23]`https://docs.readthedocs.io/en/stable/config-file/v2.html`

14.7 Software Journals

As a final step to releasing our new package, we might want to give it a **DOI** so that it can be cited by researchers. As we saw in Section 13.2.3, GitHub integrates with Zenodo[24] for precisely this purpose.

While creating a DOI using a site like Zenodo is often the end of the software publishing process, there is the option of publishing a journal paper to describe the software in detail. Some research disciplines have journals devoted to describing particular types of software (e.g., *Geoscientific Model Development*[25]), and there are also a number of generic software journals such as the *Journal of Open Research Software*[26] and the *Journal of Open Source Software*[27]. Packages submitted to these journals are typically assessed against a range of criteria relating to how easy the software is to install and how well it is documented, so the peer review process can be a great way to get critical feedback from people who have seen many research software packages come and go over the years.

Once you have obtained a DOI and possibly published with a software journal, the last step is to tell users how to cite your new software package. This is traditionally done by adding a `CITATION` file to the associated GitHub repository (alongside `README`, `LICENSE`, `CONDUCT` and similar files discussed in Section 1.1.1), containing a plain text citation that can be copied and pasted into email as well as entries formatted for various bibliographic systems like BibTeX[28].

```
$ cat CITATION.md


# Citation

If you use the pyzipf package for work/research presented in a
publication, we ask that you please cite:

Khan A and Virtanen S, 2020. pyzipf: A Python package for word
count analysis. *Journal of Important Software*, 5(51), 2317,
https://doi.org/10.21105/jois.02317
```

[24]https://guides.github.com/activities/citable-code/
[25]https://www.geoscientific-model-development.net/
[26]https://openresearchsoftware.metajnl.com/
[27]https://joss.theoj.org/
[28]http://www.bibtex.org/

```
### BibTeX entry

@article{Khan2020,
    title={pyzipf: A Python package for word count analysis.},
    author={Khan, Amira and Virtanen, Sami},
    journal={Journal of Important Software},
    volume={5},
    number={51},
    eid={2317},
    year={2020},
    doi={10.21105/jois.02317},
    url={https://doi.org/10.21105/jois.02317},
}
```

14.8 Summary

Thousands of people have helped write the software that our Zipf's Law example relies on, but their work is only useful because they packaged it and documented how to use it. Doing this is increasingly recognized as a credit-worthy activity by universities, government labs, and other organizations, particularly for research software engineers. It is also deeply satisfying to make strangers' lives better, if only in small ways.

14.9 Exercises

14.9.1 Package metadata

In a number of places on our TestPyPI webpage, it says that no project description was provided (Figure 14.1). How could we edit our `setup.py` file to include a description? What other metadata would you add?

Hint: The `setup() args` documentation[29] might be useful.

[29] `https://packaging.python.org/guides/distributing-packages-using-setuptools/#setup-args`

14.9.2 Separating requirements

As well as `requirements_docs.txt`, developers often create a `requirements_dev.txt` file to list packages that are not needed by the package's users, but are required for its development and testing. Pull `pytest` out of `requirements.txt` and put it in a new `requirements_dev.txt` file.

14.9.3 Software review

The *Journal of Open Source Software*[30] has a checklist[31] that reviewers must follow when assessing a submitted software paper. Run through the checklist (skipping the criteria related to the software paper) and see how the Zipf's Law package would rate on each criteria.

14.9.4 Packaging quotations

Each chapter in this book opens with a quote from the British author Terry Pratchett. This script `quotes.py` contains a function `random_quote` which prints a random Pratchett quote:

```
import random


quote_list = ["It's still magic even if you know how it's done.",
              "Everything starts somewhere, "\
              "though many physicists disagree.",
              "Ninety percent of most magic merely consists "\
              "of knowing one extra fact.",
              "Wisdom comes from experience. "\
              "Experience is often a result of lack of wisdom.",
              "There isn't a way things should be. "\
              "There's just what happens, and what we do.",
              "Multiple exclamation marks are a sure sign "\
              "of a diseased mind.",
              "+++ Divide By Cucumber Error. "\
              "Please Reinstall Universe And Reboot +++",
              "It's got three keyboards and a hundred extra "\
              "knobs, including twelve with '?' on them.",
             ]
```

[30] https://joss.theoj.org/

[31] https://joss.readthedocs.io/en/latest/review_checklist.html

```
def random_quote():
    """Print a random Pratchett quote."""
    print(random.choice(quote_list))
```

Create a new conda development environment called pratchett and use pip to install a new package called pratchett into that environment. The package should contain quotes.py, and once the package has been installed the user should be able to run:

```
from pratchett import quotes

quotes.random_quote()
```

14.10 Key Points

- Use setuptools[32] to build and distribute Python packages.
- Create a directory named mypackage containing a setup.py script with a subdirectory also called mypackage containing the package's source files.
- Use **semantic versioning** for software releases.
- Use a **virtual environment** to test how your package installs without disrupting your main Python installation.
- Use pip[33] to install Python packages.
- The default repository for Python packages is PyPI[34].
- Use TestPyPI[35] to test the distribution of your package.
- Use a README file for package-level documentation.
- Use Sphinx[36] to generate documentation for a package.
- Use Read the Docs[37] to host package documentation online.
- Create a **DOI** for your package using GitHub's Zenodo integration[38].
- Publish details of your package in a software journal so others can cite it.

[32] https://setuptools.readthedocs.io/
[33] https://pypi.org/project/pip/
[34] https://pypi.org/
[35] https://test.pypi.org
[36] https://www.sphinx-doc.org/en/master/
[37] https://docs.readthedocs.io/en/latest/
[38] https://guides.github.com/activities/citable-code/

15

Finale

We have come a long way since we first met Amira, Jun, and Sami in Section 0.2. Amira now has her scripts, datasets and reports organized, in version control, and on GitHub. Her work has already paid off: a colleague spotted a problem in an analysis step, that impacted five figures in one of her reports. Because Amira has embraced GitHub, the colleague could easily suggest a fix through a pull request. It took Amira some time to regenerate all of the affected figures, because she had to recall which code needed to be rerun. She recognized this as a great reason to embrace Make, and has added implementing it to her to-do list.

She used to be intimidated by the Unix shell, but now Amira finds it an essential part of her everyday work: she uses shell scripts to automate data processing tasks, she runs her own command-line tools she wrote in Python, and issues countless commands to Git. Her comfort with the shell is also helping as she learns how to run her analyses on a remote computing cluster from Sami.

Sami had experience with Git and Make in software projects from their undergraduate studies, but they have a new appreciation for their importance in research projects. They've reworked their Git and Make workshops to use examples of data pipelines, and have been getting rave reviews from participants. Sami has also gained an appreciation for the importance of provenance of both data and code in research. They've been helping their users by suggesting ways they can make their research more accessible, inspectable and reproducible.

Jun is now working on his first Python package. He's added better error handling so he and his users get better information when something goes wrong, he's implemented some testing strategies to give him and his users confidence that his code works, and he's improved his documentation. Jun has also added a license, a Code of Conduct and contributing guidelines to his project repo, and has already had a contribution that fixes some typos in the documentation.

15.1 Why We Wrote This Book

Shell scripts, branching, automated workflows, healthy team dynamics—they all take time to learn, but they enable researchers to get more done in less time and with less pain, and that's why we wrote this book. The climate crisis is the greatest threat our species has faced since civilization began. The COVID-19 pandemic has shown just how ill-prepared we are to deal with problems of that magnitude, or with the suspicion and disinformation that now poisons every public discussion online.

Every hour that a researcher *doesn't* waste wrestling with software is an hour they can spend solving a new problem; every meeting that ends early with a clear decision frees up time to explain to the public what we actually know and why it matters. We don't expect this book to change the world, but we hope that knowing how to write, test, document, package, and share your work will give you a slightly better chance of doing so. We hope you have enjoyed reading what we have written; if so, we would enjoy hearing from you.

- Damien Irving (`https://damienirving.github.io/`)
- Kate Hertweck (`https://katehertweck.com/`)
- Luke Johnston (`https://lukewjohnston.com/`)
- Joel Ostblom (`https://joelostblom.com/`)
- Charlotte Wickham (`https://www.cwick.co.nz/`)
- Greg Wilson (`https://third-bit.com`)

> So much universe, and so little time.
>
> — Terry Pratchett

A

Solutions

The exercises included in this book represent a wide variety of problems, from multiple choice questions to larger coding tasks. It's relatively straightforward to indicate a correct answer for the former, though there may be unanticipated cases in which the specific software version you're using leads to alternative answers being preferable. It's even more difficult to identify the "right" answer for the latter, since there are often many ways to accomplish the same task with code. Here we present possible solutions that the authors generally agree represent "good" code, but we encourage you to explore additional approaches.

Commits noted in a solution reference Amira's `zipf` repository on GitHub[1], which allow you to see the specific lines of a file modified to arrive at the answer.

Chapter 2

Exercise 2.10.1

The `-l` option makes `ls` use a **l**ong listing format, showing not only the file/directory names but also additional information such as the file size and the time of its last modification. If you use both the `-h` option and the `-l` option, this makes the file size "**h**uman readable," i.e., displaying something like `5.3K` instead of `5369`.

Exercise 2.10.2

The command `ls -R -t` results in the contents of each directory sorted by time of last change.

[1] `https://github.com/amira-khan/zipf`

Exercise 2.10.3

1. No: `.` stands for the current directory.
2. No: `/` stands for the root directory.
3. No: Amira's home directory is `/Users/Amira`.
4. No: This goes up two levels, i.e., ends in `/Users`.
5. Yes: `~` stands for the user's home directory, in this case `/Users/amira`.
6. No: This would navigate into a directory `home` in the current directory if it exists.
7. Yes: Starting from the home directory `~`, this command goes into `data` then back (using `..`) to the home directory.
8. Yes: Shortcut to go back to the user's home directory.
9. Yes: Goes up one level.
10. Yes: Same as the previous answer, but with an unnecessary `.` (indicating the current directory).

Exercise 2.10.4

1. No: There *is* a directory `backup` in `/Users`.
2. No: This is the content of `Users/sami/backup`, but with `..` we asked for one level further up.
3. No: Same as previous explanation, but results shown as directories (which is what the `-F` option specifies).
4. Yes: `../backup/` refers to `/Users/backup/`.

Exercise 2.10.5

1. No: `pwd` is not the name of a directory.
2. Yes: `ls` without directory argument lists files and directories in the current directory.
3. Yes: Uses the absolute path explicitly.

Exercise 2.10.6

The `touch` command updates a file's timestamp. If no file exists with the given name, `touch` will create one. Assuming you don't already have `my_file.txt` in your working directory, `touch my_file.txt` will create the file. When you inspect the file with `ls -l`, note that the size of `my_file.txt` is 0 bytes. In other words, it contains no data. If you open `my_file.txt` using your text editor, it is blank.

Some programs do not generate output files themselves, but instead require that empty files have already been generated. When the program is run, it searches for an existing file to populate with its output. The touch command allows you to efficiently generate a blank text file to be used by such programs.

Exercise 2.10.7

```
$ remove my_file.txt? y
```

The `-i` option will prompt before (every) removal (use y to confirm deletion or n to keep the file). The Unix shell doesn't have a trash bin, so all the files removed will disappear forever. By using the `-i` option, we have the chance to check that we are deleting only the files that we want to remove.

Exercise 2.10.8

```
$ mv ../data/chapter1.txt ../data/chapter2.txt .
```

Recall that `..` refers to the parent directory (i.e., one above the current directory) and that `.` refers to the current directory.

Exercise 2.10.9

1. No: While this would create a file with the correct name, the incorrectly named file still exists in the directory and would need to be deleted.
2. Yes: This would work to rename the file.
3. No: The period (`.`) indicates where to move the file, but does not provide a new filename; identical filenames cannot be created.
4. No: The period (`.`) indicates where to copy the file, but does not provide a new filename; identical filenames cannot be created.

Exercise 2.10.10

We start in the `/Users/amira/data` directory, containing a single file, `books.dat`. We create a new folder called `doc` and move (`mv`) the file `books.dat` to that new folder. Then we make a copy (`cp`) of the file we just moved named `books-saved.dat`.

The tricky part here is the location of the copied file. Recall that `..` means "go up a level," so the copied file is now in `/Users/amira`. Notice that `..` is interpreted with respect to the current working directory, **not** with respect to the location of the file being copied. So, the only thing that will show using ls (in `/Users/amira/data`) is the `doc` folder.

1. No: `books-saved.dat` is located at `/Users/amira`
2. Yes.
3. No: `books.dat` is located at `/Users/amira/data/doc`
4. No: `books-saved.dat` is located at `/Users/amira`

Exercise 2.10.11

If given more than one filename followed by a directory name (i.e., the destination directory must be the last argument), `cp` copies the files to the named directory.

If given three filenames, `cp` throws an error because it is expecting a directory name as the last argument.

Exercise 2.10.12

1. Yes: Shows all files whose names contain two different characters (`?`) followed by the letter `n`, then zero or more characters (`*`) followed by `txt`.
2. No: Shows all files whose names start with zero or more characters (`*`) followed by `e_`, zero or more characters (`*`), then `txt`. The output includes the two desired books, but also `time_machine.txt`.
3. No: Shows all files whose names start with zero or more characters (`*`) followed by `n`, zero or more characters (`*`), then `txt`. The output includes the two desired books, but also `frankenstein.txt` and `time_machine.txt`.
4. No: Shows all files whose names start with zero or more characters (`*`) followed by `n`, a single character `?`, `e`, zero or more characters (`*`), then `txt`. The output shows `frankenstein.txt` and `sense_and_sensibility.txt`.

Exercise 2.10.13

```
$ mv *.txt data
```

Amira needs to move her files `books.txt` and `titles.txt` to the `data` directory. The shell will expand `*.txt` to match all `.txt` files in the current directory. The `mv` command then moves the list of `.txt` files to the `data` directory.

Exercise 2.10.14

1. Yes: This accurately re-creates the directory structure.
2. Yes: This accurately re-creates the directory structure.
3. No: The first line of this code set gives an error:

   ```
   mkdir: 2016-05-20/data: No such file or directory
   ```

   `mkdir` won't create a subdirectory for a directory that doesn't yet exist (unless you use an option like `-p` that explicitly creates parent directories).
4. No: This creates `raw` and `processed` directories at the same level as `data`:

   ```
   2016-05-20/
           ├── data
           ├── processed
           └── raw
   ```

Exercise 2.10.15

1. A solution using two wildcard expressions:

   ```
   $ ls s*.txt
   $ ls t*.txt
   ```

2. When there are no files beginning with `s` and ending in `.txt`, or when there are no files beginning with `t` and ending in `.txt`.

Exercise 2.10.16

1. No: This would remove only `.csv` files with one-character names.
2. Yes: This removes only files ending in `.csv`.
3. No: The shell would expand `*` to match everything in the current directory, so the command would try to remove all matched files and an additional file called `.csv`.
4. No: The shell would expand `*.*` to match all files with any extension, so this command would delete all files in the current directory.

Exercise 2.10.17

`novel-????-[ab]*.{txt,pdf}` matches:

- Files whose names start with `novel-`,
- which is then followed by exactly four characters (since each `?` matches one character),
- followed by another literal `-`,
- followed by either the letter `a` or the letter `b`,
- followed by zero or more other characters (the `*`),
- followed by `.txt` or `.pdf`.

Chapter 3

Exercise 3.8.1

`echo hello > testfile01.txt` writes the string "hello" to `testfile01.txt`, but the file gets overwritten each time we run the command.

`echo hello >> testfile02.txt` writes "hello" to `testfile02.txt`, but appends the string to the file if it already exists (i.e., when we run it for the second time).

Exercise 3.8.2

1. No: This results from only running the first line of code (`head`).
2. No: This results from only running the second line of code (`tail`).
3. Yes: The first line writes the first three lines of `dracula.txt`, the second line appends the last two lines of `dracula.txt` to the same file.

4. No: We would need to pipe the commands to obtain this answer (`head -n 3 dracula.txt | tail -n 2 > extracted.txt`).

Exercise 3.8.3

Try running each line of code in the `data` directory.

1. No: This incorrectly uses redirect (`>`), and will result in an error.
2. No: The number of lines desired for `head` is reported incorrectly; this will result in an error.
3. No: This will extract the first three files from the `wc` results, which have not yet been sorted into length of lines.
4. Yes: This output correctly orders and connects each of the commands.

Exercise 3.8.4

To obtain a list of unique results from these data, we need to run:

```
$ sort genres.txt | uniq
```

It makes sense that `uniq` is almost always run after using `sort`, because that allows a computer to compare only adjacent lines. If `uniq` did not compare only adjacent lines, it would require comparing each line to all other lines. For a small set of comparisons, this doesn't matter much, but this isn't always possible for large files.

Exercise 3.8.5

When used on a single file, `cat` prints the contents of that file to the screen. In this case, the contents of `titles.txt` are sent as input to `head -n 5`, so the first five lines of `titles.txt` is output. These five lines are used as the input for `tail -n 3`, which results in lines 3–5 as output. This is used as input to the final command, which sorts them in reverse order. These results are written to the file `final.txt`, the contents of which are:

```
Sense and Sensibility,1811
Moby Dick,1851
Jane Eyre,1847
```

Exercise 3.8.6

`cut` selects substrings from a line by:

- breaking the string into pieces wherever it finds a separator (`-d ,`), which in this case is a comma, and
- keeping one or more of the resulting fields/columns (`-f 2`).

In this case, the output is only the dates from `titles.txt`, since this is in the second column.

```
$ cut -d , -f 2 titles.txt
```

```
1897
1818
1847
1851
1811
1892
1897
1895
1847
```

Exercise 3.8.7

1. No: This sorts by the book title.
2. No: This results in an error because `sort` is being used incorrectly.
3. No: There are duplicate dates in the output because they have not been sorted first.
4. Yes: This results in the output shown below.
5. No: This extracts the desired data (below), but then counts the number of lines, resulting in the incorrect answer.

```
1 1811
1 1818
2 1847
1 1851
1 1892
1 1895
2 1897
```

If you have difficulty understanding the answers above, try running the commands or sub-sections of the pipelines (e.g., the code between pipes).

Exercise 3.8.8

The difference between the versions is whether the code after `echo` is inside quotation marks.

The first version redirects the output from `echo analyze $file` to a file (`analyzed-$file`). This doesn't allow us to preview the commands, but instead creates files (`analyzed-$file`) containing the text `analyze $file`.

The second version will allow us to preview the commands. This prints to screen everything enclosed in the quotation marks, expanding the loop variable name (prefixed with `$`).

Try both versions for yourself to see the output. Be sure to open the `analyzed-*` files to view their contents.

Exercise 3.8.9

The first version gives the same output on each iteration through the loop. Bash expands the wildcard `*.txt` to match all files ending in `.txt` and then lists them using `ls`. The expanded loop would look like this (we'll only show the first two data files):

```
$ for datafile in dracula.txt  frankenstein.txt ...
> do
>   ls dracula.txt  frankenstein.txt ...
```

```
dracula.txt  frankenstein.txt ...
dracula.txt  frankenstein.txt ...
...
```

The second version lists a different file on each loop iteration. The value of the `datafile` variable is evaluated using `$datafile`, and then listed using `ls`.

```
dracula.txt
frankenstein.txt
jane_eyre.txt
moby_dick.txt
sense_and_sensibility.txt
sherlock_holmes.txt
time_machine.txt
```

Exercise 3.8.10

The first version results in only `dracula.txt` output, because it is the only file beginning in "d."

The second version results in the following, because these files all contain a "d" with zero or more characters before and after:

```
README.md
dracula.txt
moby_dick.txt
sense_and_sensibility.txt
```

Exercise 3.8.11

Both versions write the first 16 lines (`head -n 16`) of each book to a file (`headers.txt`).

The first version results in the text from each file being overwritten in each iteration because of use of `>` as a redirect.

The second version uses `>>`, which appends the lines to the existing file. This is preferable because the final `headers.txt` includes the first 16 lines from all files.

Exercise 3.8.12

If a command causes something to crash or hang, it might be useful to know what that command was, in order to investigate the problem. Were the command only be recorded after running it, we would not have a record of the last command run in the event of a crash.

Chapter 4

Exercise 4.8.1

```
$ cd ~/zipf
```

Change into the zipf directory, which is located in the home directory (designated by ~).

```
$ for file in $(find . -name "*.bak")
> do
>   rm $file
> done
```

Find all the files ending in .bak and remove them one by one.

```
$ rm bin/summarize_all_books.sh
```

Remove the summarize_all_books.sh script.

```
$ rm -r results
```

Recursively remove each file in the results directory and then remove the directory itself. (It is necessary to remove all the files first because you cannot remove a non-empty directory.)

Exercise 4.8.2

Running this script with the given parameters will print the first and last line from each file in the directory ending in .txt.

1. No: This answer misinterprets the lines printed.
2. Yes.
3. No: This answer includes the wrong files.
4. No: Leaving off the quotation marks would result in an error.

Exercise 4.8.3

One possible script (longest.sh) to accomplish this task:

```
# Shell script which takes two arguments:
#    1. a directory name
#    2. a file extension
# and prints the name of the file in that directory
# with the most lines which matches the file extension.
#
# Usage: bash longest.sh directory/ txt

wc -l $1/*.$2 | sort -n | tail -n 2 | head -n 1
```

Exercise 4.8.4

1. `script1.sh` will print the names of all files in the directory on a single line, e.g., `README.md dracula.txt frankenstein.txt jane_eyre.txt moby_dick.txt script1.sh sense_and_sensibility.txt sherlock_holmes.txt time_machine.txt`. Although `*.txt` is included when running the script, the commands run by the script do not reference `$1`.
2. `script2.sh` will print the contents of the first three files ending in `.txt`; the three variables (`$1`, `$2`, `$3`) refer to the first, second, and third argument entered after the script, respectively.
3. `script3.sh` will print the name of each file ending in `.txt`, since `$@` refers to *all* the arguments (e.g., filenames) given to a shell script. The list of files would be followed by `.txt`: `dracula.txt frankenstein.txt jane_eyre.txt moby_dick.txt sense_and_sensibility.txt sherlock_holmes.txt time_machine.txt.txt`.

Exercise 4.8.5

1. No: This command extracts any line containing "he," either as a word or within a word.
2. No: This results in the same output as the answer for #1. `-E` allows the search term to represent an extended regular expression, but the search term is simple enough that it doesn't make a difference in the result.
3. Yes: `-w` means to return only matches for the word "he."
4. No: `-i` means to invert the search result; this would return all lines *except* the one we desire.

Exercise 4.8.6

```
# Obtain unique years from multiple comma-delimited
# lists of titles and publication years
#
# Usage: bash year.sh file1.txt file2.txt ...

for filename in $*
do
  cut -d , -f 2 $filename | sort -n | uniq
done
```

Exercise 4.8.7

One possible solution:

```
for sister in Elinor Marianne
do
    echo $sister:
    grep -o -w $sister sense_and_sensibility.txt | wc -l
done
```

The `-o` option prints only the matching part of a line.

An alternative (but possibly less accurate) solution is:

```
for sister in Elinor Marianne
do
    echo $sister:
    grep -o -c -w $sister sense_and_sensibility.txt
done
```

This solution is potentially less accurate because `grep -c` only reports the number of lines matched. The total number of matches reported by this method will be lower if there is more than one match per line.

Exercise 4.8.8

1. Yes: This returns `data/jane_eyre.txt`.

2. Maybe: This option may work on your computer, but may not behave consistently across all shells because expansion of the wildcard (`*e.txt`) may prevent piping from working correctly. We recommend enclosing `*e.txt` in quotation marks, as in answer 1.
3. No: This searches the contents of files for lines matching "machine," rather than the filenames.
4. See above.

Exercise 4.8.9

1. Find all files with a `.dat` extension recursively from the current directory.
2. Count the number of lines each of these files contains.
3. Sort the output from step 2 numerically.

Exercise 4.8.10

The following command works if your working directory is `Desktop/` and you replace "username" with that of your current computer. `-mtime` needs to be negative because it is referencing a day prior to the current date.

```
$ find . -type f -mtime -1 -user username
```

Chapter 5

Exercise 5.11.1

Running a Python statement directly from the command line is useful as a basic calculator and for simple string operations, since these commands occur in one line of code. More complicated commands will require multiple statements; when run using `python -c`, statements must be separated by semi-colons:

```
$ python -c "import math; print(math.log(123))"
```

Multiple statements, therefore, quickly become more troublesome to run in this manner.

Exercise 5.11.2

The `my_ls.py` script could read as follows:

```
"""List the files in a given directory with a given suffix."""

import argparse
import glob


def main(args):
    """Run the program."""
    dir = args.dir if args.dir[-1] == '/' else args.dir + '/'
    glob_input = dir + '*.' + args.suffix
    glob_output = sorted(glob.glob(glob_input))
    for item in glob_output:
        print(item)


if __name__ == '__main__':
    parser = argparse.ArgumentParser(description=__doc__)
    parser.add_argument('dir', type=str, help='Directory')
    parser.add_argument('suffix', type=str,
                        help='File suffix (e.g. py, sh)')
    args = parser.parse_args()
    main(args)
```

Exercise 5.11.3

The `sentence_endings.py` script could read as follows:

```
"""Count the occurrence of different sentence endings."""

import argparse


def main(args):
    """Run the command line program."""
    text = args.infile.read()
    for ending in ['.', '?', '!']:
        count = text.count(ending)
```

```
        print(f'Number of {ending} is {count}')


if __name__ == '__main__':
    parser = argparse.ArgumentParser(description=__doc__)
    parser.add_argument('infile', type=argparse.FileType('r'),
                        nargs='?', default='-',
                        help='Input file name')
    args = parser.parse_args()
    main(args)
```

Exercise 5.11.4

While there may be other ways for `plotcounts.py` to meet the requirements of the exercise, we'll be using this script in subsequent chapters so we recommend that the script reads as follows:

```
"""Plot word counts."""

import argparse

import pandas as pd


def main(args):
    """Run the command line program."""
    df = pd.read_csv(args.infile, header=None,
                     names=('word', 'word_frequency'))
    df['rank'] = df['word_frequency'].rank(ascending=False,
                                           method='max')
    df['inverse_rank'] = 1 / df['rank']
    ax = df.plot.scatter(x='word_frequency',
                         y='inverse_rank',
                         figsize=[12, 6],
                         grid=True,
                         xlim=args.xlim)
    ax.figure.savefig(args.outfile)


if __name__ == '__main__':
    parser = argparse.ArgumentParser(description=__doc__)
```

```
    parser.add_argument('infile', type=argparse.FileType('r'),
                        nargs='?', default='-',
                        help='Word count csv file name')
    parser.add_argument('--outfile', type=str,
                        default='plotcounts.png',
                        help='Output image file name')
    parser.add_argument('--xlim', type=float, nargs=2,
                        metavar=('XMIN', 'XMAX'),
                        default=None, help='X-axis limits')
    args = parser.parse_args()
    main(args)
```

Chapter 6

Exercise 6.11.1

Amira does not need to make the heaps-law subdirectory a Git repository because the zipf repository will track everything inside it regardless of how deeply nested.

Amira *shouldn't* run git init in heaps-law because nested Git repositories can interfere with each other. If someone commits something in the inner repository, Git will not know whether to record the changes in that repository, the outer one, or both.

Exercise 6.11.2

git status now shows:

```
On branch master
Untracked files:
  (use "git add <file>..." to include in what will be committed)
    example.txt

nothing added to commit but untracked files present
(use "git add" to track)
```

Nothing has happened to the file; it still exists but Git no longer has it in the staging area. `git rm --cached` is equivalent to `git restore --staged`. With newer versions of Git, older commands will still work, and you may encounter references to them when reading help documentation. If you created this file in your `zipf` project, we recommend removing it before proceeding.

Exercise 6.11.3

If we make a few changes to `.gitignore` such that it now reads:

```
__pycache__ this is a change

this is another change
```

then `git diff` would show:

```
diff --git a/.gitignore b/.gitignore
index bee8a64..5c83419 100644
--- a/.gitignore
+++ b/.gitignore
@@ -1 +1,3 @@
-__pycache__
+__pycache__ this is a change
+
+this is another change
```

Whereas `git diff --word-diff` shows:

```
diff --git a/.gitignore b/.gitignore
index bee8a64..5c83419 100644
--- a/.gitignore
+++ b/.gitignore
@@ -1 +1,3 @@
__pycache__ {+this is a change+}

{+this is another change+}
```

Depending on the nature of the changes you are viewing, the latter may be easier to interpret since it shows exactly what has been changed.

Exercise 6.11.4

1. Maybe: would only create a commit if the file has already been staged.
2. No: would try to create a new repository, which results in an error if a repository already exists.
3. Yes: first adds the file to the staging area, then commits.
4. No: would result in an error, as it would try to commit a file "my recent changes" with the message "myfile.txt."

Exercise 6.11.5

1. Go into your home directory with `cd ~`.
2. Create a new folder called `bio` with `mkdir bio`.
3. Make the repository your working directory with `cd bio`.
4. Turn it into a repository with `git init`.
5. Create your biography using `nano` or another text editor.
6. Add it and commit it in a single step with `git commit -a -m "Some message"`.
7. Modify the file.
8. Use `git diff` to see the differences.

Exercise 6.11.6

1. Create `employment.txt` using an editor like Nano.
2. Add both `me.txt` and `employment.txt` to the staging area with `git add *.txt`.
3. Check that both files are there with `git status`.
4. Commit both files at once with `git commit`.

Exercise 6.11.7

GitHub displays timestamps in a human-readable relative format (i.e., "22 hours ago" or "three weeks ago"), since this makes it easy for anyone in any time zone to know what changes have been made recently. However, if we hover over the timestamp we can see the exact time at which the last change to the file occurred.

Exercise 6.11.8

The answer is 1.

The command `git add motivation.txt` adds the current version of `motivation.txt` to the staging area. The changes to the file from the second `echo` command are only applied to the working copy, not the version in the staging area.

As a result, when `git commit -m "Motivate project"` is executed, the version of `motivation.txt` committed to the repository is the content from the first `echo`.

However, the working copy still has the output from the second `echo`; `git status` would show that the file is modified. `git restore HEAD motivation.txt` therefore replaces the working copy with the most recently committed version of `motivation.txt` (the content of the first `echo`), so `cat motivation.txt` prints:

```
Zipf's Law describes the relationship between the frequency and
rarity of words.
```

Exercise 6.11.9

Add this line to `.gitignore`:

```
results/plots/
```

Exercise 6.11.10

Add the following two lines to `.gitignore`:

```
*.dat           # ignore all data files
!final.dat      # except final.data
```

The exclamation point `!` includes a previously excluded entry.

Note also that if we have previously committed `.dat` files in this repository, they will not be ignored once these rules are added to `.gitignore`. Only future `.dat` files will be ignored.

Exercise 6.11.11

The left button (with the picture of a clipboard) copies the full identifier of the commit to the clipboard. In the shell, `git log` shows the full commit identifier for each commit.

The middle button (with seven letters and numbers) shows all of the changes that were made in that particular commit; green shaded lines indicate additions and red lines indicate removals. We can show the same thing in the shell using `git diff` or `git diff FROM..TO` (where `FROM` and `TO` are commit identifiers).

The right button lets us view all of the files in the repository at the time of that commit. To do this in the shell, we would need to check out the repository as it was at that commit using `git checkout ID`, where `ID` is the tag, branch name, or commit identifier. If we do this, we need to remember to put the repository back to the right state afterward.

Exercise 6.11.12

Committing updates our local repository. Pushing sends any commits we have made locally that aren't yet in the remote repository to the remote repository.

Exercise 6.11.13

When GitHub creates a `README.md` file while setting up a new repository, it actually creates the repository and then commits the `README.md` file. When we try to pull from the remote repository to our local repository, Git detects that their histories do not share a common origin and refuses to merge them.

```
$ git pull origin master
```

```
warning: no common commits
remote: Enumerating objects: 3, done.
remote: Counting objects: 100% (3/3), done.
remote: Total 3 (delta 0), reused 0 (delta 0), pack-reused 0
Unpacking objects: 100% (3/3), done.
From https://github.com/frances/eniac
 * branch            master     -> FETCH_HEAD
 * [new branch]      master     -> origin/master
fatal: refusing to merge unrelated histories
```

We can force Git to merge the two repositories with the option `--allow-unrelated-histories`. Please check the contents of the local and remote repositories carefully before doing this.

Exercise 6.11.14

The `checkout` command restores files from the repository, overwriting the files in our working directory. `HEAD` indicates the latest version.

1. No: this can be dangerous; without a filename, `git checkout` will restore all files in the current directory (and all directories below it) to their state at the commit specified. This command will restore `data_cruncher.sh` to the latest commit version, but will also reset any other files we have changed to that version, which will erase any unsaved changes you may have made to those files.
2. Yes: this restores the latest version of only the desired file.
3. No: this gets the version of `data_cruncher.sh` from the commit before `HEAD`, which is not what we want.
4. Yes: the unique ID (identifier) of the last commit is what `HEAD` means.
5. Yes: this is equivalent to the answer to 2.
6. No: `git restore` assumes `HEAD`, so Git will assume you're trying to restore a file called `HEAD`, resulting in an error.

Exercise 6.11.15

1. Compares what has changed between the current `bin/plotcounts.py` and the same file nine commits ago.
2. It returns an error: `fatal: ambiguous argument 'HEAD~9': unknown revision or path not in the working tree.` We don't have enough commits in history for the command to properly execute.
3. It compares changes (either staged or unstaged) to the most recent commit.

Exercise 6.11.16

No, using `git checkout` on a staged file does not unstage it. The changes are in the staging area and checkout would affect the working directory.

Exercise 6.11.17

Each line of output corresponds to a line in the file, and includes the commit identifier, who last modified the line, when that change was made, and what is included on that line. Note that the edit you just committed is not present here; `git blame` only shows the current lines in the file, and doesn't report on lines that have been removed.

Chapter 7

Exercise 7.12.1

1. `--oneline` shows each commit on a single line with the **short identifier** at the start and the title of the commit beside it. `-n NUMBER` limits the number of commits to show.
2. `--since` and `--after` can be used to show commits in a range of dates or times; `--author` can be used to show commits by a particular person; and `-w` tells Git to ignore whitespace when comparing commits.

Exercise 7.12.2

An online search for "show Git branch in Bash prompt" turns up several approaches, one of the simplest of which is to add this line to our `~/.bashrc` file:

```
export PS1="\\w + \$(git branch 2>/dev/null | grep '^*' |
colrm 1 2) \$ "
```

Breaking it down:

1. Setting the `PS1` variable defines the primary shell **prompt**.
2. `\\w` in a shell prompt string means "the current directory."
3. The `+` is a literal `+` sign between the current directory and the Git branch name.
4. The command that gets the name of the current Git branch is in `$(...)`. (We need to escape the `$` as `\$` so Bash doesn't just run it once when defining the string.)
5. The `git branch` command shows *all* the branches, so we pipe that to `grep` and select the one marked with a `*`.
6. Finally, we remove the first column (i.e., the one containing the `*`) to leave just the branch name.

So what's `2>/dev/null` about? That redirects any error messages to `/dev/null`, a special "file" that consumes input without saving it. We need that because sometimes we will be in a directory that isn't inside a Git repository, and we don't want error messages showing up in our shell prompt.

None of this is obvious, and we didn't figure it out ourselves. Instead, we did a search and pasted various answers into explainshell.com[2] until we had something we understood and trusted.

Exercise 7.12.3

`https://github.com/github/gitignore/blob/master/Python.gitignore` ignores 76 files or patterns. Of those, we recognized less than half. Searching online for some of these, like `"*.pot file"`, turns up useful explanations. Searching for others like `var/` does not; in that case, we have to look at the category (in this case, "Python distribution") and set aside time to do more reading.

Exercise 7.12.4

1. `git diff master..same` does not print anything because there are no differences between the two branches.
2. `git merge same master` prints `merging` because Git combines histories even when the files themselves do not differ. After running this command, `git history` shows a commit for the merge.

Exercise 7.12.5

1. Git refuses to delete a branch with unmerged commits because it doesn't want to destroy our work.
2. Using the `-D` (capital-D) option to `git branch` will delete the branch anyway. This is dangerous because any content that exists only in that branch will be lost.
3. Even with `-D`, `git branch` will not delete the branch we are currently on.

[2] `http://explainshell.com`

Exercise 7.12.6

1. Chartreuse has repositories on GitHub and their desktop containing identical copies of `README.md` and nothing else.
2. Fuchsia has repositories on GitHub and their desktop with exactly the same content as Chartreuse's repositories.
3. `fuchsia.txt` is in both of Fuchsia's repositories but not in Chartreuse's repositories.
4. `fuchsia.txt` is still in both of Fuchsia's repositories but still not in Chartreuse's repositories.
5. `chartreuse.txt` is in both of Chartreuse's repositories but not yet in either of Fuchsia's repositories.
6. `chartreuse.txt` is in Fuchsia's desktop repository but not yet in their GitHub repository.
7. `chartreuse.txt` is in both of Fuchsia's repositories.
8. `fuchsia.txt` is in Chartreuse's GitHub repository but not in their desktop repository.
9. All four repositories contain both `fuchsia.txt` and `chartreuse.txt`.

Chapter 8

Exercise 8.14.1

- Our license is at `https://github.com/merely-useful/py-rse/blob/book/LICENSE.md`.
- Our contribution guidelines are at `https://github.com/merely-useful/py-rse/blob/book/CONTRIBUTING.md`.

Exercise 8.14.2

The `CONDUCT.md` file should have contents that mimic those given in Section 8.3.

Exercise 8.14.3

The newly created `LICENSE.md` should look something like the example MIT License shown in Section 8.4.1.

Exercise 8.14.4

The text in the `README.md` might look something like:

```
## Contributing

Interested in contributing?
Check out the [CONTRIBUTING.md](CONTRIBUTING.md)
file for guidelines on how to contribute.
Please note that this project is released with a
[Contributor Code of Conduct](CONDUCT.md).
By contributing to this project,
you agree to abide by its terms.
```

Your `CONTRIBUTING.md` file might look something like the following:

```
# Contributing

Thank you for your interest
in contributing to the Zipf's Law package!

If you are new to the package and/or
collaborative code development on GitHub,
feel free to discuss any suggested changes via issue or email.
We can then walk you through the pull request process if need be.
As the project grows,
we intend to develop more detailed guidelines for submitting
bug reports and feature requests.

We also have a code of conduct
(see [`CONDUCT.md`](CONDUCT.md)).
Please follow it in all your interactions with the project.
```

Exercise 8.14.5

Be sure to tag the new issue as a feature request to help **triage**.

Exercise 8.14.6

We often delete the `duplicate` label: when we mark an issue that way, we (almost) always add a comment saying which issue it's a duplicate *of*, in which case it's just as sensible to label the issue `wontfix`.

Exercise 8.14.7

Some solutions could be:

- Give the team member their own office space so they don't distract others.
- Buy noise-cancelling headphones for the employees that find it distracting.
- Re-arrange the work spaces so that there is a "quiet office" and a regular office space and have the team member with the attention disorder work in the regular office.

Exercise 8.14.8

Possible solutions:

- Change the rule so that anyone who contributes to the project, in any way, gets included as a co-author.
- Update the rule to include a contributor list on all projects with descriptions of duties, roles, and tasks the contributor provided for the project.

Exercise 8.14.9

We obviously can't say which description fits you best, but:

- Use **three sticky notes** and **interruption bingo** to stop *Anna* from cutting people off.

- Tell *Bao* that the devil doesn't need more advocates, and that he's only allowed one "but what about" at a time.

- *Hediyeh*'s lack of self-confidence will take a long time to remedy. Keeping a list of the times she's been right and reminding her of them frequently is a start, but the real fix is to create and maintain a supportive environment.

- Unmasking *Kenny*'s hitchhiking will feel like nit-picking, but so does the accounting required to pin down other forms of fraud. The most important thing is to have the discussion in the open so that everyone realizes he's taking credit for everyone else's work as well as theirs.

- *Melissa* needs a running partner—someone to work beside her so that she starts when she should and doesn't get distracted. If that doesn't work, the project may need to assign everything mission-critical to someone else (which will probably lead to her leaving).

- *Petra* can be managed with a one-for-one rule: each time she builds or fixes something that someone else needs, she can then work on something she thinks is cool. However, she's only allowed to add whatever it is to the project if someone else will publicly commit to maintaining it.
- Get *Frank* and *Raj* off your project as quickly as you can.

Chapter 9

Exercise 9.11.1

`make -n target` will show commands without running them.

Exercise 9.11.2

1. The `-B` option rebuilds everything, even files that aren't out of date.
2. The `-C` option tells Make to change directories before executing, so that `make -C ~/myproject` runs Make in `~/myproject` regardless of the directory it is invoked from.
3. By default, Make looks for (and runs) a file called `Makefile` or `makefile`. If you use another name for your Makefile (which is necessary if you have multiple Makefiles in the same directory), then you need to specify the name of that Makefile using the `-f` option.

Exercise 9.11.3

`mkdir -p some/path` makes one or more nested directories if they don't exist, and does nothing (without complaining) if they already exist. It is useful for creating the output directories for build rules.

Exercise 9.11.4

The build rule for generated the result for any book should now be:

```
## results/%.csv : regenerate result for any book.
results/%.csv : data/%.txt $(COUNT)
```

```
    @bash $(SUMMARY) $< Title
    @bash $(SUMMARY) $< Author
    python $(COUNT) $< > $@
```

where `SUMMARY` is defined earlier in the `Makefile` as

```
SUMMARY=bin/book_summary.sh
```

and the settings build rule now includes:

```
@echo SUMMARY: $(SUMMARY)
```

Exercise 9.11.5

Since we already have a variable `RESULTS` that contains all of the results files, all we need is a phony target that depends on them:

```
.PHONY: results # and all the other phony targets

## results : regenerate result for all books.
results : ${RESULTS}
```

Exercise 9.11.6

If we use a shell **wildcard** in a rule like this:

```
results/collated.csv : results/*.csv
    python $(COLLATE) $^ > $@
```

then if `results/collated.csv` already exists, the rule tells Make that the file depends on itself.

Exercise 9.11.7

Our rule is:

```
help :
	@grep -h -E '^##' ${MAKEFILE_LIST} | sed -e 's/## //g' \
	| column -t -s ':'
```

- The `-h` option to `grep` tells it *not* to print filenames, while the `-E` option tells it to interpret `^##` as a pattern.
- `MAKEFILE_LIST` is an automatically defined variable with the names of all the Makefiles in play. (There might be more than one because Makefiles can include other Makefiles.)
- `sed` can be used to do string substitution.
- `column` formats text nicely in columns.

Exercise 9.11.8

This strategy would be advantageous if in the future we intended to write a number of different Makefiles that all use the `countwords.py`, `collate.py` and `plotcounts.py` scripts.

We discuss configuration strategies in more detail in Chapter 10.

Chapter 10

Exercise 10.8.1

The build rule involving `plotcounts.py` should now read:

```
## results/collated.png: plot the collated results.
results/collated.png : results/collated.csv $(PARAMS)
	python $(PLOT) $< --outfile $@ --plotparams $(word 2,$^)
```

where `PARAMS` is defined earlier in the `Makefile` along with all the other variables and also included later in the settings build rule:

```
COUNT=bin/countwords.py
COLLATE=bin/collate.py
PARAMS=bin/plotparams.yml
PLOT=bin/plotcounts.py
SUMMARY=bin/book_summary.sh
DATA=$(wildcard data/*.txt)
RESULTS=$(patsubst data/%.txt,results/%.csv,$(DATA))
```

```
## settings : show variables' values.
settings :
	@echo COUNT: $(COUNT)
	@echo DATA: $(DATA)
	@echo RESULTS: $(RESULTS)
	@echo COLLATE: $(COLLATE)
	@echo PARAMS: $(PARAMS)
	@echo PLOT: $(PLOT)
	@echo SUMMARY: $(SUMMARY)
```

Exercise 10.8.2

1. Make the following additions to `plotcounts.py`:

```
import matplotlib.pyplot as plt
```

```
parser.add_argument('--style', type=str,
                    choices=plt.style.available,
                    default=None, help='matplotlib style')
```

```
def main(args):
    """Run the command line program."""
    if args.style:
        plt.style.use(args.style)
```

3. Add `nargs='*'` to the definition of the `--style` option:

```
parser.add_argument('--style', type=str, nargs='*',
                    choices=plt.style.available,
                    default=None, help='matplotlib style')
```

Exercise 10.8.3

The first step is to add a new command-line argument to tell `plotcount.py` what we want to do:

```
if __name__ == '__main__':
    parser = argparse.ArgumentParser(description=__doc__)
    # ...other options as before...
    parser.add_argument('--saveconfig', type=str, default=None,
                        help='Save configuration to file')
    args = parser.parse_args()
    main(args)
```

Next, we add three lines to `main` to act on this option *after* all of the plotting parameters have been set. For now we use `return` to exit from `main` as soon as the parameters have been saved; this lets us test our change without overwriting any of our actual plots.

```
def save_configuration(fname, params):
    """Save configuration to a file."""
    with open(fname, 'w') as reader:
        yaml.dump(params, reader)


def main(args):
    """Run the command line program."""
    if args.style:
        plt.style.use(args.style)
    set_plot_params(args.plotparams)
    if args.saveconfig:
        save_configuration(args.saveconfig, mpl.rcParams)
        return
    df = pd.read_csv(args.infile, header=None,
                     names=('word', 'word_frequency'))
    # ...carry on producing plot...
```

Finally, we add a target to `Makefile` to try out our change. We do the test this way so that we can be sure that we're testing with the same options we use with the real program; if we were to type in the whole command ourselves, we might use something different. We also save the configuration to `/tmp` rather than to our project directory to keep it out of version control's way:

```
## test-saveconfig : save plot configuration.
test-saveconfig :
	python $(PLOT) --saveconfig /tmp/test-saveconfig.yml \
	  --plotparams $(PARAMS)
```

The output is over 400 lines long, and includes settings for everything from the animation bit rate to the size of y-axis ticks:

```
!!python/object/new:matplotlib.RcParams
dictitems:
  _internal.classic_mode: false
  agg.path.chunksize: 0
  animation.avconv_args: []
  animation.avconv_path: avconv
  animation.bitrate: -1
  ...
  ytick.minor.size: 2.0
  ytick.minor.visible: false
  ytick.minor.width: 0.6
  ytick.right: false
```

The beautiful thing about this file is that the entries are automatically sorted alphabetically, which makes it easy for both human beings and the `diff` command to spot differences. This helps reproducibility because any one of these settings might change in a new release of `matplotlib`, and any of those changes might affect our plots. Saving the settings allows us to compare what we had when we did our work to what we have when we're trying to re-create it, which in turn gives us a starting point for debugging if we need to.

Exercise 10.8.4

```
import configparser


def set_plot_params(param_file):
    """Set the matplotlib parameters."""
    if param_file:
        config = configparser.ConfigParser()
        config.read(param_file)
        for section in config.sections():
            for param in config[section]:
                value = config[section][param]
                mpl.rcParams[param] = value
```

1. Most people seem to find Windows INI files easier to write and read, since it's easier to see what's a heading and what's a value.
2. However, Windows INI files only provide one level of sectioning, so complex configurations are harder to express. Thus, while YAML may be a bit more difficult to get started with, it will take us further.

Exercise 10.8.5

The answer depends on whether we are able to make changes to Program A and Program B. If we can, we can modify them to use overlay configuration and put the shared parameters in a single file that both programs load. If we can't do that, the next best thing is to create a small helper program that reads their configuration files and checks that common parameters have consistent values. The first solution prevents the problem; the second detects it, which is a lot better than nothing.

Chapter 11

Exercise 11.11.1

- The first assertion checks that the input sequence `values` is not empty. An empty sequence such as `[]` will make it fail.

- The second assertion checks that each value in the list can be turned into an integer. Input such as `[1, 2,'c', 3]` will make it fail.
- The third assertion checks that the total of the list is greater than 0. Input such as `[-10, 2, 3]` will make it fail.

Exercise 11.11.2

1. Remove the comments about inserting preconditions and add the following:

```
assert len(rect) == 4, 'Rectangles must contain 4 coordinates'
x0, y0, x1, y1 = rect
assert x0 < x1, 'Invalid X coordinates'
assert y0 < y1, 'Invalid Y coordinates'
```

2. Remove the comment about inserting postconditions and add the following

```
assert 0 < upper_x <= 1.0, \
  'Calculated upper X coordinate invalid'
assert 0 < upper_y <= 1.0, \
  'Calculated upper Y coordinate invalid'
```

3. The problem is that the following section of `normalize_rectangle` should read `float(dy) / dx`, not `float(dx) / dy`.

```
if dx > dy:
    scaled = float(dx) / dy
```

4. `test_geometry.py` should read as follows:

```
import geometry
```

```python
def test_tall_skinny():
    """Test normalization of a tall, skinny rectangle."""
    rect = [20, 15, 30, 20]
    expected_result = (0, 0, 1.0, 0.5)
    actual_result = geometry.normalize_rectangle(rect)
    assert actual_result == expected_result
```

5. Other tests might include (but are not limited to):

```python
def test_short_wide():
    """Test normalization of a short, wide rectangle."""
    rect = [2, 5, 3, 10]
    expected_result = (0, 0, 0.2, 1.0)
    actual_result = geometry.normalize_rectangle(rect)
    assert actual_result == expected_result


def test_negative_coordinates():
    """Test rectangle normalization with negative coords."""
    rect = [-2, 5, -1, 10]
    expected_result = (0, 0, 0.2, 1.0)
    actual_result = geometry.normalize_rectangle(rect)
    assert actual_result == expected_result
```

Exercise 11.11.3

There are three approaches to testing when pseudo-random numbers are involved:

1. Run the function once with a known **seed**, check and record its output, and then compare the output of subsequent runs to that saved output. (Basically, if the function does the same thing it did the first time, we trust it.)
2. Replace the **pseudo-random number generator** with a function of our own that generates a predictable series of values. For example, if we are randomly partitioning a list into two equal halves, we could instead use a function that puts odd-numbered values in one partition and even-numbered values in another (which is a legal but unlikely outcome of truly random partitioning).

3. Instead of checking for an exact result, check that the result lies within certain bounds, just as we would with the result of a physical experiment.

Exercise 11.11.4

This result seems counter-intuitive to many people because relative error is a measure of a single value, but in this case we are looking at a distribution of values: each result is off by 0.1 compared to a range of 0–2, which doesn't "feel" infinite. In this case, a better measure might be the largest **absolute error** divided by the standard deviation of the data.

Chapter 12

Exercise 12.6.1

Add a new command-line argument to `collate.py`:

```
parser.add_argument('-v', '--verbose',
                    action="store_true", default=False,
                    help="Set logging level to DEBUG")
```

and two new lines to the beginning of the `main` function:

```
log_level = logging.DEBUG if args.verbose else logging.WARNING
logging.basicConfig(level=log_level)
```

such that the full `collate.py` script now reads as follows:

```
"""
Combine multiple word count CSV-files
into a single cumulative count.
"""

import csv
```

```python
import argparse
from collections import Counter
import logging

import utilities as util


ERRORS = {
    'not_csv_suffix' : '{fname}: File must end in .csv',
    }


def update_counts(reader, word_counts):
    """Update word counts with data from another reader/file."""
    for word, count in csv.reader(reader):
        word_counts[word] += int(count)


def main(args):
    """Run the command line program."""
    log_lev = logging.DEBUG if args.verbose else logging.WARNING
    logging.basicConfig(level=log_lev)
    word_counts = Counter()
    logging.info('Processing files...')
    for fname in args.infiles:
        logging.debug(f'Reading in {fname}...')
        if fname[-4:] != '.csv':
            msg = ERRORS['not_csv_suffix'].format(fname=fname)
            raise OSError(msg)
        with open(fname, 'r') as reader:
            logging.debug('Computing word counts...')
            update_counts(reader, word_counts)
    util.collection_to_csv(word_counts, num=args.num)


if __name__ == '__main__':
    parser = argparse.ArgumentParser(description=__doc__)
    parser.add_argument('infiles', type=str, nargs='*',
                        help='Input file names')
    parser.add_argument('-n', '--num',
                        type=int, default=None,
                        help='Output n most frequent words')
    parser.add_argument('-v', '--verbose',
                        action="store_true", default=False,
```

```
                        help="Set logging level to DEBUG")
    args = parser.parse_args()
    main(args)
```

Exercise 12.6.2

Add a new command-line argument to `collate.py`:

```
parser.add_argument('-l', '--logfile',
                    type=str, default='collate.log',
                    help='Name of the log file')
```

and pass the name of the log file to `logging.basicConfig` using the `filename` argument:

```
logging.basicConfig(level=log_lev, filename=args.logfile)
```

such that the `collate.py` script now reads as follows:

```
"""
Combine multiple word count CSV-files
into a single cumulative count.
"""

import csv
import argparse
from collections import Counter
import logging

import utilities as util


ERRORS = {
    'not_csv_suffix' : '{fname}: File must end in .csv',
    }
```

```
def update_counts(reader, word_counts):
    """Update word counts with data from another reader/file."""
    for word, count in csv.reader(reader):
        word_counts[word] += int(count)


def main(args):
    """Run the command line program."""
    log_lev = logging.DEBUG if args.verbose else logging.WARNING
    logging.basicConfig(level=log_lev, filename=args.logfile)
    word_counts = Counter()
    logging.info('Processing files...')
    for fname in args.infiles:
        logging.debug(f'Reading in {fname}...')
        if fname[-4:] != '.csv':
            msg = ERRORS['not_csv_suffix'].format(fname=fname)
            raise OSError(msg)
        with open(fname, 'r') as reader:
            logging.debug('Computing word counts...')
            update_counts(reader, word_counts)
    util.collection_to_csv(word_counts, num=args.num)


if __name__ == '__main__':
    parser = argparse.ArgumentParser(description=__doc__)
    parser.add_argument('infiles', type=str, nargs='*',
                        help='Input file names')
    parser.add_argument('-n', '--num',
                        type=int, default=None,
                        help='Output n most frequent words')
    parser.add_argument('-v', '--verbose',
                        action="store_true", default=False,
                        help="Set logging level to DEBUG")
    parser.add_argument('-l', '--logfile',
                        type=str, default='collate.log',
                        help='Name of the log file')
    args = parser.parse_args()
    main(args)
```

Exercise 12.6.3

1. The loop in `collate.py` that reads/processes each input file should now read as follows:

```
for fname in args.infiles:
    try:
        logging.debug(f'Reading in {fname}...')
        if fname[-4:] != '.csv':
            msg = ERRORS['not_csv_suffix'].format(fname=fname)
            raise OSError(msg)
        with open(fname, 'r') as reader:
            logging.debug('Computing word counts...')
            update_counts(reader, word_counts)
    except Exception as error:
        logging.warning(f'{fname} not processed: {error}')
```

2. The loop in `collate.py` that reads/processes each input file should now read as follows:

```
for fname in args.infiles:
    try:
        logging.debug(f'Reading in {fname}...')
        if fname[-4:] != '.csv':
            msg = ERRORS['not_csv_suffix'].format(
                fname=fname)
            raise OSError(msg)
        with open(fname, 'r') as reader:
            logging.debug('Computing word counts...')
            update_counts(reader, word_counts)
    except FileNotFoundError:
        msg = f'{fname} not processed: File does not exist'
        logging.warning(msg)
    except PermissionError:
        msg = f'{fname} not processed: No read permission'
        logging.warning(msg)
    except Exception as error:
        msg = f'{fname} not processed: {error}'
        logging.warning(msg)
```

Exercise 12.6.4

1. The try/except block in collate.py should begin as follows:

```
try:
    process_file(fname, word_counts)
except FileNotFoundError:
# ... the other exceptions
```

2. The following additions need to be made to test_zipfs.py.

```
import collate



def test_not_csv_error():
    """Error handling test for csv check"""
    fname = 'data/dracula.txt'
    word_counts = Counter()
    with pytest.raises(OSError):
        collate.process_file(fname, word_counts)
```

3. The following unit test needs to be added to test_zipfs.py.

```
def test_missing_file_error():
    """Error handling test for missing file"""
    fname = 'fake_file.csv'
    word_counts = Counter()
    with pytest.raises(FileNotFoundError):
        collate.process_file(fname, word_counts)
```

4. The following sequence of commands is required to test the code coverage.

```
$ coverage run -m pytest
$ coverage html
```

   Open htmlcov/index.html and click on bin/collate.py to view a coverage summary. The lines of process_files that include the raise OSError and open(fname, 'r') commands should appear in green after clicking the green "run" box in the top left-hand corner of the page.

Exercise 12.6.5

1. The convention is to use `ALL_CAPS_WITH_UNDERSCORES` when defining global variables.
2. Python's f-strings interpolate variables that are in scope: there is no easy way to interpolate values from a lookup table. In contrast, `str.format` can be given any number of named keyword arguments (Appendix F), so we can look up a string and then interpolate whatever values we want.
3. Once `ERRORS` has been moved to the `utilities` module, all references to it in `collate.py` must be updated to `util.ERRORS`.

Exercise 12.6.6

A **traceback** is an object that records where an exception was raised), what **stack frames** were on the call stack when the error occurred, and other details that are helpful for debugging. Python's traceback[3] library can be used to get and print information from these objects.

Chapter 13

Exercise 13.4.1

You can get an ORCID by registering here[4]. Please add this 16-digit identifier to all of your published works and to your online profiles.

Exercise 13.4.2

If possible, compare your answers with those of a colleague who works with the same data. Where did you agree and disagree, and why?

[3]`https://docs.python.org/3/library/traceback.html`
[4]`https://orcid.org/register`

Exercise 13.4.3

1. 51 solicitors were interviewed as the participants.
2. Interview data, and data from a database on court decisions.
3. This information is not available within the documentation. Information on their jobs and opinions are there, but the participant demographics are only described within the associated article. The difficulty is that the article is not linked within the documentation or the metadata.
4. We can search the dataset name and author name trying to find this. A search for the grant information with "National Science Foundation (1228602)" finds the grant page[5]. Two articles are linked there, but both the DOI links are broken. We can search with the citation for each paper to find them. The Forced Migration article[6] uses a different subset of interviews and does not mention demographics, nor links to the deposited dataset. The Boston College Law Review article[7] has the same two problems of different data and no dataset citation.

   Searching more broadly through Meili's work, we can find Meili (2015). This lists the dataset as a footnote and reports the 51 interviews with demographic data on reported gender of the interviewees. This paper lists data collection as 2010–2014, while the other two say 2010–2013. We might come to a conclusion that this extra year is where the extra 9 interviews come in, but that difference is not explained anywhere.

Exercise 13.4.4

For `borstlab/reversephi_paper`[8]:

1. The software requirements are documented in `README.md`. In addition to the tools used in the `zipf/` project (Python, Make and Git), the project also requires ImageMagick. No information on installing ImageMagick or a required version of ImageMagick is provided.

   To re-create the `conda` environment, you would need the file `my_environment.yml`. Instructions for creating and using the environment are provided in `README.md`.

[5] https://www.nsf.gov/awardsearch/showAward?AWD_ID=1228602
[6] https://www.fmreview.org/fragilestates/meili
[7] https://lawdigitalcommons.bc.edu/cgi/viewcontent.cgi?article=3318&context=bclr
[8] https://github.com/borstlab/reversephi_paper/

2. Like zipf the data processing and analysis steps are documented in a Makefile. The README includes instructions for re-creating the results using make all.
3. There doesn't seem to be a DOI for the archived code and data, but the GitHub repo does have a release v1.0 with the description "Published manuscript (1.0)" beside it. A zip file of this release could be downloaded from GitHub.

For the figshare page[9] that accompanies the paper Irving, Wijffels, and Church (2019):

1. The figshare page includes a "Software environment" section. To re-create the conda environment, you would need the file environment.yml.
2. figure*_log.txt are log files for each figure in the paper. These files show the computational steps performed in generating the figure, in the form of a list of commands executed at the command line.

   code.zip is a version controlled (using git) file directory containing the code written to perform the analysis (i.e., it contains the scripts referred to in the log files). This code can also be found on GitHub.
3. The figshare page itself is the archive, and includes a version history for the contents.

For the GitHub repo blab/h3n2-reassortment[10]:

1. README.md includes an "Install requirements" section that describes setting up the conda environment using the file h3n2_reassortment.yaml.

   The analysis also depends on components from Nextstrain. Instructions for cloning them from GitHub are provided.
2. The code seems to be spread across the directories jupyter_notebooks, hyphy, flu_epidemiology, and src, but it isn't clear what order the code should be run in, or how the components depend on each other.
3. The data itself is not archived, but links are provided in the "Install requirements" section of README.md to documents that describe how to obtain the data. Some intermediate data is also provided in the data/ directory.

   The GitHub repo has a release with files "that are up-to-date with

[9]https://doi.org/10.6084/m9.figshare.7575830

[10]https://github.com/blab/h3n2-reassortment

the version of the manuscript that was submitted to Virus Evolution on 31 January 2019."

Exercise 13.4.5

`https://web.archive.org/web/20191105173924/https://ukhomeoffice.github.io/accessibility-posters/posters/accessibility-posters.pdf`

Exercise 13.4.6

You'll know you've completed this exercise when you have a URL that points to zip archive for a specific release of your repository on GitHub, e.g:

`https://github.com/amira-khan/zipf/archive/KhanVirtanen2020.zip`

Exercise 13.4.7

Some steps to publishing your project's code would be:

1. Upload the code on GitHub.
2. Use a standard folder and file structure as taught in this book.
3. Include `README`, `CONTRIBUTING`, `CONDUCT`, and `LICENSE` files.
4. Make sure these files explain how to install and configure the required software and tells people how to run the code in the project.
5. Include a `requirements.txt` file for Python package dependencies.

Chapter 14

Exercise 14.9.1

A `description` and `long_description` argument need to be provided when the `setup` function is called in `setup.py`. On the TestPyPI webpage, the user interface displays description in the grey banner and long_description in the section named "Project Description."

Other metadata that might be added includes the author email address, software license details and a link to the documentation at Read the Docs.

Exercise 14.9.2

The new `requirements_dev.txt` file will have this inside it:

```
pytest
```

Exercise 14.9.3

The answers to the relevant questions from the checklist are shown below.

- *Repository*: Is the source code for this software available at the repository url?
  - Yes. The source code is available at PyPI.
- *License*: Does the repository contain a plain-text LICENSE file with the contents of an OSI approved software license?
  - Yes. Our GitHub repository contains LICENSE.md (Section 8.4.1).
- *Installation*: Does installation proceed as outlined in the documentation?
  - Yes. Our README says the package can be installed via pip.
- *Functionality*: Have the functional claims of the software been confirmed?
  - Yes. The command-line programs `countwords`, `collate`, and `plotcounts` perform as described in the README.
- *A statement of need*: Do the authors clearly state what problems the software is designed to solve and who the target audience is?
  - Yes. The "Motivation" section of the README explains this.
- *Installation instructions*: Is there a clearly stated list of dependencies? Ideally these should be handled with an automated package management solution.
  - Yes. In our `setup.py` file the `install_requires` argument lists dependencies.
- *Example usage*: Do the authors include examples of how to use the software (ideally to solve real-world analysis problems).
  - Yes. There are examples in the README.
- *Functionality documentation*: Is the core functionality of the software documented to a satisfactory level (e.g., API method documentation)?
  - Yes. This information is available on Read the Docs.

- *Automated tests*: Are there automated tests or manual steps described so that the functionality of the software can be verified?
  - We have unit tests written and available (`test_zipfs.py`), but our documentation needs to be updated to tell people to run `pytest` in order to manually run those tests.
- *Community guidelines*: Are there clear guidelines for third parties wishing to 1) Contribute to the software 2) Report issues or problems with the software 3) Seek support?
  - Yes. Our CONTRIBUTING file explains this (Section 8.11).

Exercise 14.9.4

The directory tree for the `pratchett` package is:

```
pratchett
├── pratchett
│   └── quotes.py
├── README.md
└── setup.py
```

`README.md` should contain a basic description of the package and how to install/use it, while `setup.py` should contain:

```
from setuptools import setup


setup(
    name='pratchett',
    version='0.1',
    author='Amira Khan',
    packages=['pratchett'],
)
```

The following sequence of commands will create the development environment, activate it, and then install the package:

```
$ conda create -n pratchett python
$ conda activate pratchett
(pratchett)$ cd pratchett
(pratchett)$ pip install -e .
```

B

Learning Objectives

This appendix lists the learning objectives for each chapter, and is intended to help instructors who want to use this curriculum.

B.1 Getting Started

- Identify the few standard files that should be present in every research software project.
- Explain the typical directory structure used in small and medium-sized data analysis projects.
- Download the required data.
- Install the required software.

B.2 The Basics of the Unix Shell

- Explain how the shell relates to the keyboard, the screen, the operating system, and users' programs.
- Explain when and why a **command-line interface** should be used instead of **graphical user interfaces**.
- Explain the steps in the shell's **read-evaluate-print loop**.
- Identify the command, options, and filenames in a command-line call.
- Explain the similarities and differences between files and directories.
- Translate an **absolute path** into a **relative path** and vice versa.
- Construct absolute and relative paths that identify files and directories.
- Delete, copy, and move files and directories.

B.3 Building Tools with the Unix Shell

- **Redirect** a command's output to a file.
- Use redirection to process a file instead of keyboard input.
- Construct **pipelines** with two or more stages.
- Explain Unix's "small pieces, loosely joined" philosophy.
- Write a loop that applies one or more commands separately to each file in a set of files.
- Trace the values taken on by a loop variable during execution of the loop.
- Explain the difference between a variable's name and its value.
- Demonstrate how to see recently executed commands.
- Re-run recently executed commands without retyping them.

B.4 Going Further with the Unix Shell

- Write a **shell script** that uses command-line arguments.
- Create pipelines that include shell scripts as well as built-in commands.
- Create and use variables in shell scripts with correct quoting.
- Use `grep` to select lines from text files that match simple patterns.
- Use `find` to find files whose names match simple patterns.
- Edit the `.bashrc` file to change default shell variables.
- Create aliases for commonly used commands.

B.5 Building Command-Line Tools with Python

- Explain the benefits of writing Python programs that can be executed at the command line.
- Create a command-line Python program that respects **Unix shell** conventions for reading input and writing output.
- Use the `argparse`[1] library to handle command-line arguments in a program.
- Explain how to tell if a module is being run directly or being loaded by another program.
- Write **docstrings** for programs and functions.

[1] `https://docs.python.org/3/library/argparse.html`

- Explain the difference between **optional arguments** and **positional arguments**.
- Create a module that contains functions used by multiple programs and import that module.

B.6 Using Git at the Command Line

- Explain the advantages and disadvantages of using **Git** at the command line.
- Demonstrate how to configure Git on a new computer.
- Create a local Git repository at the command line.
- Demonstrate the modify-add-commit cycle for one or more files.
- Synchronize a local repository with a **remote repository**.
- Explain what the `HEAD` of a repository is and demonstrate how to use it in commands.
- Identify and use Git commit identifiers.
- Demonstrate how to compare revisions to files in a repository.
- Restore old versions of files in a repository.
- Explain how to use `.gitignore` to ignore files and identify files that are being ignored.

B.7 Going Further with Git

- Explain why **branches** are useful.
- Demonstrate how to create a branch, make changes on that branch, and **merge** those changes back into the original branch.
- Explain what **conflicts** are and demonstrate how to resolve them.
- Explain what is meant by a **branch-per-feature** workflow.
- Define the terms **fork**, **clone**, **remote**, and **pull request**.
- Demonstrate how to fork a repository and submit a pull request to the original repository.

B.8 Working in Teams

- Explain how a project lead can be a good **ally**.
- Explain the purpose of a Code of Conduct and add one to a project.
- Explain why every project should include a license and add one to a project.
- Describe different kinds of licenses for software and written material.
- Explain what an **issue tracking system** does and what it should be used for.
- Describe what a well-written issue should contain.
- Explain how to **label** issues to manage work.
- Submit an issue to a project.
- Describe common approaches to prioritizing tasks.
- Describe some common-sense rules for running meetings.
- Explain why every project should include contribution guidelines and add some to a project.
- Explain how to handle conflict between project participants.

B.9 Automating Analyses with Make

- Explain what a **build manager** is and how they aid reproducible research.
- Name and describe the three parts of a **build rule**.
- Write a Makefile that re-runs a multi-stage data analysis.
- Explain and trace how Make chooses an order in which to execute rules.
- Explain what **phony targets** are and define a phony target.
- Explain what **automatic variables** are and identify three commonly used automatic variables.
- Write Make rules that use automatic variables.
- Explain why and how to write **pattern rules** in a Makefile.
- Write Make rules that use patterns.
- Define variables in a Makefile explicitly and by using functions.
- Make a self-documenting Makefile.

B.10 Configuring Programs

- Explain what **overlay configuration** is.

- Describe the four levels of configuration typically used by robust software.
- Create a configuration file using **YAML**.

B.11 Testing Software

- Explain three different goals for testing software.
- Add **assertions** to a program to check that it is operating correctly.
- Write and run unit tests using `pytest`.
- Determine the **coverage** of those tests and identify untested portions of code.
- Explain **continuous integration** and implement it using Travis CI[2].
- Describe and contrast **test-driven development** and checking-driven development.

B.12 Handling Errors

- Explain how to use exceptions to signal and handle errors in programs.
- Write `try`/`except` blocks to **raise** and **catch** exceptions.
- Explain what is meant by "throw low, catch high."
- Describe the most common built-in exception types in Python and how they relate to each other.
- Explain what makes a useful error message.
- Create and use a lookup table for common error messages.
- Explain the advantages of using a **logging framework** rather than `print` statements.
- Describe the five standard logging levels and explain what each should be used for.
- Create, configure, and use a simple logger.

B.13 Tracking Provenance

- Explain what a **DOI** is and how to get one.

[2]`https://travis-ci.com/`

- Explain what an **ORCID** is and get one.
- Describe the FAIR Principles[3] and determine whether a dataset conforms to them.
- Explain where to archive small, medium, and large datasets.
- Describe good practices for archiving analysis code and determine whether a report conforms to them.
- Explain the difference between **reproducibility** and **inspectability**.

B.14 Creating Packages with Python

- Create a Python package using `setuptools`[4].
- Create and use a **virtual environment** to manage Python package installations.
- Install a Python package using `pip`[5].
- Distribute that package via TestPyPI[6].
- Write a README file for a Python package.
- Use Sphinx[7] to create and preview documentation for a package.
- Explain where and how to obtain a **DOI** for a software release.
- Describe some academic journals that publish software papers.

[3] `https://www.go-fair.org/fair-principles/`
[4] `https://setuptools.readthedocs.io/`
[5] `https://pypi.org/project/pip/`
[6] `https://test.pypi.org`
[7] `https://www.sphinx-doc.org/en/master/`

C

Key Points

This appendix lists the key points for each chapter.

C.1 Getting Started

- Make tidiness a habit, rather than cleaning up your project files later.
- Include a few standard files in all your projects, such as README, LICENSE, CONTRIBUTING, CONDUCT and CITATION.
- Put runnable code in a `bin/` directory.
- Put raw/original data in a `data/` directory and never modify it.
- Put results in a `results/` directory. This includes cleaned-up data and figures (i.e., everything created using what's in `bin` and `data`).
- Put documentation and manuscripts in a `docs/` directory.
- Refer to The Carpentries software installation guide[1] if you're having trouble.

C.2 The Basics of the Unix Shell

- A **shell** is a program that reads commands and runs other programs.
- The **filesystem** manages information stored on disk.
- Information is stored in files, which are located in directories (folders).
- Directories can also store other directories, which forms a directory tree.
- `pwd` prints the user's **current working directory**.
- `/` on its own is the **root directory** of the whole filesystem.
- `ls` prints a list of files and directories.
- An **absolute path** specifies a location from the root of the filesystem.

[1] `https://carpentries.github.io/workshop-template/#setup`

- A **relative path** specifies a location in the filesystem starting from the current directory.
- `cd` changes the current working directory.
- `..` means the **parent directory**.
- `.` on its own means the current directory.
- `mkdir` creates a new directory.
- `cp` copies a file.
- `rm` removes (deletes) a file.
- `mv` moves (renames) a file or directory.
- `*` matches zero or more characters in a filename.
- `?` matches any single character in a filename.
- `wc` counts lines, words, and characters in its inputs.
- `man` displays the manual page for a given command; some commands also have a `--help` option.

C.3 Building Tools with the Unix Shell

- `cat` displays the contents of its inputs.
- `head` displays the first few lines of its input.
- `tail` displays the last few lines of its input.
- `sort` sorts its inputs.
- Use the up-arrow key to scroll up through previous commands to edit and repeat them.
- Use `history` to display recent commands and `!number` to repeat a command by number.
- Every process in Unix has an input channel called **standard input** and an output channel called **standard output**.
- `>` redirects a command's output to a file, overwriting any existing content.
- `>>` appends a command's output to a file.
- `<` operator redirects input to a command.
- A **pipe** `|` sends the output of the command on the left to the input of the command on the right.
- A `for` loop repeats commands once for every thing in a list.
- Every `for` loop must have a variable to refer to the thing it is currently operating on and a **body** containing commands to execute.
- Use `$name` or `${name}` to get the value of a variable.

C.4 Going Further with the Unix Shell

- Save commands in files (usually called **shell scripts**) for re-use.
- `bash filename` runs the commands saved in a file.
- `$@` refers to all of a shell script's command-line arguments.
- `$1`, `$2`, etc., refer to the first command-line argument, the second command-line argument, etc.
- Place variables in quotes if the values might have spaces or other special characters in them.
- `find` prints a list of files with specific properties or whose names match patterns.
- `$(command)` inserts a command's output in place.
- `grep` selects lines in files that match patterns.
- Use the `.bashrc` file in your home directory to set shell variables each time the shell runs.
- Use `alias` to create shortcuts for things you type frequently.

C.5 Building Command-Line Programs in Python

- Write command-line Python programs that can be run in the **Unix shell** like other command-line tools.
- If the user does not specify any input files, read from **standard input**.
- If the user does not specify any output files, write to **standard output**.
- Place all `import` statements at the start of a module.
- Use the value of `__name__` to determine if a file is being run directly or being loaded as a module.
- Use `argparse`[2] to handle command-line arguments in standard ways.
- Use **short options** for common controls and **long options** for less common or more complicated ones.
- Use **docstrings** to document functions and scripts.
- Place functions that are used across multiple scripts in a separate file that those scripts can import.

[2]`https://docs.python.org/3/library/argparse.html`

C.6 Using Git at the Command Line

- Use `git config` with the `--global` option to configure your username, email address, and other preferences once per machine.
- `git init` initializes a **repository**.
- Git stores all repository management data in the `.git` subdirectory of the repository's root directory.
- `git status` shows the status of a repository.
- `git add` puts files in the repository's staging area.
- `git commit` saves the staged content as a new commit in the local repository.
- `git log` lists previous commits.
- `git diff` shows the difference between two versions of the repository.
- Synchronize your local repository with a **remote repository** on a **forge** such as GitHub[3].
- `git remote` manages bookmarks pointing at remote repositories.
- `git push` copies changes from a local repository to a remote repository.
- `git pull` copies changes from a remote repository to a local repository.
- `git restore` and `git checkout` recover old versions of files.
- The `.gitignore` file tells Git what files to ignore.

C.7 Going Further with Git

- Use a **branch-per-feature** workflow to develop new features while leaving the master branch in working order.
- `git branch` creates a new branch.
- `git checkout` switches between branches.
- `git merge` **merges** changes from another branch into the current branch.
- **Conflicts** occur when files or parts of files are changed in different ways on different branches.
- Version control systems do not allow people to overwrite changes silently; instead, they highlight conflicts that need to be resolved.
- **Forking** a repository makes a copy of it on a server.
- **Cloning** a repository with `git clone` creates a local copy of a remote repository.
- Create a remote called `upstream` to point to the repository a fork was derived from.

[3] `https://github.com`

- Create **pull requests** to submit changes from your fork to the upstream repository.

C.8 Working in Teams

- Welcome and nurture community members proactively.
- Create an explicit Code of Conduct for your project modeled on the Contributor Covenant[4].
- Include a license in your project so that it's clear who can do what with the material.
- Create **issues** for bugs, enhancement requests, and discussions.
- **Label issues** to identify their purpose.
- **Triage** issues regularly and group them into **milestones** to track progress.
- Include contribution guidelines in your project that specify its workflow and its expectations of participants.
- Make rules about **governance** explicit.
- Use common-sense rules to make project meetings fair and productive.
- Manage conflict between participants rather than hoping it will take care of itself.

C.9 Automating Analyses with Make

- Make[5] is a widely used build manager.
- A **build manager** re-runs commands to update files that are out of date.
- A **build rule** has **targets**, **prerequisites**, and a **recipe**.
- A target can be a file or a **phony target** that simply triggers an action.
- When a target is out of date with respect to its prerequisites, Make executes the recipe associated with its rule.
- Make executes as many rules as it needs to when updating files, but always respects prerequisite order.
- Make defines **automatic variables** such as `$@` (target), `$^` (all prerequisites), and `$<` (first prerequisite).
- **Pattern rules** can use `%` as a placeholder for parts of filenames.
- Makefiles can define variables using `NAME=value`.
- Make also has functions such as `$(wildcard...)` and `$(patsubst...)`.

[4] https://www.contributor-covenant.org
[5] https://www.gnu.org/software/make/

- Use specially formatted comments to create self-documenting Makefiles.

C.10 Configuring Programs

- **Overlay configuration** specifies settings for a program in layers, each of which overrides previous layers.
- Use a system-wide configuration file for general settings.
- Use a user-specific configuration file for personal preferences.
- Use a job-specific configuration file with settings for a particular run.
- Use command-line options to change things that commonly change.
- Use **YAML** or some other standard syntax to write configuration files.
- Save configuration information to make your research **reproducible**.

C.11 Testing Software

- Test software to convince people (including yourself) that software is correct enough and to make tolerances on "enough" explicit.
- Add **assertions** to code so that it checks itself as it runs.
- Write **unit tests** to check individual pieces of code.
- Write **integration tests** to check that those pieces work together correctly.
- Write **regression tests** to check if things that used to work no longer do.
- A **test framework** finds and runs tests written in a prescribed fashion and reports their results.
- Test **coverage** is the fraction of lines of code that are executed by a set of tests.
- **Continuous integration** re-builds and/or re-tests software every time something changes.

C.12 Handling Errors

- Signal errors by **raising exceptions**.
- Use `try`/`except` blocks to **catch** and handle exceptions.
- Python organizes its standard exceptions in a hierarchy so that programs can catch and handle them selectively.

- "Throw low, catch high," i.e., raise exceptions immediately but handle them at a higher level.
- Write error messages that help users figure out what to do to fix the problem.
- Store error messages in a lookup table to ensure consistency.
- Use a **logging framework** instead of `print` statements to report program activity.
- Separate logging messages into `DEBUG`, `INFO`, `WARNING`, `ERROR`, and `CRITICAL` levels.
- Use `logging.basicConfig` to define basic logging parameters.

C.13 Tracking Provenance

- Publish data and code as well as papers.
- Use **DOIs** to identify reports, datasets, and software releases.
- Use an **ORCID** to identify yourself as an author of a report, dataset, or software release.
- Data should be FAIR[6]: findable, accessible, interoperable, and reusable.
- Put small datasets in version control repositories; store large ones on data sharing sites.
- Describe your software environment, analysis scripts, and data processing steps in **reproducible** ways.
- Make your analyses **inspectable** as well as reproducible.

C.14 Creating Packages with Python

- Use `setuptools`[7] to build and distribute Python packages.
- Create a directory named `mypackage` containing a `setup.py` script with a subdirectory also called `mypackage` containing the package's source files.
- Use **semantic versioning** for software releases.
- Use a **virtual environment** to test how your package installs without disrupting your main Python installation.
- Use `pip`[8] to install Python packages.
- The default repository for Python packages is PyPI[9].

[6] `https://www.go-fair.org/fair-principles/`
[7] `https://setuptools.readthedocs.io/`
[8] `https://pypi.org/project/pip/`
[9] `https://pypi.org/`

- Use TestPyPI[10] to test the distribution of your package.
- Use a README file for package-level documentation.
- Use Sphinx[11] to generate documentation for a package.
- Use Read the Docs[12] to host package documentation online.
- Create a **DOI** for your package using GitHub's Zenodo integration[13].
- Publish details of your package in a software journal so others can cite it.

[10] `https://test.pypi.org`
[11] `https://www.sphinx-doc.org/en/master/`
[12] `https://docs.readthedocs.io/en/latest/`
[13] `https://guides.github.com/activities/citable-code/`

D

Project Tree

The final directory tree for the Zipf's Law project looks like the following:

```
pyzipf/
├── .gitignore
├── CITATION.md
├── CONDUCT.md
├── CONTRIBUTING.md
├── KhanVirtanen2020.md
├── LICENSE.md
├── Makefile
├── README.rst
├── environment.yml
├── requirements.txt
├── requirements_docs.txt
├── setup.py
├── data
│   ├── README.md
│   ├── dracula.txt
│   ├── frankenstein.txt
│   ├── jane_eyre.txt
│   ├── moby_dick.txt
│   ├── sense_and_sensibility.txt
│   ├── sherlock_holmes.txt
│   └── time_machine.txt
├── docs
│   ├── Makefile
│   ├── conf.py
│   ├── index.rst
│   └── source
│       ├── collate.rst
│       ├── countwords.rst
│       ├── modules.rst
│       ├── plotcounts.rst
│       ├── test_zipfs.rst
│       └── utilities.rst
├── results
```

```
│   ├── collated.csv
│   ├── collated.png
│   ├── dracula.csv
│   ├── dracula.png
│   ├── frankenstein.csv
│   ├── jane_eyre.csv
│   ├── jane_eyre.png
│   ├── moby_dick.csv
│   ├── sense_and_sensibility.csv
│   ├── sherlock_holmes.csv
│   └── time_machine.csv
├── test_data
│   ├── random_words.txt
│   └── risk.txt
└── pyzipf
    ├── book_summary.sh
    ├── collate.py
    ├── countwords.py
    ├── plotcounts.py
    ├── plotparams.yml
    ├── script_template.py
    ├── test_zipfs.py
    └── utilities.py
```

You can view the complete project, including the version history, in Amira's `zipf` repository on GitHub[1].

Each file was introduced and subsequently modified in the following chapters, sections and exercises:

- `pyzipf/`: Introduced as `zipf/` in Section 1.2 and changed name to `pyzipf/` in Section 14.1.
- `pyzipf/.gitignore`: Introduced in Section 6.9, and updated in various other chapters following GitHub's `.gitignore` templates[2].
- `pyzipf/CITATION.md`: Introduced in Section 14.7.
- `pyzipf/CONDUCT.md`: Introduced in Section 8.3 and committed to the repository in Exercise 8.14.2.
- `pyzipf/CONTRIBUTING.md`: Introduced in Section 8.11 and committed to the repository in Exercise 8.14.4.
- `pyzipf/KhanVirtanen2020.md`: Introduced in Section 13.2.2.

[1] https://github.com/amira-khan/zipf
[2] https://github.com/github/gitignore

- `pyzipf/LICENSE.md`: Introduced in Section 8.4.1 and committed to the repository in Exercise 8.14.3.
- `pyzipf/Makefile`: Introduced and updated throughout Chapter 9. Updated again in Exercise 10.8.1.
- `pyzipf/README.rst`: Introduced as a `.md` file in Section 7.6, updated in Section 7.8 and then converted to a `.rst` file with further updates in Section 14.6.1.
- `pyzipf/environment.yml`: Introduced in Section 13.2.1.
- `pyzipf/requirements.txt`: Introduced in Section 11.8.
- `pyzipf/requirements_docs.txt`: Introduced in Section 14.6.2.
- `pyzipf/setup.py`: Introduced and updated throughout Chapter 14.
- `pyzipf/data/*` : Downloaded as part of the setup instructions (Section 1.2).
- `pyzipf/docs/*`: Introduced in Section 14.6.2.
- `pyzipf/results/collated.*`: Generated in Section 9.9.
- `pyzipf/results/dracula.csv`: Generated in Section 5.7.
- `pyzipf/results/dracula.png`: Generated in Section 6.5 and updated in Section 7.4.
- `pyzipf/results/jane_eyre.csv`: Generated in Section 5.7.
- `pyzipf/results/jane_eyre.png`: Generated in Section 5.9.
- `pyzipf/results/moby_dick.csv`: Generated in Section 5.7.
- `pyzipf/results/frankenstein.csv`: Generated in Section 9.7.
- `pyzipf/results/sense_and_sensibility.csv`: Generated in Section 9.7.
- `pyzipf/results/sherlock_holmes.csv`: Generated in Section 9.7.
- `pyzipf/results/time_machine.csv`: Generated in Section 9.7.
- `pyzipf/test_data/random_words.txt`: Generated in Section 11.5.
- `pyzipf/test_data/risk.txt`: Introduced in Section 11.2.
- `pyzipf/pyzipf/`: Introduced as `bin/` in Section 1.1.2 and changed name to `pyzipf/` in Section 14.1.
- `pyzipf/pyzipf/book_summary.sh`: Introduced and updated throughout Chapter 4.

- `pyzipf/pyzipf/collate.py`: Introduced in Section 5.7 and updated in Section 5.8, throughout Chapter 12 and in Section 14.1.
- `pyzipf/pyzipf/countwords.py`: Introduced in Section 5.4 and updated in Sections 5.8 and 14.1.
- `pyzipf/pyzipf/plotcounts.py`: Introduced in Exercise 5.11.4 and updated throughout Chapters 6, 7 and 10.
- `pyzipf/pyzipf/plotparams.yml`: Introduced in Section 10.6.
- `pyzipf/pyzipf/script_template.py`: Introduced in Section 5.2 and updated in Section 5.3.
- `pyzipf/pyzipf/test_zipfs.py`: Introduced and updated throughout Chapter 11.
- `pyzipf/pyzipf/utilities.py`: Introduced in Section 5.8.

E

Working Remotely

When the Internet was young, people didn't encrypt anything except the most sensitive information when sending it over a network. However, this meant that villains could steal usernames and passwords. The **SSH protocol** was invented to prevent this (or at least slow it down). It uses several sophisticated (and heavily tested) encryption protocols to ensure that outsiders can't see what's in the messages going back and forth between different computers.

To understand how it works, let's take a closer look at what happens when we use the shell on a desktop or laptop computer. The first step is to log in so that the operating system knows who we are and what we're allowed to do. We do this by typing our username and password; the operating system checks those values against its records, and if they match, runs a shell for us.

As we type commands, characters are sent from our keyboard to the shell. It displays those characters on the screen to represent what we type, and then executes the command and displays its output (if any). If we want to run commands on another machine, such as the server in the basement that manages our database of experimental results, we have to log in to that machine so that our commands will go to it instead of to our laptop. We call this a **remote login**.

E.1 Logging In

In order for us to be able to log in, the remote computer must run a **remote login server** and we must run a program that can talk to that server. The client program passes our login credentials to the remote login server; if we are allowed to log in, that server then runs a shell for us on the remote computer (Figure E.1).

Once our local client is connected to the remote server, everything we type into the client is passed on, by the server, to the shell running on the remote computer. That remote shell runs those commands on our behalf, just as a

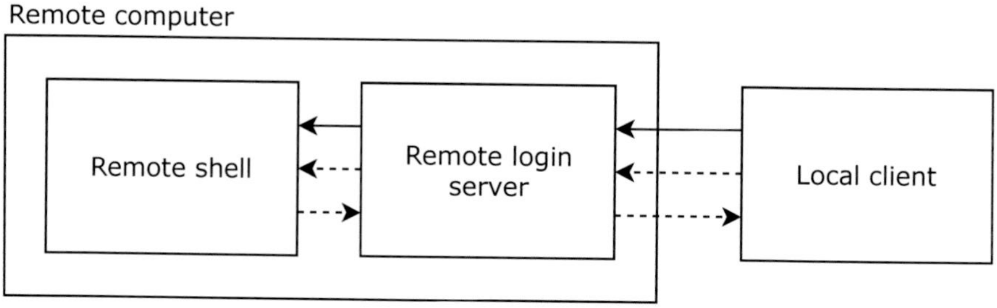

FIGURE E.1: The local client connects to the remote login server, which starts a remote shell (solid lines). Commands from the local client are passed to the remote shell and output is passed back (dashed lines).

local shell would, then sends back output, via the server, to our client, for our computer to display.

The remote login server which accepts connections from client programs is known as the **SSH daemon**, or `sshd`. The client program we use to log in remotely is the **secure shell**, or `ssh`. It has a companion program called `scp` that allows us to copy files to or from a remote computer using the same kind of encrypted connection.

We issue the command `ssh username@computer` to log in remotely. This command tries to make a connection to the SSH daemon running on the remote computer we have specified. After we log in, we can use the remote shell to use the remote computer's files and directories. Typing `exit` or Control-D terminates the remote shell, and the local client program, and returns us to our previous shell.

In the example below, the remote machine's command prompt is `moon>` instead of `$` to make it clearer which machine is doing what.

```
$ pwd
```

```
/Users/amira
```

```
$ ssh amira@moon.euphoric.edu
Password: ********
```

```
moon> hostname
```

```
moon
```

```
moon> pwd
```

```
/Users/amira
```

```
moon> ls -F
```

```
bin/     cheese.txt   dark_side/   rocks.cfg
```

```
moon> exit
```

```
$ pwd
```

```
/Users/amira
```

E.2 Copying Files

To copy a file, we specify the source and destination paths, either of which may include computer names. If we leave out a computer name, `scp` assumes we mean the machine we're running on. For example, this command copies our latest results to the backup server in the basement, printing out its progress as it does so:

```
$ scp results.dat amira@backup:backups/results-2019-11-11.dat
Password: ********
```

```
results.dat                100%  9   1.0 MB/s 00:00
```

Note the colon :, separating the hostname of the server and the pathname of the file we are copying to. It is this character that informs scp that the source or target of the copy is on the remote machine and the reason it is needed can be explained as follows:

In the same way that the default directory into which we are placed when running a shell on a remote machine is our home directory on that machine, the default target, for a remote copy, is also the home directory.

This means that:

```
$ scp results.dat amira@backup:
```

would copy results.dat into our home directory on backup, however, if we did not have the colon to inform scp of the remote machine, we would still have a valid command:

```
$ scp results.dat amira@backup
```

but now we have merely created a file called amira@backup on our local machine, as we would have done with cp.

```
$ cp results.dat amira@backup
```

Copying a whole directory between remote machines uses the same syntax as the cp command: we just use the -r option to signal that we want copying to be recursive. For example, this command copies all of our results from the backup server to our laptop:

```
$ scp -r amira@backup:backups ./backups
Password: ********
```

```
results-2019-09-18.dat                100%  7   1.0 MB/s 00:00
results-2019-10-04.dat                100%  9   1.0 MB/s 00:00
results-2019-10-28.dat                100%  8   1.0 MB/s 00:00
results-2019-11-11.dat                100%  9   1.0 MB/s 00:00
```

E.3 Running Commands

Here's one more thing the `ssh` client program can do for us. Suppose we want to check whether we have already created the file `backups/results-2019-11-12.dat` on the backup server. Instead of logging in and then typing `ls`, we could do this:

```
$ ssh amira@backup "ls results*"
Password: ********
```

```
results-2019-09-18.dat  results-2019-10-28.dat
results-2019-10-04.dat  results-2019-11-11.dat
```

Here, `ssh` takes the argument after our remote username and passes it to the shell on the remote computer. (`ls results` has multiple words, so we have to put quotes around it to make it look like one value.) Since those arguments are a legal command, the remote shell runs `ls results` for us and sends the output back to our local shell for display.

E.4 Creating Keys

Typing our password over and over again is annoying, especially if the commands we want to run remotely are in a loop. To remove the need to do this, we can create an **SSH key** to tell the remote machine that it should always trust us.

SSH keys come in pairs, a public key that gets shared with services like GitHub, and a private key that is stored only on our computer. If the keys match, we are granted access. The cryptography behind SSH keys ensures that no one can reverse-engineer our private key from the public one.

We might already have an SSH key pair on our machine. We can check by moving to our `.ssh` directory and listing the contents.

```
$ cd ~/.ssh
$ ls
```

If we see id_rsa.pub, we already have a key pair and don't need to create a new one.

If we don't see id_rsa.pub, this command will generate a new key pair. (Make sure to replace your@email.com with your own email address.)

```
$ ssh-keygen -t rsa -C "your@email.com"
```

When asked where to save the new key, press enter to accept the default location.

```
Generating public/private rsa key pair.
Enter file in which to save the key
(/Users/username/.ssh/id_rsa):
```

We will then be asked to provide an optional passphrase. This can be used to make your key even more secure, but if we want to avoid typing our password every time, we can skip it by pressing enter twice:

```
Enter passphrase (empty for no passphrase):
Enter same passphrase again:
```

When key generation is complete, we should see the following confirmation:

```
Your identification has been saved in
/Users/username/.ssh/id_rsa.
Your public key has been saved in
/Users/username/.ssh/id_rsa.pub.
The key fingerprint is:
01:0f:f4:3b:ca:85:d6:17:a1:7d:f0:68:9d:f0:a2:db your@email.com
The key's randomart image is:
+--[ RSA 2048]----+
|                 |
|                 |
|        . E +    |
|       . o = .   |
|      . S =    o |
|       o.O . o   |
|       o .+ .    |
|      . o+..     |
|       .+=o      |
+-----------------+
```

(The random art image is an alternate way to match keys.) We now need to place a copy of our public key on any servers we would like to connect to. Display the contents of our public key file with `cat`:

```
$ cat ~/.ssh/id_rsa.pub
```

```
ssh-rsa AAAAB3NzaC1yc2EAAAABIwAAAQEA879BJGYlPTLIuc9/R5MYiN4yc/
YiCLcdBpSdzgK9Dt0Bkfe3rSz5cPm4wmehdE7GkVFXrBJ2YHqPLuM1yx1AUxIe
bpwlIl9f/aUHOts9eVnVh4NztPy0iSU/Sv0b2ODQQvcy2vYcujlorscl8JjAgf
WsO3W4iGEe6QwBpVomcME8IU35v5VbylM9ORQa6wvZMVrPECBvwItTY8cPWH3M
GZiK/74eHbSLKA4PY3gM4GHI450Nie16yggEg2aTQfWA1rry9JYWEoHS9pJ1dn
LqZU3k/8OWgqJrilwSoC5rGjgp93iu0H8T6+mEHGRQe84Nk1y5lESSWIbn6P63
6Bl3uQ== your@email.com
```

Copy the contents of the output, then log in to the remote server as usual:

```
$ ssh amira@moon.euphoric.edu
Password: ********
```

Paste the copied content at the end of `~/.ssh/authorized_keys`.

```
moon> nano ~/.ssh/authorized_keys
```

After appending the content, log out of the remote machine and try to log in again. If we set up the SSH key correctly, we won't need to type our password:

```
moon> exit
```

```
$ ssh amira@moon.euphoric.edu
```

E.5 Dependencies

The example of copying our public key to a remote machine, so that it can then be used when we next SSH into that remote machine, assumed that we already had a directory `~/.ssh/`.

While a remote server may support the use of SSH to log in, your home directory there may not contain a `.ssh` directory by default.

We have already seen that we can use SSH to run commands on remote machines, so we can ensure that everything is set up as required before we place the copy of our public key on a remote machine.

Walking through this process allows us to highlight some of the typical requirements of the SSH protocol itself, as documented in the man page for the `ssh` command.

Firstly, we check that we have a `.ssh/` directory on another remote machine, `comet`

```
$ ssh amira@comet "ls -ld ~/.ssh"
Password: ********
```

```
ls: cannot access /Users/amira/.ssh: No such file or directory
```

Oops: we should create the directory and check that it's there:

```
$ ssh amira@comet "mkdir ~/.ssh"
Password: ********
```

```
$ ssh amira@comet "ls -ld ~/.ssh"
Password: ********
```

```
drwxr-xr-x 2 amira amira 512 Jan 01 09:09 /Users/amira/.ssh
```

Now we have a `.ssh` directory, into which to place SSH-related files, but we can see that the default permissions allow anyone to inspect the files within that directory. This is not considered a good thing for a protocol that is supposed to be secure, so the recommended permissions are read/write/execute for the user, and not accessible by others.

Let's alter the permissions on the directory:

```
$ ssh amira@comet "chmod 700 ~/.ssh; ls -ld ~/.ssh"
Password: ********
```

```
drwx------ 2 amira amira 512 Jan 01 09:09 /Users/amira/.ssh
```

That looks much better.

In the above example, it was suggested that we paste the content of our public key at the end of `~/.ssh/authorized_keys`, however as we didn't have a `~/.ssh/` on this remote machine, we can simply copy our public key over as the initial `~/.ssh/authorized_keys`, and of course, we will use `scp` to do this, even though we don't yet have passwordless SSH access set up.

```
$ scp ~/.ssh/id_rsa.pub amira@comet:.ssh/authorized_keys
Password: ********
```

Note that the default target for the `scp` command on a remote machine is the home directory, so we have not needed to use the shorthand `~/.ssh/` or even the full path `/Users/amira/.ssh/` to our home directory there.

Checking the permissions of the file we have just created on the remote machine, also serves to indicate that we no longer need to use our password, because we now have what's needed to use SSH without it.

```
$ ssh amira@comet "ls -l ~/.ssh"
```

```
-rw-r--r-- 2 amira amira 512 Jan 01 09:11
/Users/amira/.ssh/authorized_keys
```

While the authorized keys file is not considered to be highly sensitive (after all, it contains public keys), we alter the permissions to match the man page's recommendations.

```
$ ssh amira@comet "chmod go-r ~/.ssh/authorized_keys; ls -l
~/.ssh"
```

```
-rw------- 2 amira amira 512 Jan 01 09:11
/Users/amira/.ssh/authorized_keys
```

F

Writing Readable Code

Nothing in biology makes sense except in light of evolution (Dobzhansky 1973). Similarly, nothing in software development makes sense except in light of human psychology. This is particularly true when we look at programming style. Computers don't need to understand programs in order to execute them, but people do if they are to create, debug, and extend them.

Throughout this book we have written code to analyze word counts in classic novels using good Python style. This appendix discusses the style choices we made, presents guidelines for good Python programming style, and introduces some language features that can make programs more flexible and more readable.

F.1 Python Style

The single most important rule of style is to be consistent, both internally and with other programs (Kernighan and Pike 1999). Python's standard style is called PEP-8[1]; the acronym "PEP" is short for "Python Enhancement Proposal," and PEP-8 lays out the rules that Python's own libraries use. Some of its rules are listed below, along with others borrowed from "Code Smells and Feels[2]."

F.1.1 Spacing

Always indent code blocks using 4 spaces, and use spaces instead of tabs.

Python doesn't actually require consistent indentation so long as each block is indented the same amount, which means that this is legal:

[1] `https://www.python.org/dev/peps/pep-0008/`
[2] `https://github.com/jennybc/code-smells-and-feels`

```
def transpose(original):
  result = Matrix(original.numRow, original.numCol)
  for row in range(original.numRow):
            for col in range(original.numCol):
             result[row, col] = original[col, row]
  return result
```

The same block of code is much more readable when written as:

```
def transpose(original):
    result = Matrix(original.numRow, original.numCol)
    for row in range(original.numRow):
        for col in range(original.numCol):
            result[row, col] = original[col, row]
    return result
```

The use of 4 spaces is a compromise between 2 (which we find perfectly readable, but some people find too crowded) and 8 (which most people agree uses up too much horizontal space). As for the use of spaces rather than tabs, the original reason was that the most common interpretation of tabs by the editors of the 1980s was 8 spaces, which again was more than most people felt necessary. Today, almost all editors will auto-indent or auto-complete when the tab key is pressed (or insert spaces, if configured to do so), but the legacy of those ancient times lives on.

Do not put spaces inside parentheses.

Write `(1+2)` instead of `( 1+2 )`. This applies to function calls as well: write `max(a, b)` rather than `max( a, b )`. (We will see a related rule when we discuss default parameter values in Section F.6.)

Always use spaces around comparisons like `>` and `<=`.

Python automatically interprets `a+b<c+d` as `(a+b)<(c+d)`, but that's a lot of punctuation crowded together. Using spaces around comparison operators makes it easier to see what's being compared to what. However, we should use our own judgment for spacing around arithmetic operators like `+` and `/`. For example, `a+b+c` is perfectly readable, but

```
substrate[i, j] + overlay[i, j]
```

is easier for the eye to follow than the spaceless:

```
substrate[i, j]+overlay[i, j]
```

Most programmers would also write:

```
a*b + c*d
```

instead of:

```
a*b+c*d
```

or:

```
(a*b)+(c*d)
```

Adding spaces makes simple expressions more readable, but does not change the way Python interprets them—when it encounters `a * b+c`, for example, Python still does the multiplication before the addition.

Put two blank lines between each function definition.

This helps the eye see where one ends and the next begins, though the fact that functions always start in the first column helps as well.

Add an empty line at the end of the script.

Ending a file in a newline character is required for some other programming languages. Although it's not required for Python code to function, it does make it easier to view and edit code.

F.1.2 Naming

Use `ALL_CAPS_WITH_UNDERSCORES` for global variables.

This convention is inherited from C, which was used to write the first version of Python. In that language, upper case was used to indicate a constant whose value couldn't be modified; Python doesn't enforce that rule, but `SHOUTING_AT_PROGRAMMERS` helps remind them that some things shouldn't be messed with.

Use `lower_case_with_underscores` for the names of functions and variables.

Research on naming conventions has produced mixed results (Binkley et al. 2012; Schankin et al. 2018) but Python has (mostly) settled on underscored names for most things. This style is called **snake case** or **pothole case**; we should only use **CamelCase** for classes, which are outside the scope of this lesson.

Avoid abbreviations in function and variable names.

Abbreviations and acronyms can be ambiguous (does `xcl` mean "Excel," "exclude," or "excellent?"), and can be hard for non-native speakers to understand. Following this rule doesn't necessarily require more typing: a good programming editor will **auto-complete** names for us.

Use short names for short-lived local variables and longer names for things with wider scope.

Using `i` and `j` for loop indices is perfectly readable provided the loop is only a few lines long (Beniamini et al. 2017). Anything that is used at a greater distance or whose purpose isn't immediately clear (such as a function) should have a longer name.

Do not comment and uncomment sections of code to change behavior.

If we need to do something in some runs of the program and not in others, use an `if` statement to enable or disable that block of code: it eliminates the risk of accidentally commenting out one too many lines. If the lines we were removing or commenting out print debugging information, we should replace them with logging calls (Section 12.4). If they are operations that we want to execute, we can add a configuration option (Chapter 10), and if we are sure we don't need the code, we should take it out completely: we can always get it back from version control (Section 6.11.14).

F.2 Order

The order of items in each file should be:

- The **shebang** line (because it has to be first to work).
- The file's documentation string (Section 5.3).
- All of the `import` statements, one per line.
- Global variable definitions (especially things that would be constants in languages that support them).
- Function definitions.

- If the file can be run as a program, the `if __name__ == '__main__'` statement discussed in Section 5.1.

That much is clear, but programmers disagree (strongly) on whether high-level functions should come first or last, i.e., whether `main` should be the first function in the file or the last one. Our scripts put it last, so that it is immediately before the check on `__name__`. Wherever it goes, `main` tends to follow one of three patterns:

1. Figure out what the user has asked it to do (Chapter 10).
2. Read all input data.
3. Process it.
4. Write output.

or:

1. Figure out what the user has asked for.
2. For each input file:
    1. Read.
    2. Process.
    3. Write file-specific output (if any).
3. Write summary output (if any).

or:

1. Figure out what the user has asked for.
2. Repeatedly:
    1. Wait for user input.
    2. Do what the user has asked.
3. Exit when a "stop" command of some sort is received.

Each step in each of the outlines above usually becomes a function. Those functions depend on others, some of which are written to break code into comprehensible chunks and are then called just once, others of which are utilities that may be called many times from many different places.

We put all of the single-use functions in the first half of the file in the order in which they are likely to be called, and then put all of the multi-use utility functions in the bottom of the file in alphabetical order. If any of those utility functions are used by other scripts or programs, they should go in a file of their own (Section 5.8).

In fact, this is a good practice even if those functions are only used by one program, since it signals even more clearly which are specific to this program and which are likely to be reused elsewhere. This is why we create `collate.py` in Section 5.7: we could have kept all of our code in `countwords.py`, but collating felt like something we might want to do separately.

F.3 Checking Style

Checking that code conforms to guidelines like PEP-8 would be time consuming if we had to do it manually, but most languages have tools that will check style rules for us. These tools are often called **linters**, after an early tool called `lint`[3] that found lint (or fluff) in C code.

Python's linter used to be called `pep8` and is now called `pycodestyle`. To see how it works, let's look at this program, which is supposed to count the number of **stop words** in a document:

```
stops = ['a', 'A', 'the', 'The', 'and']

def count(ln):
    n = 0
    for i in range(len(ln)):
        line = ln[i]
        stuff = line.split()
        for word in stuff:
            # print(word)
            j = stops.count(word)
            if (j > 0) == True:
                n = n + 1
    return n

import sys

lines = sys.stdin.readlines()
# print('number of lines', len(lines))
n = count(lines)
print('number', n)
```

When we run:

[3] `https://en.wikipedia.org/wiki/Lint_(software)`

```
$ pycodestyle count_stops.py
```

it prints:

```
src/style/count_stops_before.py:3:1:
E302 expected 2 blank lines, found 1
src/style/count_stops_before.py:11:24:
E712 comparison to True should be
'if cond is True:' or 'if cond:'
src/style/count_stops_before.py:12:13:
E101 indentation contains mixed spaces and tabs
src/style/count_stops_before.py:12:13:
W191 indentation contains tabs
src/style/count_stops_before.py:15:1:
E305 expected 2 blank lines after class or function definition,
  found 1
src/style/count_stops_before.py:15:1:
E402 module level import not at top of file
```

which tells us that:

- We should use two blank lines before the function definition on line 3 and after it on line 15.
- Using `== True` or `== False` is redundant (because `x == True` is the same as `x` and `x == False` is the same as `not x`).
- Line 12 uses tabs instead of just spaces.
- The `import` on line 15 should be at the top of the file.

Fixing these issues gives us:

```
import sys


stops = ['a', 'A', 'the', 'The', 'and']


def count(ln):
    n = 0
    for i in range(len(ln)):
        line = ln[i]
        stuff = line.split()
```

```
        for word in stuff:
            # print(word)
            j = stops.count(word)
            if j > 0:
                n = n + 1
    return n


lines = sys.stdin.readlines()
# print('number of lines', len(lines))
n = count(lines)
print('number', n)
```

F.4 Refactoring

Once a program gets a clean bill of health from `pycodestyle`, it's worth having a human being look it over and suggest improvements. To **refactor** code means to change its structure without changing what it does, like simplifying an equation. It is just as much a part of programming as writing code in the first place: nobody gets things right the first time (Brand 1995), and needs or insights can change over time.

Most discussions of refactoring focus on **object-oriented programming**, but many patterns can and should be used to clean up **procedural** code. Knowing a few of these patterns helps us create better software and makes it easier to communicate with our peers.

F.4.1 Do not repeat values

The first and simplest refactoring is "replace value with name." It tells us to replace magic numbers with names, i.e., to define constants. This can seem ridiculous in simple cases (why define and use `inches_per_foot` instead of just writing 12?). However, what may be obvious to us when we're writing code won't be obvious to the next person, particularly if they are working in a different context (most of the world uses the metric system and doesn't know how many inches are in a foot). It is also a matter of habit: if we write numbers without explanation in our code for simple cases, we are more likely to do so in complex cases, and more likely to regret it afterward.

Using names instead of raw values also makes it easier to understand code when we read it aloud, which is always a good test of its style. Finally, a single value defined in one place is much easier to change than a bunch of numbers scattered throughout our program. We may not think we will have to change it, but then people want to use our software on Mars and we discover that constants aren't (Mak 2006).

```
# ...before...
seconds_elapsed = num_days * 24 * 60 * 60
```

```
# ...after...
SECONDS_PER_DAY = 24 * 60 * 60
# ...other code...
seconds_elapsed = num_days * SECONDS_PER_DAY
```

F.4.2 Do not repeat calculations in loops

It's inefficient to calculate the same value over and over again. It also makes code less readable: if a calculation is inside a loop or a function, readers will assume that it might change each time the code is executed.

Our second refactoring, "hoist repeated calculation out of loop," tells us to move the repeated calculation out of the loop or function. Doing this signals that its value is always the same. And naming that common value helps readers understand what its purpose is.

```
# ...before...
for sample in signals:
    output.append(2 * pi * sample / weight)
```

```
# ...after...
scaling = 2 * pi / weight
for sample in signals:
    output.append(sample * scaling)
```

F.4.3 Replace tests with flags to clarify repeated tests

Novice programmers frequently write conditional tests like this:

```
if (a > b) == True:
    # ...do something...
```

The comparison to `True` is unnecessary because `a > b` is a Boolean value that is itself either `True` or `False`. Like any other value, Booleans can be assigned to variables, and those variables can then be used directly in tests:

```
was_greater = estimate > 0.0
# ...other code that might change estimate...
if was_greater:
    # ...do something...
```

This refactoring is "replace repeated test with flag." Again, there is no need to write `if was_greater == True`: that always produces the same result as `if was_greater`. Similarly, the equality tests in `if was_greater == False` is redundant: the expression can simply be written `if not was_greater`. Creating and using a **flag** instead of repeating the test is therefore like moving a calculation out of a loop: even if that value is only used once, it makes our intention clearer.

```
# ...before...
def process_data(data, scaling):
    if len(data) > THRESHOLD:
        scaling = sqrt(scaling)
    # ...process data to create score...
    if len(data) > THRESHOLD:
        score = score ** 2
```

```
# ...after...
def process_data(data, scaling):
    is_large_data = len(data) > THRESHOLD
    if is_large_data:
        scaling = sqrt(scaling)
    # ...process data to create score...
    if is_large_data:
        score = score ** 2
```

If it takes many lines of code to process data and create a score, and the test then needs to change from > to >=, we are more likely to get the refactored version right the first time, since the test only appears in one place and its result is given a name.

F.4.4 Use in-place operators to avoid duplicating expression

An **in-place operator**, sometimes called an **update operator**, does a calculation with two values and overwrites one of the values. For example, instead of writing:

```
step = step + 1
```

we can write:

```
step += 1
```

In-place operators save us some typing. They also make the intention clearer, and most importantly, they make it harder to get complex assignments wrong. For example:

```
samples[least_factor,
        max(current_offset, offset_limit)] *= scaling_factor
```

is less difficult to read than the equivalent expression:

```
samples[least_factor, max(current_offset, offset_limit)] = \
    scaling_factor * samples[least_factor,
                             max(current_limit,
                                 offset_limit)]
```

(The proof of this claim is that you probably didn't notice that the long form uses different expressions to index `samples` on the left and right of the assignment.) The refactoring "use in-place operator" does what its name suggests: converts normal assignments into their briefer equivalents.

```
# ...before...
for least_factor in all_factors:
    samples[least_factor] = \
        samples[least_factor] * bayesian_scaling
```

```
# ...after...
for least_factor in all_factors:
    samples[least_factor] *= bayesian_scaling
```

F.4.5 Handle special cases first

A **short circuit test** is a quick check to handle a special case, such as checking the length of a list of values and returning `math.nan` for the average if the list is empty. "Place short circuits early" tells us to put short-circuit tests near the start of functions so that readers can mentally remove special cases from their thinking while reading the code that handles the usual case.

```
# ...before...
def rescale_by_average(values, factors, weights):
    a = 0.0
    for (f, w) in zip(factors, weights):
        a += f * w
    if a == 0.0:
        return
    a /= len(f)
    if not values:
        return
    else:
        for (i, v) in enumerate(values):
            values[i] = v / a
```

```
# ...after...
def rescale_by_average(values, factors, weights):
    if (not values) or (not factors) or (not weights):
        return
    a = 0.0
    for (f, w) in zip(factors, weights):
```

```
    a += f * w
a /= len(f)
for (i, v) in enumerate(values):
    values[i] = v / a
```

Return Consistently

PEP-8 says, "Be consistent in **return** statements," and goes on to say that either all **return** statements in a function should return a value, or none of them should. If a function contains any explicit **return** statements at all, it should end with one as well.

A related refactoring pattern is "default and override." To use it, assign a default or most common value to a variable unconditionally, and then override it in a special case. The result is fewer lines of code and clearer control flow; however, it does mean executing two assignments instead of one, so it shouldn't be used if the common case is expensive (e.g., involves a database lookup or a web request).

```
# ...before..
if configuration['threshold'] > UPPER_BOUND:
    scale = 0.8
else:
    scale = 1.0
```

```
# ...after...
scale = 1.0
if configuration['threshold'] > UPPER_BOUND:
    scale = 0.8
```

In simple cases, people will sometimes put the test and assignment on a single line:

```
scale = 1.0
if configuration['threshold'] > UPPER_BOUND: scale = 0.8
```

Some programmers take this even further and use a **conditional expression**:

```
scale = 0.8 if configuration['threshold'] > UPPER_BOUND else 1.0
```

However, this puts the default last instead of first, which is less clear.

A Little Jargon

`X if test else Y` is called a **ternary expression**. Just as a binary expression like `A + B` has two parts, a ternary expression has three. Conditional expressions are the only ternary expressions in most programming languages.

F.4.6 Use functions to make code more comprehensible

Functions were created so that programmers could re-use common operations, but moving code into functions also reduces **cognitive load** by reducing the number of things that have to be understood simultaneously.

A common rule of thumb is that no function should be longer than a printed page (about 80 lines) or have more than four levels of indentation because of nested loops and conditionals. Anything longer or more deeply nested is hard for readers to understand, so we should move pieces of long functions into small ones.

```
# ...before...
def check_neighbors(grid, point):
    if (0 < point.x) and (point.x < grid.width-1) and \
       (0 < point.y) and (point.y < grid.height-1):
        # ...look at all four neighbors
```

```
# ...after..
def check_neighbors(grid, point):
    if in_interior(grid, point):
        # ...look at all four neighbors...

def in_interior(grid, point):
    return \
    (0 < point.x) and (point.x < grid.width-1) and \
    (0 < point.y) and (point.y < grid.height-1)
```

We should *always* extract functions when code can be re-used. Even if they are only used once, multi-part conditionals, long equations, and the bodies of loops are good candidates for extraction. If we can't think of a plausible name, or if a lot of data has to be passed into the function after it's extracted, the code should probably be left where it is. Finally, it's often helpful to keep using the original variable names as parameter names during refactoring to reduce typing.

F.4.7 Combine operations in functions

"Combine functions" is the opposite of "extract function." If operations are always done together, it can sometimes be be more efficient to do them together, and might be easier to understand. However, combining functions often reduces their reusability and readability. (One sign that functions shouldn't have been combined is people using the combination and throwing some of the result away.)

The fragment below shows how two functions can be combined:

```
# ...before...
def count_vowels(text):
    num = 0
    for char in text:
        if char in VOWELS:
            num += 1
    return num

def count_consonants(text):
    num = 0
    for char in text:
        if char in CONSONANTS:
            num += 1
    return num
```

```
# ...after...
def count_vowels_and_consonants(text):
    num_vowels = 0
    num_consonants = 0
    for char in text:
        if char in VOWELS:
            num_vowels += 1
```

```
        elif char in CONSONANTS:
            num_consonants += 1
    return num_vowels, num_consonants
```

F.4.8 Replace code with data

It is easier to understand and maintain lookup tables than complicated conditionals, so the "create lookup table" refactoring tells us to turn the latter into the former:

```
# ...before..
def count_vowels_and_consonants(text):
    num_vowels = 0
    num_consonants = 0
    for char in text:
        if char in VOWELS:
            num_vowels += 1
        elif char in CONSONANTS:
            num_consonants += 1
    return num_vowels, num_consonants
```

```
# ...after...
IS_VOWEL = {'a' : 1, 'b' : 0, 'c' : 0, ... }
IS_CONSONANT = {'a' : 0, 'b' : 1, 'c' : 1, ... }

def count_vowels_and_consonants(text):
    num_vowels = num_consonants = 0
    for char in text:
        num_vowels += IS_VOWEL[char]
        num_consonants += IS_CONSONANT[char]
    return num_vowels, num_consonants
```

The more cases there are, the greater the advantage lookup tables have over multi-part conditionals. Those advantages multiply when items can belong to more than one category, in which case the table is often best written as a dictionary with items as keys and sets of categories as values:

```
LETTERS = {
    'A' : {'vowel', 'upper_case'},
    'B' : {'consonant', 'upper_case'},
    # ...other upper-case letters...
    'a' : {'vowel', 'lower_case'},
    'b' : {'consonant', 'lower_case'},
    # ...other lower-case letters...
    '+' : {'punctuation'},
    '@' : {'punctuation'},
    # ...other punctuation...
}

def count_vowels_and_consonants(text):
    num_vowels = num_consonants = 0
    for char in text:
        num_vowels += int('vowel' in LETTERS[char])
        num_consonants += int('consonant' in LETTERS[char])
    return num_vowels, num_consonants
```

The expressions used to update `num_vowels` and `num_consonants` make use of the fact that `in` produces either `True` or `False`, which the function `int` converts to either 1 or 0. We will explore ways of making this code more readable in the exercises.

F.5 Code Reviews

At the end of Section F.3, our stop-word program looked like this:

```
import sys


stops = ['a', 'A', 'the', 'The', 'and']


def count(ln):
    n = 0
    for i in range(len(ln)):
        line = ln[i]
```

```
        stuff = line.split()
        for word in stuff:
            # print(word)
            j = stops.count(word)
            if j > 0:
                n = n + 1
    return n


lines = sys.stdin.readlines()
# print('number of lines', len(lines))
n = count(lines)
print('number', n)
```

This passes a PEP-8 style check, but based on our coding guidelines and our discussion of refactoring, these things should be changed:

- The commented-out `print` statements should either be removed or turned into logging statements (Section 12.4).
- The variables `ln`, `i`, and `j` should be given clearer names.
- The outer loop in `count` loops over the indices of the line list rather than over the lines. It should do the latter (which will allow us to get rid of the variable `i`).
- Rather than counting how often a word occurs in the list of stop words with `stops.count`, we can turn the stop words into a set and use `in` to check words. This will be more readable *and* more efficient.
- There's no reason to store the result of `line.split` in a temporary variable: the inner loop of `count` can use it directly.
- Since the set of stop words is a global variable, it should be written in upper case.
- We should use `+=` to increment the counter `n`.
- Rather than reading the input into a list of lines and then looping over that, we can give `count` a stream and have it process the lines one by one.
- Since we might want to use `count` in other programs someday, we should put the two lines at the bottom that handle input into a conditional so that they aren't executed when this script is imported.

After making all these changes, our program looks like this:

```
import sys


STOPS = {'a', 'A', 'the', 'The', 'and'}


def count(reader):
    n = 0
    for line in reader:
        for word in line.split():
            if word in STOPS:
                n += 1
    return n


if __name__ == '__main__':
    n = count(sys.stdin)
    print('number', n)
```

Reading code in order to find bugs and suggest improvements like these is called **code review**. Multiple studies over more than 40 years have shown that code review is the most effective way to find bugs in software (Fagan 1976, 1986; Cohen 2010; Bacchelli and Bird 2013). It is also a great way to transfer knowledge between programmers: reading someone else's code critically will give us lots of ideas about what we could do better, and highlight things that we should probably stop doing as well.

Despite this, code review still isn't common in research software development. This is partly a chicken-and-egg problem: people don't do it because other people don't do it (Segal 2005). Code review is also more difficult to do in specialized scientific fields: in order for review to be useful, reviewers need to understand the problem domain well enough to comment on algorithms and design choices rather than indentation and variable naming, and the number of people who can do that for a research project is often very small (Petre and Wilson 2014).

Section 7.9 explained how to create and merge pull requests. How we review these is just as important as what we look for: being dismissive or combative are good ways to ensure that people don't pay attention to our reviews, or avoid having us review their work (Bernhardt 2018). Equally, being defensive when someone offers suggestions politely and sincerely is very human, but can stunt our development as a programmer.

Lots of people have written guidelines for doing reviews that avoid these traps (Quenneville 2018; Sankarram 2018). A few common points are:

Work in small increments. As Cohen (2010) and others have found, code review is most effective when done in short bursts. That means that change requests should also be short: anything that's more than a couple of screens long should be broken into smaller pieces.

Look for algorithmic problems first. Code review isn't just (or even primarily) about style: its real purpose is to find bugs before they can affect anyone. The first pass over any change should therefore look for algorithmic problems. Are the calculations right? Are any rare cases going to be missed? Are errors being caught and handled (Chapter 12)? Using a consistent style helps reviewers focus on these issues.

Use a checklist. Linters are great, but can't decide when someone should have used a lookup table instead of conditionals. A list of things to check for can make review faster and more comprehensible, especially when we can copy-and-paste or drag-and-drop specific comments onto specific lines (something that GitHub unfortunately doesn't yet support).

Ask for clarification. If we don't understand something, or don't understand why the author did it, we should ask. (When the author explains it, we might suggest that the explanation should be documented somewhere.)

Offer alternatives. Telling authors that something is wrong is helpful; telling them what they might do instead is more so.

Don't be sarcastic or disparaging. "Did you maybe think about *testing* this garbage?" is a Code of Conduct violation in any well-run project.

Don't present opinions as facts. "Nobody uses X anymore" might be true. If it is, the person making the claim ought to be able to point at download statistics or a Google Trends search; if they can't, they should say, "I don't think we use X anymore" and explain why they think that.

Don't feign surprise or pass judgment. "Gosh, didn't you know [some obscure fact]?" isn't helpful; neither is, "Geez, why don't you [some clever trick] here?"

Don't overwhelm people with details. If someone has used the letter `x` as a variable name in several places, and they shouldn't have, comment on the first two or three and simply put a check beside the others—the reader won't need the comment repeated.

Don't try to sneak in feature requests. Nobody enjoys fixing bugs and style violations. Asking them to add entirely new functionality while they're at it is rude.

How we respond to reviews is just as important:

Be specific in replies to reviewers. If someone has suggested a better variable name, we can probably simply fix it. If someone has suggested a major overhaul to an algorithm, we should reply to their comment to point at the commit that includes the fix.

Thank our reviewers. If someone has taken the time to read our code carefully, thank them for doing it.

And finally:

Don't let anyone break these rules just because they're frequent contributors or in positions of power. As Gruenert and Whitaker (2015) says, the culture of any organization is shaped by the worst behavior it is willing to tolerate. The main figures in a project should be *more* respectful than everyone else in order to show what standards everyone else is expected to meet.

F.6 Python Features

Working memory can only hold a few items at once: initial estimates in the 1950s put the number at 7±2 (G. A. Miller 1956), and more recent estimates put it as low as 4 or 5. High-level languages from Fortran to Python are essentially a way to reduce the number of things programmers have to think about at once so that they can fit what the computer is doing into this limited space. The sections below describe some of these features; as we become more comfortable with Python we will find and use others.

But beware: the things that make programs more compact and comprehensible for experienced programmers can make them less comprehensible for novices. For example, suppose we want to create this matrix as a list of lists:

```
[[0, 1, 2, 3, 4],
 [1, 2, 3, 4, 5],
 [2, 3, 4, 5, 6],
 [3, 4, 5, 6, 7],
 [4, 5, 6, 7, 8]]
```

One way is to use loops:

```
matrix = []
for i in range(5):
    row = []
```

```
    for j in range(5):
        row.append(i+j)
    matrix.append(row)
```

Another is to use a nested **list comprehension**:

```
[[i+j for j in range(5)] for i in range(5)]
```

An experienced programmer might recognize what the latter is doing; the rest of us are probably better off reading and writing the more verbose solution.

F.6.1 Provide default values for parameters

If our function requires two dozen parameters, the odds are very good that users will frequently forget them or put them in the wrong order. One solution is to bundle parameters together so that (for example) people pass three `point` objects instead of nine separate `x`, `y`, and `z` values.

A second approach (which can be combined with the previous one) is to specify **default values** for some of the parameters. Doing this gives users control over everything while also allowing them to ignore details; it also indicates what we consider "normal" for the function.

For example, suppose we are comparing images to see if they are the same or different. We can specify two kinds of tolerance: how large a difference in color value to notice, and how many differences above that threshold to tolerate as a percentage of the total number of pixels. By default, any color difference is considered significant, and only 1% of pixels are allowed to differ:

```
def image_diff(left, right, per_pixel=0, fraction=0.01):
    # ...implementation...
```

When this function is called using `image_diff(old, new)`, those default values apply. However, it can also be called like this:

- `image_diff(old, new, per_pixel=2)` allows pixels to differ slightly without those differences being significant.
- `image_diff(old, new, fraction=0.05)` allows more pixels to differ.
- `image_diff(old, new, per_pixel=1, fraction=0.005)` raises the per-pixel threshold but decreases number of allowed differences.

Note that we do not put spaces around the = when defining a default parameter value. This is consistent with PEP-8's rules about spacing in function definitions and calls (Section F.1).

Default parameter values make code easier to understand and use, but there is a subtle trap. When Python executes a function definition like this:

```
def collect(new_value, accumulator=set()):
    accumulator.add(new_value)
    return accumulator
```

it calls `set()` to create a new empty set *when it is reading the function definition*, and then uses that set as the default value for `accumulator` every time the function is called. It does *not* call `set()` once for each call, so all calls using the default will share the same set:

```
>>> collect('first')
{'first'}
>>> collect('second')
{'first', 'second'}
```

A common way to avoid this is to pass `None` to the function to signal that the user didn't provide a value:

```
def collect(new_value, accumulator=None):
    if accumulator is None:
        accumulator = set()
    accumulator.add(new_value)
    return accumulator
```

F.6.2 Handle a variable number of arguments

We can often make programs simpler by writing functions that take **a variable number of arguments**, just like `print` and `max`. One way is to require users to stuff those arguments into a list, e.g., to write `find_limits([a, b, c, d])`. However, Python can do this for us. If we declare a single argument whose name starts with a single `*`, Python will put all "extra" arguments into a **tuple** and pass that as the argument. By convention, this argument is called `args`:

```
def find_limits(*args):
    print(args)

find_limits(1, 3, 5, 2, 4)
```

```
(1, 3, 5, 2, 4)
```

This catch-all parameter can be used with regular parameters, but must come last in the parameter list to avoid ambiguity:

```
def select_outside(low, high, *values):
    result = []
    for v in values:
        if (v < low) or (v > high):
            result.add(v)
    return result

print(select_outside(0, 1.0, 0.3, -0.2, -0.5, 0.4, 1.7))
```

```
[-0.2, -0.5, 1.7]
```

An equivalent special form exists for **keyword arguments**: the catch-all variable's name is prefixed with `**` (i.e., two asterisks instead of one), and it is conventionally called `kwargs` (for "keyword arguments"). When this is used, the function is given a **dictionary** of names and values rather than a list:

```
def set_options(tag, **kwargs):
    result = f'<{tag}'
    for key in kwargs:
        result += f' {key}="{kwargs[key]}"'
    result += '/>'
    return result

print(set_options('h1', color='blue'))
print(set_options('p', align='center', size='150%'))
```

```
<h1 color="blue"/>
<p align="center" size="150%"/>
```

Notice that the names of parameters are not quoted: we pass `color='blue'` to the function, not `'color'='blue'`.

F.6.3 Unpacking variable arguments

We can use the inverse of `*args` and `**kwargs` to match a list of values to arguments. In this case, we put the `*` in front of a list and `**` in front of a dictionary when *calling* the function, rather than in front of the parameter when *defining* it:

```
def trim_value(data, low, high):
    print(data, "with", low, "and", high)

parameters = ['some matrix', 'lower bound']
named_parameters = {'high': 'upper bound'}
trim_value(*parameters, **named_parameters)
```

```
some matrix with lower bound and upper bound
```

F.6.4 Use destructuring to assign multiple values at once

One last feature of Python is **destructuring assignment**. Suppose we have a nested list such as `[1, [2, 3]]`, and we want to assign its numbers to three variables called `first`, `second`, and `third`. Instead of writing this:

```
first = values[0]
second = values[1][0]
third = values[1][1]
```

we can write this:

```
[first, [second, third]] = [1, [2, 3]]
```

In general, if the variables on the left are arranged in the same way as the values on the right, Python will automatically unpack the values and assign them correctly. This is particularly useful when looping over lists of structured values:

```
people = [
    [['Kay', 'McNulty'], 'mcnulty@eniac.org'],
    [['Betty', 'Jennings'], 'jennings@eniac.org'],
    [['Marlyn', 'Wescoff'], 'mwescoff@eniac.org']
]
for [[first, last], email] in people:
    print('{first} {last} <{email}>')
```

```
Kay McNulty <mcnulty@eniac.org>
Betty Jennings <jennings@eniac.org>
Marlyn Wescoff <mwescoff@eniac.org>
```

F.7 Summary

George Orwell laid out six rules for good writing[4], the last and most important of which is, "Break any of these rules sooner than say anything outright barbarous." PEP-8 conveys the same message[5]: there will always be cases where your code will be easier to understand if you *don't* do the things described in this lesson, but there are probably fewer of them than you think.

[4] https://en.wikipedia.org/wiki/Politics_and_the_English_Language#Remedy_of_Six_Rules

[5] https://www.python.org/dev/peps/pep-0008/#a-foolish-consistency-is-the-hobgoblin-of-little-minds

G

Documenting Programs

An old proverb says, "Trust, but verify." The equivalent in programming is, "Be clear, but document." No matter how well software is written, it always embodies decisions that aren't explicit in the final code or accommodates complications that aren't going to be obvious to the next reader. Putting it another way, the best function names in the world aren't going to answer the questions "Why does the software do this?" and "Why doesn't it do this in a simpler way?"

In this appendix we address some issues that commonly arise when people start to think about documenting their software. We start with tips for writing good docstrings, before considering more detailed documentation like tutorials, cookbooks and frequently asked question (**FAQ**) lists. Clearly defining your audience can make it easier to determine which types of documentation you need to provide, and there are a number of things you can do to reduce the work involved in providing support for FAQs.

G.1 Writing Good Docstrings

If we are doing **exploratory programming**, a short **docstring** to remind ourselves of each function's purpose is probably as much documentation as we need. (In fact, it's probably better than what most people do.) That one- or two-liner should begin with an active verb and describe either how inputs are turned into outputs, or what side effects the function has; as we discuss below, if we need to describe both, we should probably rewrite our function.

An active verb is something like "extract," "normalize," or "plot." For example, here's a function that calculates someone's age from their birthday, which uses a one-line docstring beginning with the active verb "find":

```
import re
from datetime import date
```

```
def calculate_age(birthday):
    """Find current age from birth date."""
    valid_date = '([0-9]{4})-([0-9]{2})-([0-9]{2})$'
    if not bool(re.search(valid_date, birthday)):
        message = 'Birthday must be in YYYY-MM-DD format'
        raise ValueError(message)
    today = date.today()
    born = date.fromisoformat(birthday)
    no_bday_this_year_yet = (today.month,
                             today.day) < (born.month,
                                           born.day)
    age = today.year - born.year - no_bday_this_year_yet
    return age
```

Other examples of good one-line docstrings include:

- "Create a list of capital cities from a list of countries."
- "Clip signals to lie in [0...1]."
- "Reduce the red component of each pixel."

We can tell our one-liners are useful if we can read them aloud in the order the functions are called in place of the function's name and parameters.

Once we start writing code for other people (or our future selves) our docstrings should include:

1. The name and purpose of every public class, function, and constant in our code.
2. The name, purpose, and default value (if any) of every parameter to every function.
3. Any side effects the function has.
4. The type of value returned by every function.
5. What exceptions those functions can raise and when.

The word "public" in the first rule is important. We don't have to write full documentation for helper functions that are only used inside our package and aren't meant to be called by users, but these should still have at least a comment explaining their purpose.

Here's the previous function with a more complete docstring:

```
def calculate_age(birthday):
    """Find current age from birth date.

    :param birthday str: birth date
    :returns: age in years
    :rtype: int
    :raises ValueError: if birthday not in YYYY-MM-DD format
    """
```

We could format and organize the information in the docstring any way we like, but here we've decided to use **reStructuredText**, which is the default plain-text markup format supported by Sphinx[1] (Section 14.6). The Sphinx documentation[2] describes the precise syntax for parameters, returns, exceptions and other items that are typically included in a docstring.

While the reStructuredText docstring format suggested in the Sphinx documentation looks nice once Sphinx parses and converts it to HTML for the web, the code itself is somewhat dense and hard to read. To address this issue, a number of different formats have been proposed. Two of the most prominent are Google style[3]:

```
def calculate_age(birthday):
    """Find current age from birth date.

    Args:
        birthday (str): birth date.

    Returns:
        Age in years.

    Raises:
        ValueError: If birthday not in YYYY-MM-DD format.
    """
```

and `numpydoc` style[4]:

[1] `https://www.sphinx-doc.org/en/master/`
[2] `https://www.sphinx-doc.org/en/master/usage/restructuredtext/domains.html#info-field-lists`
[3] `https://github.com/google/styleguide/blob/gh-pages/pyguide.md#38-comments-and-docstrings`
[4] `https://numpydoc.readthedocs.io/en/latest/`

```python
def calculate_age(birthday):
    """Find current age from birth date.

    Parameters
    ----------
    birthday : string
        birth date

    Returns
    -------
    integer
        age in years

    Raises
    ------
    ValueError
        if birthday not in YYYY-MM-DD format
    """
```

These two formats have become so popular that a Sphinx extension called Napoleon[5] has been released. It parses `numpydoc` and Google style docstrings and converts them to reStructuredText before Sphinx attempts to parse them. This happens in an intermediate step while Sphinx is processing the documentation, so it doesn't modify any of the docstrings in your actual source code files. The choice between the `numpydoc` and Google styles is largely aesthetic, but the two styles should not be mixed. Choose one style for your project and be consistent with it.

G.2 Defining Your Audience

It's important to consider who documentation is for. There are three kinds of people in any domain: **novices**, **competent practitioners**, and **experts** (Wilson 2019a). A novice doesn't yet have a **mental model** of the domain: they don't know what the key terms are, how they relate, what the causes of their problems are, or how to tell whether a solution to their problem is appropriate or not.

Competent practitioners know enough to accomplish routine tasks with rou-

[5] `https://sphinxcontrib-napoleon.readthedocs.io/en/latest/`

tine effort: they may need to check Stack Overflow[6] every few minutes, but they know what to search for and what "done" looks like.

Finally, experts have such a deep and broad understanding of the domain that they can solve routine problems at a glance and are able to handle the one-in-a-thousand cases that would baffle the merely competent.

Each of these three groups needs a different kind of documentation:

- A novice needs a tutorial that introduces her to key ideas one by one and shows how they fit together.
- A competent practitioner needs reference guides, cookbooks, and Q&A sites; these give her solutions close enough to what she needs that she can tweak them the rest of the way.
- Experts need this material as well—nobody's memory is perfect—but they may also paradoxically want tutorials. The difference between them and novices is that experts want tutorials on how things work and why they were designed that way.

The first thing to decide when writing documentation is therefore to decide which of these needs we are trying to meet. Tutorials like this book should be long-form prose that contain code samples and diagrams. They should use **authentic tasks** to motivate ideas, i.e., show people things they actually want to do rather than printing the numbers from 1 to 10, and should include regular check-ins so that learners and instructors alike can tell if they're making progress.

Tutorials help novices build a mental model, but competent practitioners and experts will be frustrated by their slow pace and low information density. They will want single-point solutions to specific problems, like how to find cells in a spreadsheet that contain a certain string or how to configure the web server to load an access control module. They can make use of an alphabetical list of the functions in a library, but are much happier if they can search by keyword to find what they need; one of the signs that someone is no longer a novice is that they're able to compose useful queries and tell if the results are on the right track or not.

[6] `https://stackoverflow.com/`

False Beginners

A **false beginner** is someone who appears not to know anything, but who has enough prior experience in other domains to be able to piece things together much more quickly than a genuine novice. Someone who is proficient with MATLAB, for example, will speed through a tutorial on Python's numerical libraries much more quickly than someone who has never programmed before. Creating documentation for false beginners is especially challenging; if resources permit, the best option is often a translation guide that shows them how they would do a task with the system they know well and then how to do the equivalent task with the new system.

In an ideal world, we would satisfy these needs with a chorus of explanations (Caulfield 2016), some long and detailed, others short and to the point. In our world, though, time and resources are limited, so all but the most popular packages must make do with single explanations.

G.3 Creating an FAQ

As projects grow, documentation within functions alone may be insufficient for users to apply code to their own problems. One strategy to assist other people with understanding a project is with an **FAQ**: a list of frequently asked questions and corresponding answers. A good FAQ uses the terms and concepts that people bring to the software rather than the vocabulary of its authors; putting it another way, the questions should be things that people might search for online, and the answers should give them enough information to solve their problem.

Creating and maintaining an FAQ is a lot of work, and unless the community is large and active, a lot of that effort may turn out to be wasted, because it's hard for the authors or maintainers of a piece of software to anticipate what newcomers will be mystified by. A better approach is to leverage sites like Stack Overflow[7], which is where most programmers are going to look for answers anyway:

1. Post every question that someone actually asks us, whether it's online, by email, or in person. Be sure to include the name of the software package in the question so that it's findable.

[7] `https://stackoverflow.com/`

2. Answer the question, making sure to mention which version of the software we're talking about (so that people can easily spot and discard stale answers in the future).

The Stack Overflow[8] guide to asking a good question[9] has been refined over many years, and is a good guide for any project:

Write the most specific title we can. "Why does division sometimes give a different result in Python 2.7 and Python 3.5?" is much better than, "Help! Math in Python!!"

Give context before giving sample code. A few sentences to explain what we are trying to do and why it will help people determine if their question is a close match to ours or not.

Provide a minimal reprex. Section 8.6 explains the value of a **reproducible example**, and why reprexes should be as short as possible. Readers will have a much easier time figuring out if this question and its answers are for them if they can see *and understand* a few lines of code.

Tag, tag, tag. Keywords make everything more findable, from scientific papers to left-handed musical instruments.

Use "I" and question words (how/what/when/where/why). Writing this way forces us to think more clearly about what someone might actually be thinking when they need help.

Keep each item short. The "minimal manual" approach to instructional design (Carroll 2014) breaks everything down into single-page steps, with half of that page devoted to troubleshooting. This may feel trivializing to the person doing the writing, but is often as much as a person searching and reading can handle. It also helps writers realize just how much implicit knowledge they are assuming.

Allow for a chorus of explanations. As discussed earlier, users are all different from one another, and are therefore best served by a chorus of explanations. Do not be afraid of providing multiple explanations to a single question that suggest different approaches or are written for different prior levels of understanding.

---

[8] `https://stackoverflow.com/`

[9] `https://stackoverflow.com/help/how-to-ask`

H

YAML

YAML is a way to write nested data structures in plain text that is often used to specify configuration options for software. The acronym stands for "YAML Ain't Markup Language," but that's misleading: YAML doesn't use tags like HTML, but can still be quite fussy about what is allowed to appear where.

Throughout this book we use YAML to configure our plotting script (Chapter 10), continuous integration with Travis CI (Section 11.8), software environment with `conda` (Section 13.2.1), and a documentation website with Read the Docs (Section 14.6.3). While you don't need to be an expert in YAML to complete tasks like these, this appendix outlines some basics that can help make things easier.

A simple YAML file has one key-value pair on each line with a colon separating the key from the value:

```
project-name: planet earth
purpose: science fair
moons: 1
```

Here, the keys are `"project-name"`, `"purpose"`, and `"moons"`, and the values are `"planet earth"`, `"science fair"`, and (hopefully) the number 1, since most YAML implementations try to guess the type of data.

If we want to create a list of values without keys, we can write it either using square brackets (like a Python array) or dashed items (like a Markdown list), so:

```
rotation-time: ["1 year", "12 months", "365.25 days"]
```

and:

```
rotation-time:
    - 1 year
    - 12 months
    - 365.25 days
```

are equivalent. (The indentation isn't absolutely required in this case, but helps make the intention clear.) If we want to write entire paragraphs, we can use a marker to show that a value spans multiple lines:

```
feedback: |
    Neat molten core concept.
    Too much water.
    Could have used more imaginative ending.
```

We can also add comments using # just as we do in many programming languages.

YAML is easy to understand when used this way, but it starts to get tricky as soon as sub-lists and sub-keys appear. For example, this is part of the YAML configuration file for formatting this book:

```
bookdown::gitbook:
  highlight: tango
  config:
    download: ["pdf", "epub"]
    toc:
      collapse: section
      before: |
        <li><a href="./">Merely Useful</a></li>
    sharing: no
```

It corresponds to the following Python data structure:

```
{
  'bookdown::gitbook': {
    'highlight': 'tango',
    'config': {
      'download': [
        'pdf',
```

```
        'epub'
      ],
      'toc': {
        'collapse': 'section',
        'before': '<li><a href="./">Merely Useful</a></li>\n'
      }
      'sharing': False
    }
  }
}
```

I

Anaconda

When people first started using Python for data science, installing the relevant libraries could be difficult. The main problem was that the Python package installer (`pip`[1]) only worked for libraries written in pure Python. Many scientific Python libraries have C and/or Fortran dependencies, so it was left up to data scientists (who often do not have a background in system administration) to figure out how to install those dependencies themselves. To overcome this problem, a number of scientific Python **distributions** have been released over the years. These come with the most popular data science libraries and their dependencies pre-installed, and some also come with a package manager to assist with installing additional libraries that weren't pre-installed. Today the most popular distribution for data science is Anaconda[2], which comes with a package (and environment) manager called `conda`[3]. In this appendix we look at some of the most important features of `conda` for research software engineers.

I.1 Package Management with conda

According to the latest documentation[4], Anaconda comes with over 250 of the most widely used data science libraries (and their dependencies) pre-installed. In addition, there are several thousand libraries available via the `conda install` command, which can be executed at the command line or by using the Anaconda Navigator graphical user interface. A package manager like conda greatly simplifies the software installation process by identifying and installing compatible versions of software and all required dependencies. It also handles the process of updating software as more recent versions become available. If you don't want to install the entire Anaconda distribution, you

[1] `https://pypi.org/project/pip/`
[2] `https://www.anaconda.com/`
[3] `https://conda.io/`
[4] `https://docs.anaconda.com/anaconda/`

can install Miniconda[5] instead. It essentially comes with conda and nothing else.

I.1.1 Anaconda Cloud

What happens if we want to install a Python package that isn't on the list of the few thousand or so most popular data science packages (i.e., the ones that are automatically available via the `conda install` command)? The answer is the Anaconda Cloud[6] website, where the community can post conda installation packages.

The utility of the Anaconda Cloud for research software engineers is best illustrated by an example. A few years ago, an atmospheric scientist by the name of Andrew Dawson wrote a Python package called windspharm[7] for performing computations on global wind fields in spherical geometry. While many of Andrew's colleagues routinely process global wind fields, atmospheric science is a relatively small field and thus the windspharm package will never have a big enough user base to make the list of popular data science packages supported by Anaconda. Andrew has therefore posted a conda installation package to Anaconda Cloud (Figure I.1) so that users can install windspharm using conda:

```
$ conda install -c ajdawson windspharm
```

The conda documentation has instructions[8] for quickly building a conda package for a Python module that is already available on PyPI[9] (Section 14.5).

I.1.2 conda-forge

It turns out there are often multiple installation packages for the same library up on Anaconda Cloud (e.g., Figure I.2). To try and address this duplication problem, conda-forge[10] was launched. It aims to be a central repository that contains just a single, up-to-date (and working) version of each installation package on Anaconda Cloud. You can therefore expand the selection of packages available via `conda install` beyond the chosen few thousand by adding the conda-forge channel to your conda configuration:

[5] https://docs.conda.io/en/latest/miniconda.html
[6] https://anaconda.org/
[7] https://ajdawson.github.io/windspharm/latest/
[8] https://docs.conda.io/projects/conda-build/en/latest/user-guide/tutorials/build-pkgs-skeleton.html
[9] https://pypi.org/
[10] https://conda-forge.org/

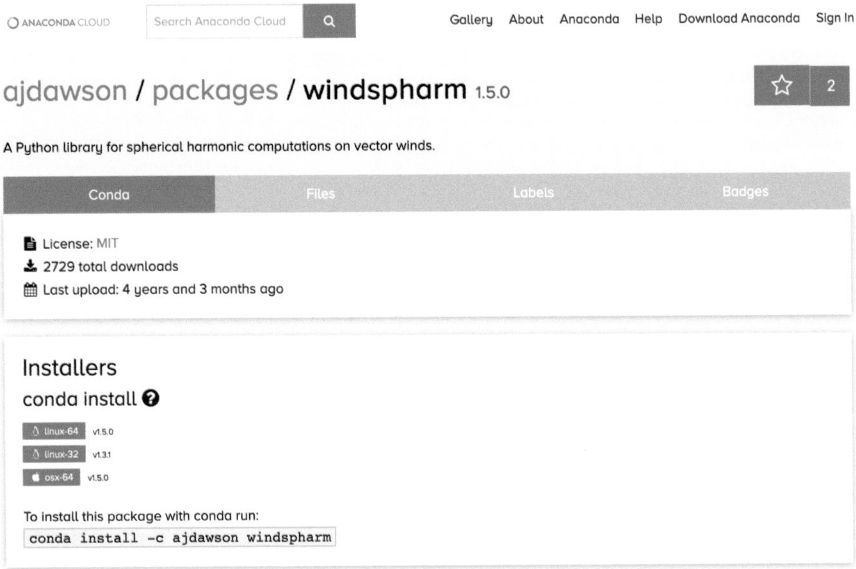

FIGURE I.1: The windspharm conda installation package on Anaconda Cloud.

```
$ conda config --add channels conda-forge
```

The conda-forge website has instructions[11] for adding a `conda` installation package to the conda-forge repository.

I.2 Environment Management with conda

If you are working on several data science projects at once, installing all the libraries you need in the same place (i.e., the system default location) can become problematic. This is especially true if the projects rely on different versions of the same package, or if you are developing a new package and need to try new things. The way to avoid these issues is to create different **virtual environments** for different projects/tasks. The original environment manager for Python development was `virtualenv`[12], which has been more re-

[11] `https://conda-forge.org/#add_recipe`
[12] `https://virtualenv.pypa.io/`

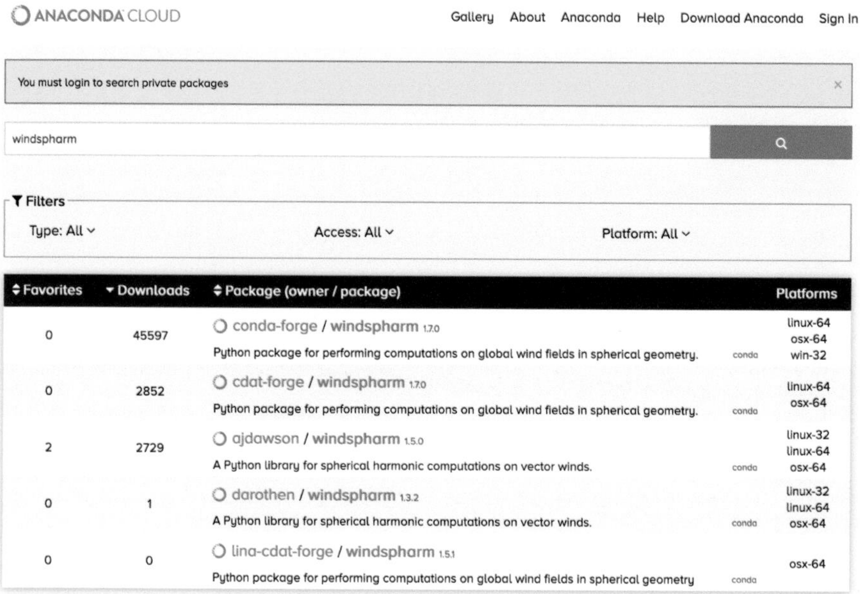

FIGURE I.2: Search results for the windspharm package on Anaconda Cloud.

cently superseded by `pipenv`[13]. The advantage that `conda` has over these options is that it is language agnostic (i.e., you can isolate non-Python packages in your environments too) and supports binary packages (i.e., you don't need to compile the source code after installing), so it has become the environment manager of choice in data science. In this book `conda` is used to export the details of an environment when documenting the computational methodology for a report (Section 13.2) and to test how a new package installs without disturbing anything in our main Python installation (Section 14.2).

[13]`https://docs.pipenv.org/`

J

Glossary

abandonware Software that is no longer being maintained.

absolute error The absolute value of the difference between the observed and the correct value. Absolute error is usually less useful than **relative error**.

absolute import In Python, an import that specifies the full location of the file to be imported.

absolute path A path that points to the same location in the **filesystem** regardless of where it is evaluated. An absolute path is the equivalent of latitude and longitude in geography. See also: relative path

actual result (of test) The value generated by running code in a test. If this matches the **expected result**, the test **passes**; if the two are different, the test **fails**.

agile development A software development methodology that emphasizes lots of small steps and continuous feedback instead of up-front planning and long-term scheduling. **Exploratory programming** is often agile.

ally Someone who actively promotes and supports inclusivity.

append mode To add data to the end of an existing file instead of overwriting the previous contents of that file. Overwriting is the default, so most programming languages require programs to be explicit about wanting to append instead.

assertion A **Boolean** expression that must be **true** at a certain point in a program. Assertions may be built into the language (e.g., **Python**'s `assert` statement) or provided as functions (e.g., **R**'s `stopifnot`). They are often used in testing, but are also put in **production code** to check that it is behaving correctly. In many languages, assertions should not be used to perform data validation as they may be silently dropped by compilers and interpreters under optimization conditions. Using assertions for data validation can therefore introduce security risks. Unlike many languages, R does not have an `assert` statement which can be disabled, and so use of a **package** such as `assertr` for data validation does not create security holes.

authentic task A task which contains important elements of things that learners would do in real life (non-classroom situations).

auto-completion A feature that allows the user to finish a word or code quickly by pressing the TAB key to list possible words or code from which the user can select.

automatic variable A variable that is automatically given a value in a **build**

rule. For example, Make automatically assigns the name of a rule's **target** to the automatic variable `$@`. Automatic variables are frequently used when writing **pattern rules**. See also: Makefile

boilerplate Standard text that is included in legal contracts, licenses, and so on.

branch-per-feature workflow A common strategy for managing work with **Git** and other **version control systems** in which a separate **branch** is created for work on each new feature or each **bug** fix and merged when that work is completed. This isolates changes from one another until they are completed.

bug report A collection of files, **logs**, or related information that describes either an unexpected output of some code or program, or an unexpected error or warning. This information is used to help find and fix a **bug** in the program or code.

bug tracker A system that tracks and manages **reported bugs** for a software program, to make it easier to address and fix the **bugs**.

build manager A program that keeps track of how files depend on one another and runs commands to update any files that are out-of-date. Build managers were invented to **compile** only those parts of programs that had changed, but are now often used to implement workflows in which plots depend on results files, which in turn depend on raw data files or configuration files. See also: build rule, Makefile

build recipe The part of a **build rule** that describes how to update something that has fallen out-of-date.

build rule A specification for a **build manager** that describes how some files depend on others and what to do if those files are out-of-date.

build target The file(s) that a **build rule** will update if they are out-of-date compared to their **dependencies**. See also: Makefile, default target

byte code A set of instructions designed to be executed efficiently by an **interpreter**.

call stack A data structure that stores information about the active subroutines executed.

camel case A style of writing code that involves naming variables and objects with no space, underscore (`_`), dot (`.`), or dash (`-`) characters, with each word being capitalized. Examples include `CalculateSum` and `findPattern`. See also: kebab case, pothole case

catch (an exception) To accept responsibility for handling an error or other unexpected event. **R** prefers "handling a condition" to "catching an exception." **Python**, on the other hand, encourages raising and catching exceptions, and in some situations, requires it.

Creative Commons license A set of **licenses** that can be applied to published work. Each license is formed by concatenating one or more of `-BY` (Attribution): users must cite the original source; `-SA` (ShareAlike): users must share their own work under a similar license; `-NC` (NonCommercial): work may not be used for commercial purposes without the creator's permission; `-ND` (NoDerivatives): no derivative works (e.g., translations) can be created without the creator's permission. Thus, `CC-BY-NC` means "users must give attribution and cannot use commercially without permission." The term `CC-0` (zero, not letter 'O') is sometimes used to mean "no restrictions," i.e., the work is in the public domain.

checklist A list of things to be checked or completed when doing a task.

command-line interface A user interface that relies solely on text for commands and output, typically running in a **shell**.

code coverage (in testing) How much of a **library** or program is executed when tests run. This is normally reported as a percentage of lines of code: for example, if 40 out of 50 lines in a file are run during testing, those tests have 80% code coverage.

code review To check a program or a change to a program by inspecting its source code.

cognitive load The amount of working memory needed to accomplish a set of simultaneous tasks.

command history An automatically created list of previously executed commands. Most read-eval-print loops (**REPLs**), including the **Unix shell**, record history and allow users to play back recent commands.

command-line argument A filename or control flag given to a command-line program when it is run.

command-line flag See **command-line argument**

command-line option See **command-line argument**

command-line switch See **command-line argument**

comment Text written in a script that is not treated as code to be run, but rather as text that describes what the code is doing. These are usually short notes, often beginning with a `#` (in many programming languages).

commit As a verb, the act of saving a set of changes to a database or **version control repository**. As a noun, the changes saved.

commit message A comment attached to a **commit** that explains what was done and why.

commons Something managed jointly by a community according to rules they themselves have evolved and adopted.

competent practitioner Someone who can do normal tasks with normal effort under normal circumstances. See also: novice, expert

computational notebook A combination of a document format that allows users to mix prose and code in a single file, and an application that executes that code interactively and in place. The **Jupyter Notebook** and **R Markdown** files are both examples of computational notebooks.

conditional expression A **ternary expression** that serves the role of an if/else statement. For example, C and similar languages use the syntax `test : ifTrue ? ifFalse` to mean "choose the value `ifTrue` if `test` is true, or the value `ifFalse` if it is not."

confirmation bias The tendency to analyze information or make decisions in ways that reinforce existing beliefs.

continuation prompt A **prompt** that indicates that the command currently being typed is not yet complete, and will not be run until it is.

continuous integration A software development practice in which changes are automatically merged as soon as they become available.

current working directory The **folder** or **directory** location in which the program operates. Any action taken by the program occurs relative to this directory.

data package A software package that, mostly, contains only data. Is used to make it simpler to disseminate data for easier use.

default target The **build target** that is used when none is specified explicitly.

default value A value assigned to a function **parameter** when the caller does not specify a value. Default values are specified as part of the function's definition.

defensive programming A set of programming practices that assumes mistakes will happen and either reports or corrects them, such as inserting **assertions** to report situations that are not ever supposed to occur.

destructuring assignment Unpacking values from data structures and assigning them to multiple variables in a single statement.

dictionary A data structure that allows items to be looked up by value, sometimes called an **associative array**. Dictionaries are often implemented using **hash tables**.

docstring Short for "documentation string," a string appearing at the start of a module, class, or function in **Python** that automatically becomes that object's documentation.

documentation generator A software tool that extracts specially formatted comments or **dostrings** from code and generates cross-referenced developer documentation.

Digital Object Identifier A unique persistent identifier for a book, paper, report, dataset, software release, or other digital artifact. See also: ORCID

down-vote A vote against something. See also: up-vote

entry point Where a program or function starts executing, or the first commands in a file that run.

exception An object that stores information about an error or other unusual event in a program. One part of a program will create and **raise an exception** to signal that something unexpected has happened; another part will **catch** it.

expected result (of test) The value that a piece of software is supposed to produce when tested in a certain way, or the state in which it is supposed to leave the system. See also: actual result (of test)

expert Someone who can diagnose and handle unusual situations, knows when the usual rules do not apply, and tends to recognize solutions rather than reasoning to them. See also: competent practitioner, novice

explicit relative import In Python, an import that specifies a path relative to the current location.

exploratory programming A software development methodology in which requirements emerge or change as the software is being written, often in response to results from early runs.

export a variable To make a variable defined inside a **shell script** available outside that script.

external error An error caused by something outside a program, such as trying to open a file that doesn't exist.

false beginner Someone whose previous knowledge allows them to learn (or re-learn) something more quickly. False beginners start at the same point as true beginners (i.e., a pre-test will show the same proficiency) but can move much more quickly.

Frequently Asked Questions A curated list of questions commonly asked about a subject, along with answers.

feature request A request to the maintainers or developers of a software program to add a specific functionality (a feature) to that program.

filename extension The last part of a filename, usually following the '.' symbol. Filename extensions are commonly used to indicate the type of content in the file, though there is no guarantee that this is correct.

filename stem The part of the filename that does not include the **extension**. For example, the stem of `glossary.yml` is `glossary`.

filesystem The part of the **operating system** that manages how files are stored and retrieved. Also used to refer to all of those files and **directories** or the specific way they are stored (as in "the Unix filesystem").

filter As a verb, to choose a set of **records** (i.e., rows of a table) based on the values they contain. As a noun, a command-line program that reads lines of text from files or **standard input**, performs some operation on them (such as filtering), and writes to a file or **stdout**.

fixture The thing on which a test is run, such as the **parameters** to the function being tested or the file being processed.

flag variable A variable that changes state exactly once to show that something has happened that needs to be dealt with later.

folder Another term for a **directory**.

forge A website that integrates **version control**, **issue tracking**, and other tools for software development.

full identifier (of a commit) A unique 160-bit identifier for a **commit** in a **Git repository**, usually written as a 20-character **hexadecimal** character **string**.

Git A **version control tool** to record and manage changes to a project.

Git branch A snapshot of a version of a **Git repository**. Multiple branches can capture multiple versions of the same repository.

Git clone Copies (and usually downloads) of a **Git remote repository** on a local computer.

Git conflict A situation in which incompatible or overlapping changes have been made on different **branches** that are now being **merged**.

Git fork To make a new copy of a **Git repository** on a **server**, or the copy that is made. See also: Git clone

Git merge Merging branches in **Git** incorporates development histories of two **branches** in one. If changes are made to similar parts of the branches on both branches, a **conflict** will occur and this must be resolved before the merge will be completed.

Git pull Downloads and synchronizes changes between a **remote repository** and a local **repository**.

Git push Uploads and synchronizes changes between a local **repository** and a **remote repository**.

Git remote A short name for a **remote repository** (like a bookmark).

Git stage To put changes in a "holding area" from which they can be **committed**.

governance The process by which an organization manages itself, or the rules used to do so.

GNU Public License A **license** that allows people to re-use software as long as they distribute the source of their changes.

graphical user interface A user interface that relies on windows, menus, pointers, and other graphical elements, as opposed to a **command-line interface** or voice-driven interface.

hitchhiker Someone who is part of a project but does not actually do any work on it.

home directory A directory that contains a user's files. Each user on a multi-user computer will have their own home directory; a personal computer will often only have one home directory.

impact/effort matrix A tool for prioritizing work in which every task is placed according to its importance and the effort required to complete it.

implicit relative import In Python, an import that does not specify a path (and hence may be ambiguous).

impostor syndrome The **false** belief that one's successes are a result of accident or luck rather than ability.

in-place operator An operator that updates one of its operands. For example, the expression `x += 2` uses the in-place operator `+=` to add 2 to the current value of `x` and assign the result back to `x`.

inspectability The degree to which a third party can figure out what was done and why. Work can be **reproducible** without being inspectable.

integration test A test that checks whether the parts of a system work properly when put together. See also: unit test

internal error An error caused by a fault in a program, such as trying to access elements beyond the end of an array.

interruption bingo A technique for managing interruptions in meetings. Everyone's name is placed on each row and each column of a grid; each time person A interrupts person B, a mark is added to the appropriate grid cell.

invariant Something that must be **true** at all times inside of a program or during the **lifecycle** of an **object**. Invariants are often expressed using

assertions. If an invariant expression is not true, this is indicative of a problem, and may result in failure or early termination of the program.

issue A **bug report**, feature request, or other to-do item associated with a project. Also called a **ticket**.

label (an issue) A short textual tag associated with an **issue** to categorize it. Common labels include `bug` and `feature request`.

issue tracking system Similar to a **bug tracking system** in that it tracks **issues** made to a **repository**, usually in the form of **feature requests**, **bug reports**, or some other to-do item.

JavaScript Object Notation A way to represent data by combining basic values like numbers and character strings in **lists** and **key/value** structures. The acronym stands for "JavaScript Object Notation"; unlike better-defined standards like **XML**, it is unencumbered by a syntax for comments or ways to define a **schema**.

kebab case A naming convention in which the parts of a name are separated with dashes, as in `first-second-third`. See also: camel case, pothole case

LaTeX A typesetting system for document preparation that uses a specialized **markup language** to define a document structure (e.g., headings), stylize text, insert mathematical equations, and manage citations and cross-references. LaTeX is widely used in academia, in particular for scientific papers and theses in mathematics, physics, engineering, and computer science.

linter A program that checks for common problems in software, such as violations of indentation rules or variable naming conventions. The name comes from the first tool of its kind, called `lint`.

list comprehension In **Python**, an expression that creates a new list in place. For example, `[2*x for x in values]` creates a new list whose items are the doubles of those in `values`.

logging framework A software **library** that manages internal reporting for programs.

logging level A setting that controls how much information is generated by a **logging framework**. Typical logging levels include `DEBUG`, `WARNING`, and `ERROR`.

long option A full-word identifier for a **command-line argument**. While most common flags are a single letter preceded by a dash, such as `-v`, long options typically use two dashes and a readable name, such as `--verbose`. See also: short option

loop body The statement or statements executed by a loop.

magic number An unnamed numerical constant that appears in a program without explanation.

Makefile A file containing commands for **Make**, often actually called `Makefile`.

Martha's Rules A simple set of rules for making decisions in small groups.

maximum likelihood estimation To choose the **parameters** for a **prob-**

ability distribution in order to maximize the likelihood of obtaining observed data.

mental model A simplified representation of the key elements and relationships of some problem domain that is good enough to support problem solving.

milestone A target that a project is trying to meet, often represented as a set of **issues** that all have to be resolved by a certain time.

MIT License A **license** that allows people to re-use software with no restrictions.

Nano (editor) A very simple text editor found on most Unix systems.

non-governmental organization An organization that is not affiliated with the government, but does the sorts of public service work that governments often do.

novice Someone who has not yet built a usable mental model of a domain. See also: competent practitioner, expert

object-oriented programming A style of programming in which functions and data are bound together in **objects** that only interact with each other through well-defined interfaces.

open license A **license** that permits general re-use, such as the **MIT License** or **GPL** for software and **CC-BY** or **CC-0** for data, prose, or other creative outputs.

open science A generic term for making scientific software, data, and publications generally available.

operating system A program that provides a standard interface to whatever hardware it is running on. Theoretically, any program that only interacts with the operating system should run on any computer that operating system runs on.

oppression A form of injustice in which one social group is marginalized or deprived while another is **privileged**.

optional argument An argument to a function or a command that may be omitted.

ORCID An Open Researcher and Contributor ID that uniquely and persistently identifies an author of scholarly works. ORCIDs are for people what **DOIs** are for documents.

orthogonality The ability to use various features of software in any order or combination. Orthogonal systems tend to be easier to understand, since features can be combined without worrying about unexpected interactions.

overlay configuration A technique for configuring programs in which several layers of configuration are used, each overriding settings in the ones before.

pager A program that displays a few lines of text at a time.

parent directory The **directory** that contains another directory of interest. Going from a directory to its parent, then its parent, and so on eventually leads to the **root directory** of the **filesystem**. See also: subdirectory

patch A single file containing a set of changes to a set of files, separated by markers that indicate where each individual change should be applied.

path (in filesystem) A **string** that specifies a location in a **filesystem**. In Unix, the **directories** in a path are joined using `/`. See also: absolute path, relative path

path coverage The fraction of possible execution paths in a piece of software that have been executed by tests. Software can have complete **code coverage** without having complete path coverage.

pattern rule A generic **build rule** that describes how to update any file whose name matches a pattern. Pattern rules often use **automatic variables** to represent the actual filenames.

phony target A **build target** that does not correspond to an actual file. Phony targets are often used to store commonly used commands in a **Makefile**.

pipe (in the Unix shell) The `|` used to make the output of one command the input of the next.

positional argument An argument to a function that gets its value according to its place in the function's definition, as opposed to a named argument that is explicitly matched by name.

postcondition Something that is guaranteed to be true after a piece of software finishes executing. See also: invariant, precondition

pothole case A naming style that separates the parts of a name with underscores, as in `first_second_third`. See also: camel case, kebab case

power law A mathematical relationship in which one quantity changes in proportion to a constant raised to the power of another quantity.

precondition Something that must be true before a piece of software runs in order for that software to run correctly. See also: invariant, postcondition

prerequisite Something that a **build target** depends on.

privilege An unearned advantage, typically as a result of belonging to a dominant social class or group.

pseudo-random number generator A function that can generate **pseudo-random numbers**. See also: seed

procedural programming A style of programming in which functions operate on data that is passed into them. The term is used in contrast to other programming styles, such as **object-oriented programming** and **functional programming**.

process An **operating system**'s representation of a running program. A process typically has some memory, the identity of the user who is running it, and a set of connections to open files.

product manager The person responsible for defining what features a product should have.

project manager The person responsible for ensuring that a project moves forward.

prompt The text printed by an **REPL** or **shell** that indicates it is ready to accept another command. The default prompt in the Unix shell is usually $, while in **Python** it is >>>, and in **R** it is >. See also: continuation prompt

provenance A record of where data originally came from and what was done to process it.

pull request The request to merge a new feature or correction created on a user's **fork** of a **Git repository** into the **upstream repository**. The developer will be notified of the change, review it, make or suggest changes, and potentially **merge** it.

raise (an exception) To signal that something unexpected or unusual has happened in a program by creating an **exception** and handing it to the **error-handling** system, which then tries to find a point in the program that will **catch** it.

raster image An image stored as a matrix of pixels.

recursion Calling a function from within a call to that function, or defining a term using a simpler version of the same term.

redirection To send a request for a web page or web service to a different page or service.

refactoring Reorganizing software without changing its behavior.

regression testing Testing software to ensure that things which used to work have not been broken.

regular expression A pattern for matching text, written as text itself. Regular expressions are sometimes called "regexp," "regex," or "RE," and are powerful tools for working with text.

relative error The absolute value of the difference between the actual and correct value divided by the correct value. For example, if the actual value is 9 and the correct value is 10, the relative error is 0.1. Relative error is usually more useful than **absolute error**.

relative path A path whose destination is interpreted relative to some other location, such as the **current working directory**. A relative path is the equivalent of giving directions using terms like "straight" and "left." See also: absolute path

remote login Starting an interactive session on one computer from another computer, e.g., by using **SSH**.

remote login server A process that handles requests to log in to a computer from other computers. See also: ssh daemon

remote repository A **repository** located on another computer. Tools such as **Git** are designed to synchronize changes between local and remote repositories in order to share work.

read-eval-print loop An interactive program that reads a command typed in by a user, executes it, prints the result, and then waits patiently for the next command. REPLs are often used to explore new ideas, or for debugging.

repository A place where a **version control system** stores the files that make up a project and the metadata that describes their history. See also: Git

reprex A reproducible example. When asking questions about coding problems online or filing issues on **GitHub**, you should always include a reprex so others can reproduce your problem and help. The reprex[1] package can help!

reproducible research The practice of describing and documenting research results in such a way that another researcher or person can re-run the analysis code on the same data to obtain the same result.

reStructured Text A plaintext **markup** format used primarily in **Python** documentation.

revision See **commit**.

root directory The **directory** that contains everything else, either directly or indirectly. The root directory is written `/` (a bare forward slash).

rotating file A set of files used to store recent information. For example, there might be one file with results for each day of the week, so that results from last Tuesday are overwritten this Tuesday.

research software engineer Someone whose primary responsibility is to build the specialized software that other researchers depend on.

script Originally, a program written in a language too user-friendly for "real" programmers to take seriously; the term is now synonymous with program.

search path The list of directories that a program searches to find something. For example, the Unix **shell** uses the search path stored in the `PATH` variable when trying to find a program whose name it has been given.

seed A value used to initialize a **pseudo-random number generator**.

semantic versioning A standard for identifying software releases. In the version identifier `major.minor.patch`, `major` changes when a new version of software is incompatible with old versions, `minor` changes when new features are added to an existing version, and `patch` changes when small **bugs** are fixed.

sense vote A preliminary vote used to determine whether further discussion is needed in a meeting. See also: Martha's Rules

shebang In Unix, a character sequence such as `#!/usr/bin/python` in the first line of an executable file that tells the **shell** what program to use to run that file.

shell A **command-line interface** that allows a user to interact with the **operating system**, such as Bash (for Unix and MacOS) or PowerShell (for Windows).

shell script A set of commands for the **shell** stored in a file so that they can be re-executed. A shell script is effectively a program.

shell variable A variable set and used in the **Unix shell**. Commonly used shell variables include `HOME` (the user's home directory) and `PATH` (their **search path**).

short circuit test A logical test that only evaluates as many arguments as it needs to. For example, if `A` is **false**, then most languages never evaluate `B` in the expression `A and B`.

[1]`https://github.com/tidyverse/reprex`

short identifier (of commit) The first few characters of a **full identifier**. Short identifiers are easy for people to type and say aloud, and are usually unique within a **repository's** recent history.

short option A single-letter identifier for a **command-line argument**. Most common flags are a single letter preceded by a dash, such as `-v`. See also: long option

snake case See **pothole case**. See also: camel case, kebab case

software distribution A set of programs that are built, tested, and distributed as a collection so that they can run together.

source distribution A **software distribution** that includes the source code, typically so that programs can be recompiled on the target computer when they are installed.

sprint A short, intense period of work on a project.

ssh daemon A **remote login server** that handles **SSH** connections.

SSH key A string of random bits stored in a file that is used to identify a user for **SSH**. Each SSH key has separate public and private parts; the public part can safely be shared, but if the private part becomes known, the key is compromised.

SSH protocol A formal standard for exchanging encrypted messages between computers and for managing **remote logins**.

stack frame A section of the **call stack** that records details of a single call to a specific function.

standard error A predefined communication channel for a **process**, typically used for error messages. See also: standard input, standard output

standard input A predefined communication channel for a **process**, typically used to read input from the keyboard or from the previous process in a **pipe**. See also: standard error, standard output

standard output A predefined communication channel for a **process**, typically used to send output to the screen or to the next process in a **pipe**. See also: standard error, standard input

stop word Common words that are filtered out of text before processing it, such as "the" and "an."

subcommand A command that is part of a larger family of commands. For example, `git commit` is a subcommand of **Git**.

subdirectory A directory that is below another directory. See also: parent directory

sustainable software Software that its users can afford to keep up to date. Sustainability depends on the quality of the software, the skills of the potential maintainers, and how much the community is willing to invest.

tag (in version control) A readable label attached to a specific **commit** so that it can easily be referred to later.

test-driven development A programming practice in which tests are written before a new feature is added or a **bug** is fixed in order to clarify the goal.

ternary expression An expression that has three parts. **Conditional expressions** are the only ternary expressions in most languages.

test framework See **test runner**.

test runner A program that finds and runs software tests and reports their results.

three stickies A technique for ensuring that everyone in a meeting gets a chance to speak. Everyone is given three sticky notes (or other tokens). Each time someone speaks, it costs them a sticky; when they are out of stickies they cannot speak until everyone has used at least one, at which point everyone gets all of their stickies back.

ticket See **issue**.

ticketing system See **issue tracking system**.

timestamp A digital identifier showing the time at which something was created or accessed. Timestamps should use **ISO date format** for portability.

tolerance How closely the **actual result** of a test must agree with the **expected result** in order for the test to pass. Tolerances are usually expressed in terms of **relative error**.

traceback In Python, an object that records where an **exception** was **raised**, what **stack frames** were on the **call stack**, and other details.

transitive dependency If A depends on B and B depends on C, C is a transitive dependency of A.

triage To go through the **issues** associated with a project and decide which are currently priorities. Triage is one of the key responsibilities of a **project manager**.

tuple A data type that has a fixed number of parts, such as the three color components of a red-green-blue color specification. Tuples are immutable (their values cannot be reset.)

unit test A test that exercises one function or feature of a piece of software and produces **pass**, **fail**, or **error**. See also: integration test

up-vote A vote in favor of something. See also: down-vote

update operator See **in-place operator**.

validation Checking that a piece of software does what its users want, i.e., "are we building the right thing?" See also: verification

variable arguments In a function, the ability to take any number of arguments. **R** uses `...` to capture the "extra" arguments. **Python** uses `*args` and `**kwargs` to capture unnamed, and named, "extra" arguments, respectively.

verification Checking that a piece of software works as intended, i.e., "did we build the thing right?" See also: validation

version control system A system for managing changes made to software during its development. See also: Git

virtual environment In **Python**, the `virtualenv` **package** allows you to create virtual, disposable, **Python** software environments containing only the packages and versions of packages you want to use for a particular project or task, and to install new packages into the environment without affecting other virtual environments, or the system-wide default environment.

virtual machine A program that pretends to be a computer. This may seem a bit redundant, but VMs are quick to create and start up, and changes made inside the virtual machine are contained within that VM so we can install new **packages** or run a completely different operating system without affecting the underlying computer.

whitespace The space, newline, carriage return, and horizontal and vertical tab characters that take up space but do not create a visible mark. The name comes from their appearance on a printed page in the era of typewriters.

wildcard A character expression that can match text, such as the `*` in `*.csv` (which matches any filename whose name ends with `.csv`).

working memory The part of memory that briefly stores information that can be directly accessed by consciousness.

K

References

Almeida, Daniel A., Gail C. Murphy, Greg Wilson, and Mike Hoye. 2017. "Do Software Developers Understand Open Source Licenses?" In *Proceedings of the 25th International Conference on Program Comprehension*, 1–11. IEEE Press. https://doi.org/10.1109/ICPC.2017.7.

Aurora, Valerie, and Mary Gardiner. 2018. *How to Respond to Code of Conduct Reports.* Frame Shift Consulting LLC.

Bacchelli, Alberto, and Christian Bird. 2013. "Expectations, Outcomes, and Challenges of Modern Code Review." In *Proc. International Conference on Software Engineering.* http://research.microsoft.com/apps/pubs/default.aspx?id=180283.

Becker, Brett A., Graham Glanville, Ricardo Iwashima, Claire McDonnell, Kyle Goslin, and Catherine Mooney. 2016. "Effective Compiler Error Message Enhancement for Novice Programming Students." *Computer Science Education* 26 (2-3): 148–75. https://doi.org/10.1080/08993408.2016.1225464.

Beniamini, Gal, Sarah Gingichashvili, Alon Klein Orbach, and Dror G. Feitelson. 2017. "Meaningful Identifier Names: The Case of Single-Letter Variables." In *Proc. 2017 International Conference on Program Comprehension (ICPC'17).* Institute of Electrical and Electronics Engineers (IEEE). https://doi.org/10.1109/icpc.2017.18.

Bernhardt, Gary. 2018. "A Case Study in Not Being a Jerk in Open Source." https://www.destroyallsoftware.com/blog/2018/a-case-study-in-not-being-a-jerk-in-open-source.

Bettenburg, Nicolas, Sascha Just, Adrian Schröter, Cathrin Weiss, Rahul Premraj, and Thomas Zimmermann. 2008. "What Makes a Good Bug Report?" In *Proc. 16th ACM SIGSOFT International Symposium on Foundations of Software Engineering - (SIGSOFT'08/FSE'16).* ACM Press. https://doi.org/10.1145/1453101.1453146.

Binkley, Dave, Marcia Davis, Dawn Lawrie, Jonathan I. Maletic, Christopher Morrell, and Bonita Sharif. 2012. "The Impact of Identifier Style on Effort and Comprehension." *Empirical Software Engineering* 18 (2): 219–76. https://doi.org/10.1007/s10664-012-9201-4.

Bollier, David. 2014. *Think Like a Commoner: A Short Introduction to the Life of the Commons*. New Society Publishers.

Borwein, Jonathan, and David H. Bailey. 2013. "The Reinhart-Rogoff Error—Or, How not to Excel at Economics." https://theconversation.com/the-reinhart-rogoff-error-or-how-not-to-excel-at-economics-13646.

Braiek, Houssem Ben, and Foutse Khomh. 2018. "On Testing Machine Learning Programs." https://arxiv.org/abs/1812.02257.

Brand, Stewart. 1995. *How Buildings Learn: What Happens after They're Built*. Penguin USA.

Brock, Jon. 2019. "'A Love Letter to Your Future Self': What Scientists Need to Know about FAIR Data." https://www.natureindex.com/news-blog/what-scientists-need-to-know-about-fair-data.

Brookfield, Stephen D., and Stephen Preskill. 2016. *The Discussion Book: 50 Great Ways to Get People Talking*. Jossey-Bass.

Brown, Neil C. C., and Greg Wilson. 2018. "Ten Quick Tips for Teaching Programming." *PLOS Computational Biology* 14 (4): e1006023. https://doi.org/10.1371/journal.pcbi.1006023.

Brown, Titus. 2017. "How I Learned to Stop Worrying and Love the Coming Archivability Crisis in Scientific Software." http://ivory.idyll.org/blog/tag/futurepaper.html.

Buckheit, Jonathan B., and David L. Donoho. 1995. "WaveLab and Reproducible Research." In *Wavelets and Statistics*, 55–81. Springer New York. https://doi.org/10.1007/978-1-4612-2544-7_5.

Carroll, John. 2014. "Creating Minimalist Instruction." *International Journal of Designs for Learning* 5 (2): 56–65. https://doi.org/10.14434/ijdl.v5i2.12887.

Caulfield, Mike. 2016. "Choral Explanations." https://hapgood.us/2016/05/13/choral-explanations/.

Cohen, Jason. 2010. "Modern Code Review." In *Making Software*, edited by Andy Oram and Greg Wilson. O'Reilly.

Devenyi, Gabriel A., Rémi Emonet, Rayna M. Harris, Kate L. Hertweck, Damien Irving, Ian Milligan, and Greg Wilson. 2018. "Ten Simple Rules for Collaborative Lesson Development." *PLOS Computational Biology* 14 (3): e1005963. https://doi.org/10.1371/journal.pcbi.1005963.

Dobzhansky, Theodosius. 1973. "Nothing in Biology Makes Sense Except in the Light of Evolution." *The American Biology Teacher* 35 (3): 125–29. https://doi.org/10.2307/4444260.

Fagan, Michael E. 1976. "Design and Code Inspections to Reduce Errors in Program Development." *IBM Systems Journal* 15 (3): 182–211. https://doi.org/10.1147/sj.153.0182.

———. 1986. "Advances in Software Inspections." *IEEE Transactions on Software Engineering* 12 (7): 744–51. https://doi.org/10.1109/TSE.1986.6312976.

Fogel, Karl. 2005. *Producing Open Source Software: How to Run a Successful Free Software Project*. O'Reilly Media.

Freeman, Jo. 1972. "The Tyranny of Structurelessness." *The Second Wave* 2 (1): 20–33.

Fucci, Davide, Hakan Erdogmus, Burak Turhan, Markku Oivo, and Natalia Juristo. 2017. "A Dissection of the Test-Driven Development Process: Does It Really Matter to Test-First or to Test-Last?" *IEEE Transactions on Software Engineering* 43 (7): 597–614. https://doi.org/10.1109/tse.2016.2616877.

Fucci, Davide, Giuseppe Scanniello, Simone Romano, Martin Shepperd, Boyce Sigweni, Fernando Uyaguari, Burak Turhan, Natalia Juristo, and Markku Oivo. 2016. "An External Replication on the Effects of Test-driven Development Using a Multi-site Blind Analysis Approach." In *Proc. 10th ACM/IEEE International Symposium on Empirical Software Engineering and Measurement (ESEM'16)*. ACM Press. https://doi.org/10.1145/2961111.2962592.

Gil, Yolanda, Cédric H. David, Ibrahim Demir, Bakinam T. Essawy, Robinson W. Fulweiler, Jonathan L. Goodall, Leif Karlstrom, et al. 2016. "Toward the Geoscience Paper of the Future: Best Practices for Documenting and Sharing Research from Data to Software to Provenance." *Earth and Space Science* 3 (10): 388–415. https://doi.org/10.1002/2015EA000136.

Goldberg, David. 1991. "What Every Computer Scientist Should Know about Floating-Point Arithmetic." *ACM Computing Surveys* 23 (1): 5–48. https://doi.org/10.1145/103162.103163.

Goodman, Alyssa, Alberto Pepe, Alexander W. Blocker, Christine L. Borgman, Kyle Cranmer, Merce Crosas, Rosanne Di Stefano, et al. 2014. "Ten Simple Rules for the Care and Feeding of Scientific Data." *PLoS Computational Biology* 10 (4): e1003542. https://doi.org/10.1371/journal.pcbi.1003542.

Goyvaerts, Jan, and Steven Levithan. 2012. *Regular Expressions Cookbook*. 2nd ed. O'Reilly Media.

Gruenert, Steve, and Todd Whitaker. 2015. *School Culture Rewired: How to Define, Assess, and Transform It*. ASCD.

Haddock, Steven, and Casey Dunn. 2010. *Practical Computing for Biologists.* Sinauer Associates.

Hart, Edmund M., Pauline Barmby, David LeBauer, François Michonneau, Sarah Mount, Patrick Mulrooney, Timothée Poisot, Kara H. Woo, Naupaka B. Zimmerman, and Jeffrey W. Hollister. 2016. "Ten Simple Rules for Digital Data Storage." *PLOS Computational Biology* 12 (10): e1005097. `https://doi.org/10.1371/journal.pcbi.1005097`.

Irving, D. B., S. Wijffels, and J. A. Church. 2019. "Anthropogenic Aerosols, Greenhouse Gases, and the Uptake, Transport, and Storage of Excess Heat in the Climate System." *Geophysical Research Letters* 46 (9): 4894–4903. `https://doi.org/10.1029/2019GL082015`.

Janssens, Jeroen. 2014. *Data Science at the Command Line.* O'Reilly Media.

Kernighan, Brian W., and Rob Pike. 1999. *The Practice of Programming.* Addison-Wesley.

Lamprecht, Anna-Lena, Leyla Garcia, Mateusz Kuzak, Carlos Martinez, Ricardo Arcila, Eva Martin Del Pico, Victoria Dominguez Del Angel, et al. 2020. "Towards FAIR Principles for Research Software." *Data Science* 3 (1): 37–59. `https://doi.org/10.3233/DS-190026`.

Lee, Stan. 1962. *Amazing Fantasy #15.* Marvel.

Leonhardt, Aljoscha, Matthias Meier, Etienne Serbe, Hubert Eichner, and Alexander Borst. 2017. "Neural Mechanisms Underlying Sensitivity to Reverse-phi Motion in the Fly." *PLoS ONE* 12 (12): 1–25. `https://doi.org/10.1371/journal.pone.0189019`.

Lin, Sarah, Ibraheem Ali, and Greg Wilson. in press. "Ten Quick Tips for Making Things Findable." *PLOS Computational Biology*, in press.

Lindberg, Van. 2008. *Intellectual Property and Open Source: A Practical Guide to Protecting Code.* O'Reilly Media.

Majumder, Suvodeep, Joymallya Chakraborty, Amritanshu Agrawal, and Tim Menzies. 2019. "Why Software Projects Need Heroes (Lessons Learned from 1000+ Projects)." `https://arxiv.org/abs/1904.09954`.

Mak, Ronald. 2006. *The Martian Principles.* Wiley.

Meili, Stephen. 2015. "Do Human Rights Treaties Help Asylum-Seekers?: Lessons from the United Kingdom." *Minnesota Legal Studies Research Paper*, no. 15-41. `https://doi.org/10.2139/ssrn.2668259`.

———. 2016. "Do Human Rights Treaties Help Asylum-Seekers: Findings from the U.K." ICPSR - Interuniversity Consortium for Political; Social Research. `https://doi.org/10.3886/E17507V2`.

Michener, William K. 2015. "Ten Simple Rules for Creating a Good Data

Management Plan." *PLOS Computational Biology* 11 (10): e1004525. https://doi.org/10.1371/journal.pcbi.1004525.

Miller, George A. 1956. "The Magical Number Seven, Plus or Minus Two: Some Limits on Our Capacity for Processing Information." *Psychological Review* 63 (2): 81–97. https://doi.org/10.1037/h0043158.

Miller, Greg. 2006. "A Scientist's Nightmare: Software Problem Leads to Five Retractions." *Science* 314 (5807): 1856–57. https://doi.org/10.1126/science.314.5807.1856.

Minahan, Anne. 1986. "Martha's Rules." *Affilia* 1 (2): 53–56. https://doi.org/10.1177/088610998600100206.

Moreno-Sánchez, Isabel, Francesc Font-Clos, and Álvaro Corral. 2016. "Large-scale Analysis of Zipf's Law in English Texts." *PLoS ONE* 11 (1): e0147073. https://doi.org/10.1371/journal.pone.0147073.

Morin, Andrew, Jennifer Urban, and Piotr Sliz. 2012. "A Quick Guide to Software Licensing for the Scientist-Programmer." *PLoS Computational Biology* 8 (7): e1002598. https://doi.org/10.1371/journal.pcbi.1002598.

Noble, William Stafford. 2009. "A Quick Guide to Organizing Computational Biology Projects." *PLoS Computational Biology* 5 (7): e1000424. https://doi.org/10.1371/journal.pcbi.1000424.

Nüst, Daniel, Vanessa Sochat, Ben Marwick, Stephen Eglen, Tim Head, Tony Hirst, and Benjamin Evans. 2020. "Ten Simple Rules for Writing Dockerfiles for Reproducible Data Science." OSF Preprints. https://doi.org/10.31219/osf.io/fsd7t.

Perez De Rosso, Santiago, and Daniel Jackson. 2013. "What's Wrong With Git?" In *Proc. 2013 ACM International Symposium on New Ideas, New Paradigms, and Reflections on Programming and Software (Onward!'13).* https://doi.org/10.1145/2509578.2509584.

Petre, Marian, and Greg Wilson. 2014. "Code Review For and By Scientists." In *Proc. Second Workshop on Sustainable Software for Science: Practice and Experience.* https://doi.org/arXiv:1407.5648.

Potter, Barney I., Rebecca Garten, James Hadfield, John Huddleston, John Barnes, Thomas Rowe, Lizheng Guo, et al. 2019. "Evolution and Rapid Spread of a Reassortant A(H3N2) Virus that Predominated the 2017–2018 Influenza Season." *Virus Evolution* 5 (2): vez046. https://doi.org/10.1093/ve/vez046.

Quenneville, Joël. 2018. "Code Review." https://github.com/thoughtbot/guides/tree/master/code-review.

Ray, Eric J., and Deborah S. Ray. 2014. *Unix and Linux: Visual QuickStart Guide.* Peachpit Press.

Sankarram, Sandya. 2018. "Unlearning Toxic Behaviors in a Code Review Culture." https://medium.freecodecamp.org/unlearning-toxic-behaviors-in-a-code-review-culture-b7c295452a3c.

Scalzi, John. 2012. "Straight White Male: The Lowest Difficulty Setting There Is." https://whatever.scalzi.com/2012/05/15/straight-white-male-the-lowest-difficulty-setting-there-is/.

Schankin, Andrea, Annika Berger, Daniel V. Holt, Johannes C. Hofmeister, Till Riedel, and Michael Beigl. 2018. "Descriptive Compound Identifier Names Improve Source Code Comprehension." In *Proc. 26th Conference on Program Comprehension (ICPC'18)*. ACM Press. https://doi.org/10.1145/3196321.3196332.

Scopatz, Anthony, and Kathryn D. Huff. 2015. *Effective Computation in Physics*. O'Reilly Media.

Segal, Judith. 2005. "When Software Engineers Met Research Scientists: A Case Study." *Empirical Software Engineering* 10 (4): 517–36. https://doi.org/10.1007/s10664-005-3865-y.

Sholler, Dan, Igor Steinmacher, Denae Ford, Mara Averick, Mike Hoye, and Greg Wilson. 2019. "Ten Simple Rules for Helping Newcomers Become Contributors to Open Projects." Edited by Scott Markel. *PLOS Computational Biology* 15 (9): e1007296. https://doi.org/10.1371/journal.pcbi.1007296.

Smith, Peter. 2011. *Software Build Systems: Principles and Experience*. Addison-Wesley Professional.

Steinmacher, Igor, Igor Scaliante Wiese, Tayana Conte, Marco Aurélio Gerosa, and David Redmiles. 2014. "The Hard Life of Open Source Software Project Newcomers." In *Proc. 7th International Workshop on Cooperative and Human Aspects of Software Engineering (CHASE/14)*. https://doi.org/10.1145/2593702.2593704.

Taschuk, Morgan, and Greg Wilson. 2017. "Ten Simple Rules for Making Research Software More Robust." *PLoS Computational Biology* 13 (4): e1005412. https://doi.org/10.1371/journal.pcbi.1005412.

Tierney, Nicholas J, and Karthik Ram. 2020. "A Realistic Guide to Making Data Available alongside Code to Improve Reproducibility." https://arxiv.org/abs/2002.11626.

Troy, Chelsea. 2018. "Why do Remote Meetings Suck so Much?" https://chelseatroy.com/2018/03/29/why-do-remote-meetings-suck-so-much/.

VanderPlas, Jake. 2014. "The Whys and Hows of Licensing Scientific Code." https://www.astrobetter.com/blog/2014/03/10/the-whys-and-hows-of-licensing-scientific-code/.

Wickes, Elizabeth, and Ayla Stein. 2016. "Data Documentation Material." http://hdl.handle.net/2142/91611.

Wilson, Greg. 2019a. *Teaching Tech Together*. Taylor & Francis.

———. 2019b. "Ten Quick Tips for Creating an Effective Lesson." Edited by Francis Ouellette. *PLOS Computational Biology* 15 (4): e1006915. https://doi.org/10.1371/journal.pcbi.1006915.

Wilson, Greg, D. A. Aruliah, C. Titus Brown, Neil P. Chue Hong, Matt Davis, Richard T. Guy, Steven H. D. Haddock, et al. 2014. "Best Practices for Scientific Computing." *PLoS Biology* 12 (1): e1001745. https://doi.org/10.1371/journal.pbio.1001745.

Wilson, Greg, Jennifer Bryan, Karen Cranston, Justin Kitzes, Lex Nederbragt, and Tracy K. Teal. 2017. "Good Enough Practices in Scientific Computing." *PLoS Computational Biology* 13 (6): e1005510. https://doi.org/10.1371/journal.pcbi.1005510.

Xu, Tianyin, Long Jin, Xuepeng Fan, Yuanyuan Zhou, Shankar Pasupathy, and Rukma Talwadker. 2015. "Hey, You Have Given Me Too Many Knobs!: Understanding and Dealing with Over-Designed Configuration in System Software." In *Proc. 10th Joint Meeting on Foundations of Software Engineering (FSE'2015)*. ACM Press. https://doi.org/10.1145/2786805.2786852.

Zampetti, Fiorella, Carmine Vassallo, Sebastiano Panichella, Gerardo Canfora, Harald Gall, and Massimiliano Di Penta. 2020. "An Empirical Characterization of Bad Practices in Continuous Integration." *Empirical Software Engineering* 25 (2): 1095–135. https://doi.org/10.1007/s10664-019-09785-8.

Zhang, Letian. 2020. "An Institutional Approach to Gender Diversity and Firm Performance." *Organization Science* 31 (2): 439–57. https://doi.org/10.1287/orsc.2019.1297.

Index